Tribology and Sustainability

Tribology and Sustainability

Edited by
Jitendra Kumar Katiyar
Mir Irfan Ul Haq
Ankush Raina
S. Jayalakshmi
R. Arvind Singh

CRC Press
Taylor & Francis Group
Boca Raton London New York

CRC Press is an imprint of the
Taylor & Francis Group, an **informa** business

First edition published 2022
by CRC Press
6000 Broken Sound Parkway NW, Suite 300, Boca Raton, FL 33487-2742

and by CRC Press
2 Park Square, Milton Park, Abingdon, Oxon, OX14 4RN

© 2022 selection and editorial matter, Jitendra Kumar Katiyar, Mir Irfan Ul Haq, Ankush Raina, S. Jayalakshmi, and R. Arvind Singh; individual chapters, the contributors

CRC Press is an imprint of Taylor & Francis Group, LLC

ISBN: 978-0-367-55146-9 (hbk)
ISBN: 978-0-367-55148-3 (pbk)
ISBN: 978-1-003-09216-2 (ebk)

Typeset in Times
by codeMantra

Contents

SECTION I Materials Tribology

SECTION II Sustainable Lubrication

SECTION III   Biotribology

Preface

Tribology is an interdisciplinary subject that addresses the study of interactions between solids as well as interactions between solids and fluids under relative motion. Tribological studies encompass various aspects of friction, wear and lubrication of engineering materials under varying environmental conditions. The tribological performance of any system depends upon several parameters such as relative speed, contact pressure, temperature, humidity and surface roughness, the nature of surfaces (i.e. hydrophobic or hydrophilic) and mechanical properties such as hardness of materials that undergo relative mechanical motion. Given the fact that tribological issues lead to considerable loss in economy, the application of better tribological practices will enhance the life of engineering equipment and reduce component failures and energy consumption, thereby leading to significant cost savings.

Growing environmental consciousness amongst end-users, stringent environmental regulations and depleting energy resources have propelled tribologists to identify sustainable futuristic solutions. Given this context, some of the initiatives taken by tribologists include (1) self-lubricating materials, (2) green additives in lubricants, (3) natural fibre-reinforced materials and (4) biomimetic approaches. All these together promote sustainability and are known by the term green tribology.

With this aim, this book highlights the fact that sustainability can be achieved with new tribological approaches and such approaches are necessary for overall well-being of human life and for the growth of economy.

The chapters in the book have been classified under three broad categories of tribology: (1) materials tribology, (2) lubricants and (3) biotribology.

The book starts with some chapters focused on sustainable materials for tribological applications such as lightweight materials, nanocomposites, self-lubricating materials and natural fibre-reinforced composites. Apart from throwing light on introductory concepts, the chapters also present the recent developments in these areas. Afterwards, the book includes chapters related to sustainable lubrication with a focus on bio-based lubricants, rheological aspects of lubricants, and minimum quantity lubrication in the case of machining. The last section of the book includes chapters related to biotribology with a focus on tribology of artificial implants for orthopaedic implants. A chapter on the tribological aspects of additive manufactured medical implants has also been included.

As the editors, we hope that the book will be of interest to students, researchers, materials scientists and industry professionals as the chapters highlight recent developments in the area of materials, green lubricants and biotribology towards achieving sustainability.

Editors

Jitendra Kumar Katiyar is a Research Assistant Professor, Department of Mechanical Engineering, SRM Institute of Science and Technology (formerly SRM University), Kattankulathur, Chennai. He is associated with the Tribology and Surface Interaction Research Laboratory, an advanced research lab dedicated to the promotion, development and understanding of tribology in different applications such as bearing, lubrication and polymer composites. He earned a PhD at the Indian Institute of Technology Kanpur in 2017. He earned a master's at the same institution in 2010. He earned a bachelor's in technology at UPTU Lucknow with honours in 2007. He has several life professional memberships, such as the Tribology Society of India, the Malaysian Society of Tribology and the Indian Society for Technical Education (ISTE). He has been an invited expert lecturer at various institutes. He has published 20+ journal papers in reputed journals and has published or attended 25+ international and national conferences. He has published more than a dozen book chapters in CRC Press, Springer and Elsevier publications. He has published two books in the field of tribology with Springer, Singapore: *Tribology in Materials and Applications* and *Automotive Tribology*. He published *Engineering Thermodynamics* in Khanna Publication, New Delhi, which is a national-level book for undergraduates. He is an active reviewer in *Tribology International, Wear, Colloid and Interface Science Communications, Friction, Materials Research Express, Defence Technology, Journal of Materials Research and Technology* and *Inderscience journals.* He received grants from various governmental organizations, such as MHRD and SERB. He has conducted several events related to tribology, such as Industrial Tribology and Tribology in Materials and Manufacturing. He is the organizer of TRIBOINDIA 2020 and the second international conference under the aegis of the Tribology Society of India in December 2020.

Mir Irfan Ul Haq is an Assistant Professor at the School of Mechanical Engineering, Shri Mata Vaishno Devi University. He has previously worked at the R&D wing of Mahindra and Mahindra. He earned a bachelor's of technology in mechanical engineering at SMVD University and a master's of engineering in mechanical system design (gold medallist) at the National Institute of Technology Srinagar. He earned a PhD in tribology at SMVD University. He is actively involved in teaching and research in the field of materials, tribology and 3D printing. His research interests include lightweight materials, new product development and additive manufacturing, development of green lubricants, self-lubricating materials and cutting fluids, mechanical testing of composites, friction and wear of materials and surface engineering.

He has published 30 research papers in SCI and Scopus indexed journals. Moreover, he has been awarded research grants from DST and NPIU-AICTE. He has attended several conferences and workshops both in India and abroad. He is a member of review boards of various international journals, apart from chairing technical sessions in various international and national conferences. Dr. Haq is actively involved in organizing conferences and workshops. He has also coordinated various student events, such as SAEBAJA, ECOKART and TEDX. He has supervised 20+ master's- and bachelor's-level projects. He has served as a member of various committees at the national and state levels.

Ankush Raina is an Assistant Professor at the School of Mechanical Engineering, Shri Mata Vaishno Devi University, Jammu and Kashmir. His area of interest includes wear and lubrication, additive manufacturing, aerodynamics and the rheological properties of lubricating oils. He earned a PhD in industrial lubrication. He earned an MTech in mechanical system design at NIT Srinagar, Kashmir. He was awarded a gold medal for securing the first position in MTech. He has published more than 25 articles in SCI/SCIE/Scopus indexed journals and attended more than 15 international conferences.

S. Jayalakshmi is a Professor at the College of Mechanical and Electrical Engineering, Wenzhou University, Wenzhou, China. She earned a PhD at the Indian Institute of Science (IISc), Bangalore, India. Her has been a Professor in India and has 10 years of experience abroad as a Research Fellow and an Academic Visitor at the National University of Singapore (NUS), Singapore, and as a Scientist at the Korea Institute of Science and Technology (KIST), Seoul, South Korea. In her research career, she is actively involved in the design and development of new Mg-based alloys/nanocomposites. She has extensive research experience in investigating the processing, characterization and structure–property relation of metallic materials, including light metal alloys/composites and amorphous alloys/bulk metallic glasses. Her current research interests include additive manufacturing of non-ferrous metals and high-entropy alloys. She received the Best Master's Thesis Award at the national level, given by the Indian National Academy of Engineering (INAE). She was selected for the prestigious INAE Research Scheme at the national level in 2018. She has two best paper awards and a best reviewer recognition by Elsevier. She has published two books and seven book chapters. She has more than 150 publications in international journals and conferences, with 1,100+ citations and h-index of 20.

R. Arvind Singh is a Professor at the College of Mechanical and Electrical Engineering, Wenzhou University, Wenzhou, China. He earned a PhD at the Indian Institute of Science (IISc), Bangalore, India. His expertise includes tribology and surface engineering related to aerospace, automotive, renewable energy, biomedical and MEMS applications. He has extensive experience in both academic and corporate R&D environments. His academic career profile includes a professorship in India and more than 10 years of experience abroad. He was a Senior Research Engineer at the Energy Research Institute (ERIAN) at the Nanyang Technological University (NTU), Singapore; Lead Engineer at the Vestas Global R&D (Singapore); Research Fellow at the National University of Singapore (NUS); Scientist at the Korea Institute of Science and Technology (KIST), Seoul, South Korea; and Materials Scientist at the John. F. Welch Technology Center (GE India). Dr. Arvind received the First Surface Engineering Best Paper Award from the Society of Tribologists and Lubrication Engineers (STLE), USA, in 2007 for his pioneering work on biomimetic tribological solutions to micro-/nanoscale devices. He received the Best Scientific Presentation Award from the Energy Research Institute (ERIAN), Nanyang Technological University (NTU), Singapore. He received the KSTLE Best Poster Award from the Korean Society of Tribologists and Lubrication Engineers (KSTLE). His current research interests include tribological behaviour of light alloys/composites and additive manufacturing of non-ferrous metals. He has seven e-disclosures, three best paper awards and three highly cited papers and has written one book and ten book chapters. He has about 150 international journals and conference publications, with 1,000+ citations and h-index of 19.

Contributors

Ahmed Abdelbary
Chief Engineering and Tribology
    Consultant
Egyptian Government
Alexandria, Egypt

Mohd Fadzli Bin Abdollah
Fakulti Kejuruteraan Mekanikal
Universiti Teknikal Malaysia Melaka
Melaka, Malaysia

A. Vivek Anand
Department of Aeronautical Engineering
MLR Institute of Technology
Hyderabad, India

Ankush Anand
School of Mechanical Engineering
Shri Mata Vaishno Devi University
Katra, India

Rahul Anand
School of Mechanical Engineering
Shri Mata Vaishno Devi University
Katra, India

V. Anandakrishnan
Department of Production Engineering
National Institute of Technology
Tiruchirappalli, India

Mokhtar Awang
Department of Mechanical Engineering
Universiti Teknologi Petronas
Perak, Malaysia

Suresh Babu
Department of Orthopaedics
Indraprastha Apollo Hospital, Mathura
    Road
New Delhi, India

Summèra Banday
Mechanical Engineering Department
Institute of Technology, University of
    Kashmir
Zakura, Srinagar, India

Hiralal Bhowmick
Mechanical Engineering Department
Thapar Institute of Engineering and
    Technology
Patiala, India

Rob Brittain
Institute of Functional Surfaces
University of Leeds
Leeds, United Kingdom

Cassie Castorena
Department of Civil Construction and
    Environmental Engineering
North Carolina State University
Raleigh, North Carolina

M.S. Charoo
Department of Mechanical
    Engineering
National Institute of Technology
Hazratbal, India

Xizhang Chen
School of Mechanical and Electrical
    Engineering
Wenzhou University
Wenzhou, China

Carsten Gachot
TU Wien-Institut für
    Konstruktionswissenschaften und
    Produktentwicklung (IKP)
Tribology Research Group
Vienna, Austria

Shraddha Gondane
Department of Mechanical
   Engineering
Visvesvaraya National Institute of
   Technology
Nagpur, India

Marcel Graf
Virtual Production Engineering
Chemnitz University of Technology
Chemnitz, Germany

Ovais Gulzar
Department of Mechanical
   Engineering
Islamic University of Science and
   Technology
Awantipora, India

Saqib Gulzar
Department of Civil Construction and
   Environmental Engineering
North Carolina State University
Raleigh, North Carolina

Manoj Gupta
Department of Mechanical
   Engineering
National University of Singapore (NUS)
Singapore

Vivudh Gupta
School of Mechanical Engineering
Shri Mata Vaishno Devi University
Katra, India

Abid Haleem
Department of Mechanical
   Engineering
Jamia Millia Islamia
New Delhi, India

Ali Sabae Hammood
Department of Materials Engineering
University of Kufa
Kufa, Iraq

M. Hanief
Department of Mechanical Engineering
National Institute of Technology
Hazratbal, India

Mir Irfan Ul Haq
School of Mechanical Engineering
Shri Mata Vaishno Devi University
Katra, India

A. P. Harsha
Department of Mechanical Engineering
Indian Institute of Technology (Banaras
   Hindu University)
Varanasi, India

Fakhruldin Mohd Hashim
Department of Mechanical Engineering
Universiti Teknologi Petronas
Perak, Malaysia

Mohd Javaid
Department of Mechanical
   Engineering
Jamia Millia Islamia
New Delhi, India

S. Jayalakshmi
School of Mechanical and Electrical
   Engineering
Wenzhou University
Wenzhou, China

Homender Kumar
Department of Mechanical Engineering
Indian Institute of Technology (Banaras
   Hindu University)
Varanasi, India

Shahira Liza
Department of Mechanical Precision
   Engineering
Malaysia-Japan International Institute
   Technology, Universiti Teknologi
   Malaysia
Kuala Lumpur, Malaysia

M. Junaid Mir
Mechanical Engineering Department
Islamic University of Science and
    Technology
Awantipora, India

R.K. Mishra
School of Mechanical Engineering
Shri Mata Vaishno Devi University
Katra, India

Norani Muti Mohamed
Department of Fundamental and
    Applied Sciences
Universiti Teknologi Petronas
Perak, Malaysia

Sanjay Mohan
School of Mechanical Engineering
Shri Mata Vaishno Devi University
Katra, India

T. Mohanraj
Department of Mechanical Engineering
Amrita School of Engineering
Amrita Vishwa Vidyapeetham
Coimbatore, India

Shuhaib Mushtaq
School of Engineering
Islamic University of Science and
    Technology
Awantipora, India

Mohd Nadeem Bhat
Materials Science and Engineering
Indian Institute of Technology Kanpur
Kanpur, India

Sharifah Khadijah Syed Mud Puad
Department of Mechanical Precision
    Engineering
Malaysia-Japan International Institute
    Technology, Universiti Teknologi
    Malaysia
Kuala Lumpur, Malaysia

Shanay Rab
Department of Mechanical
    Engineering
Jamia Millia Islamia
New Delhi, India

N. Radhika
Department of Mechanical
    Engineering
Amrita School of Engineering
Coimbatore, India

Ankush Raina
School of Mechanical Engineering
Shri Mata Vaishno Devi University
Katra, India

Ahmad Majdi Abdul Rani
Department of Mechanical
    Engineering
Universiti Teknologi Petronas
Perak, Malaysia

K. Srinivas Rao
Department of Computer Science and
    Engineering
MLR Institute of Technology
Hyderabad, India

T.V.V.L.N. Rao
Department of Mechanical
    Engineering
SRM Institute of Science and
    Technology
Kattankulathur, India

Sooraj Singh Rawat
Department of Mechanical Engineering
Indian Institute of Technology (Banaras
    Hindu University)
Varanasi, India

S. Sathish
Department of Production Engineering
National Institute of Technology
Tiruchirappalli, India

Wani Khalid Shafi
Department of Mechanical Engineering
National Institute of Technology
Hazratbal, India

Nur Hidayah Shahemi
Department of Mechanical Precision
    Engineering
Malaysia-Japan International Institute
    Technology, Universiti Teknologi
    Malaysia
Kuala Lumpur, Malaysia

Arun K. Singh
Department of Mechanical Engineering
Visvesvaraya National Institute of
    Technology
Nagpur, India

Balbir Singh
School of Mechanical Engineering
Shri Mata Vaishno Devi University
Katra, India

Harpreet Singh
Mechanical Engineering Department
Thapar Institute of Engineering and
    Technology
Patiala, India

Manjesh Kumar Singh
Department of Mechanical Engineering
Indian Institute of Technology
Kanpur, India

Pawandeep Singh
School of Mechanical Engineering
Shri Mata Vaishno Devi University
Katra, India

R. Arvind Singh
School of Mechanical and Electrical
    Engineering
Wenzhou University
Wenzhou, China

Nitish Sinha
Department of Mechanical
    Engineering
G. H. Raisoni Institute of Business
    Management
Jalgaon, India

Mahendra Kumar Soni
Department of Mechanical
    Engineering
Islamic University of Science and
    Technology
Jammu and Kashmir, India

Shane Underwood
Department of Civil Construction and
    Environmental Engineering
North Carolina State University
Raleigh, North Carolina

Raju Vaishya
Department of Orthopaedics
Indraprastha Apollo Hospital
    Mathura Road
New Delhi, India

M.F. Wani
Tribology Laboratory
National Institute of Technology
Hazratbal, India

Hamdan Haji Ya
Department of Mechanical
    Engineering
Universiti Teknologi Petronas
Perak, Malaysia

Tan Mean Yee
Department of Mechanical
    Engineering
University of Malaya
Kuala Lumpur, Malaysia

Section I

Materials Tribology

1 Materials for Tribological Applications
An Overview

Ankush Raina, Mir Irfan Ul Haq,
Sanjay Mohan, and Ankush Anand
Shri Mata Vaishno Devi University

Marcel Graf
Chemnitz University of Technology

CONTENTS

1.1   INTRODUCTION

Owing to the diverse industrial applications, a drastic increase in the energy requirements has been observed in the past few years. This ever-increasing demand has almost doubled in the last 50 years (Holmberg et al., 2017), and a considerable increase in the energy requirements has been predicted to occur in the next few decades. The major utilization of the energy is in the industrial and transportation sectors (Baba et al., 2019). The transportation sector alone accounts for 27% of the total energy consumptions (Holmberg & Erdemir, 2015). Around 80% of the total energy utilizations in the world are met by the non-renewable sources of energy. Only 20% of energy is available from the different renewable sources, which include solar, hydroelectric, wind, geothermal and tidal power. The use of non-renewable sources in such a large amount accounts for Co_x and No_x emissions, thereby causing the greenhouse effect. Thus, there is a need to look for sustainable solutions. In the recent past, different studies have also been performed in this regard (Shafi et al., 2018a). Globally, transportation industry contributes a total of 18% to the environmental degradation (Holmberg et al., 2012). According to a survey, it has been reported that energy requirements in the transportation sector increased by 37% in 1990–2005. This essentially means a further contamination of the environment by the poisonous gases (Smith, 2008). This forces the researchers to look for alternate sources of energy and at the same time encourages them to make the judicious use of available energy resources (Anand et al., 2017). Most of the losses in mechanical industry are due to the dissipation of energy in the form of heat. In the automobile sector itself, a total of 33% of energy is utilized to overcome friction itself. This includes the energy losses mainly in internal combustion engines, gears, cams, bearings and seals. As proposed by Jost (2005), the use of efficient tribological systems will enhance the economy by a substantial amount. Moreover, a saving of 22% in the energy consumption has also been proposed by reducing wear.

1.1.1 Lubrication

Worldwide, various efforts have been made in the recent past to minimize the friction and wear in the moving elements (Enomoto and Yamamoto, 1998; Tung and McMillan, 2004), and it has been proposed that in the next 20–25 years, more than 50% frictional losses can be reduced by the use of best tribopairs (Holmberg et al., 2012). In this regard, various solutions have been developed, which include (1) application of surface coatings (Holmberg et al., 1998), (2) use of lubrication (Lansdown, 2013; Raina & Anand, 2017), (3) development of self-lubricating composites (Sharma & Anand, 2017; Singh et al., 2018a), (4) surface texturing (Kovalchenko et al., 2015; Wakuda et al., 2003; Aziz et al., 2020) and (5) use of nanoparticles (Raina & Anand, 2018a). Amongst these, the use of lubrication is a commonly practised method over a number of centuries, for both scientific and industrial applications (Shafi et al., 2018b; Raina & Anand, 2018b).

The objective of lubrication is to reduce friction which is primarily due to the normal and shear stresses between solid surfaces in contact. These adverse effects can be reduced by minimizing the variations in surface stress between the two surfaces. It has been observed that every surface is rough and is composed of various peaks and valleys. These irregularities promote friction and wear at the interaction of two sliding surfaces. Moreover, the friction and wear are a function of other operating characteristics that include load, sliding speed, temperature and most importantly the geometry of the substrate (Raina & Anand, 2018c). Friction and wear are reduced by applying/generating a layer of lubricating material between two rubbing surfaces. It is the process of reducing friction and wear by interposing a layer of lubricating material between the two sliding surfaces. Therefore, any material which is used to reduce friction between the sliding contacts is known as a lubricant. A good lubricant is expected to perform the following functions:

1. It should be capable of keeping the moving surfaces apart.
2. It should have good thermal properties.
3. It must be capable of reducing friction.
4. It should act as a corrosion inhibitor.
5. It should have high wear resistance.
6. It should be capable of carrying wear debris and other contaminants.
7. It should be able to transmit desired amount of power.
8. It should act as a sealing agent against dirt and dust.
9. It should act as a water repellent.

At present, over 600 products with the above-mentioned properties are available on the market which are capable of operating under different conditions. It is believed that further advancements in this field will widen the scope of these lubricants and will lead to the enormous amount of energy and material savings (Shafi et al., 2018c).

1.1.1.1 Types of Lubricants

Different studies have been carried out to find the suitability of lubricants for various applications. These research studies include the use of lubricants in different forms, *viz.* solid lubricants, semisolid lubricants and liquid lubricants.

1.1.1.1.1 Liquid Lubricants

The liquid lubricants are generally introduced to reduce the solid–solid contact for the surfaces subjected to high relative motion. Liquid lubricants minimize wear and friction between the moving or sliding metallic surfaces by providing a protective fluid film between them. Moreover, they also act as a cooling medium, corrosion preventer and sealing agent. Liquid lubricants are classified into many types, depending on the type of base oil used.

1.1.1.1.2 Semisolid Lubricants

Semisolid lubricants are obtained by emulsification of oils or fats with metallic soap and water at 204°C–316°C. These lubricants find applications in the areas where lubricating oils are unable to perform the cooling process. In addition to this, low cost and no space constraints also make semisolid lubricants more useful. Most of the semisolid lubricants are composed of petroleum oils in which certain additives such as PTFE, silica, clay and carbon black are used along with synthetic oils. Amongst various semisolid lubricants, grease is being used widely in numerous applications.

1.1.1.1.3 Solid Lubricants

Solid lubricants are the materials in the solid form that possess substantially low friction and wear without the external supply of a lubricant. Solid lubricants find a variety of applications in case of sliding contacts where it is difficult to lubricate the surfaces by using liquid lubricants. The weak van der Waals forces between the layers of the structure of solid lubricants are the basic reason for the lowest coefficient of friction.

Solid lubricants adhere to the surface, thereby preventing the failure caused due to the insufficient or no supply of the liquid lubricants. The most important advantage of using solid lubricants is that they eliminate the use of seals and other packing materials, which otherwise is required in case of liquid lubrication. Moreover, the issues associated with liquid lubricants such as high cost, environmental degradation and continuous monitoring are also reduced.

1.1.2 LIQUID LUBRICATION

It is a fact that the most adopted method of reducing friction between the two solid surfaces is the application of liquid lubricant at the interface. Although the concept was introduced several centuries ago, researchers are still proposing the advancements, particularly in the field of industrial tribology (Shafi et al., 2018d; Haq et al., 2018). The reason for these improvements is obviously the energy considerations, as explained above. Overall performance and reliability of various tribopairs extensively depend on the properties of the lubricating oil. The selection of particular type of lubricant depends on certain factors which include understanding of lubrication regime (Shafi et al., 2019; Kerni et al., 2019), different lubricant properties, lubricant classifications and the use of various lubricant additives (Lansdown, 2013).

1.1.2.1 Properties of Liquid Lubricants

1.1.2.1.1 Viscosity

It is the property of liquid by virtue of which it offers resistance to the flow. The operating characteristics of a lubricant are determined by this property only. For low-viscosity lubricants, it is difficult to maintain the liquid film between two sliding surfaces, whereas for high-viscosity lubricants, high shear stress between the layers leads to the excessive friction. The viscosity of liquids is also a function of temperature. With the increase in temperature, oil becomes thinner, whereas at low temperatures, oil becomes thick. It is desirable that the viscosity of a good lubricant oil should be least affected by the change in temperature.

1.1.2.1.2 Flash Point and Fire Point

The minimum temperature at which the lubricating oil catches fire when its vapours are exposed to a tiny flame is known as flash point. The minimum temperature at which the vapours of the oil continue to burn at least for 5 s is called as fire point. The fire point is an important parameter for the lubricants performing at a high temperature, and it should be more than the temperature to which the oil is subjected.

1.1.2.1.3 Cloud Point and Pour Point

During the cooling process, the temperature at which the lubricating oil becomes hazy or cloudy is called cloud point. The pour point can be defined as the minimum temperature at which the oil stops flowing. Pour point and cloud point are important when the oil is subjected to cold conditions. The improper selection of the lubricant may cause inadequate supply to the machinery, thereby jamming different moving components.

1.1.2.2 Classification of Liquid Lubricants

The use of liquid lubricants to reduce the friction has been there since ancient times; however, the researchers are still proposing the advancements, particularly in the field of industrial tribology. The boosters for these improvements are the energy losses and increased frequency of failures happening due to the friction between the two mating surfaces. Thus, there is a need to reduce such losses to the maximum possible extent. At present, there are a variety of liquid lubricants available on the market, and they can be broadly classified as follows:

a. Vegetable oils
b. Animal oils
c. Mineral oils
d. Synthetic oils

Figure 1.1 describes the classification of liquid lubricants.

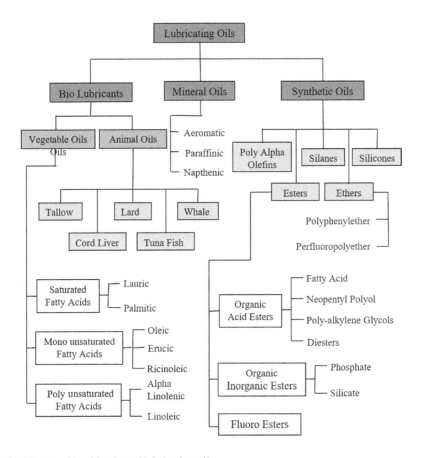

FIGURE 1.1 Classification of lubricating oils.

1.1.2.3 Lubrication Regimes

In lubricated conditions, the presence of fluid film controls the friction and wear between the surfaces in contact. The ratio of lubricant film thickness to the composite surface roughness classifies the different lubrication regimes. Different types of regimes in lubrication are discussed in the following.

1.1.2.3.1 Boundary Lubrication

Boundary lubrication is a condition wherein the two interacting surfaces are so close to each other that a very thin layer of lubricant exists between the contact surfaces. The boundary lubrication is obtained at low speeds and at high loads during start and stop conditions. This regime is not desirable due to a high coefficient of friction, increased wear and non-uniform load distribution. In this regime, the bulk properties of lubricant have a very less effect on the friction and wear behaviour. Most of the bearing failures are caused by boundary lubrication and are predominantly due to adhesive and chemical (corrosive) wear. As per the research carried out in this regard, it has been observed that nearly 70% of wear occurs during start and stop operations.

1.1.2.3.2 Mixed Lubrication

In mixed lubrication regime, the thickness of the fluid film is more as compared to that of the boundary lubrication regime and the contact between the asperities is also less. The contact between the asperities is reduced due to the increased velocity which ultimately leads to the decrease in coefficient of friction. In this regime, the contact between the asperities is not completely eliminated and the load is shared by both oil and asperities.

1.1.2.3.3 Hydrodynamic Lubrication

In hydrodynamic lubrication, a thick film of the lubricant is maintained between the contact surfaces. The thickness of the lubricating film in hydrodynamic lubrication is greater as compared to the height of asperities. The tribological behaviour of the contacting surfaces is governed by the physical properties of the lubricant. In this regime, load is supported by the film and is characterized by the formation of oil wedge which reduces the risk of asperity-to-asperity contact.

In hydrodynamic lubrication, adhesive wear occurs at the start–stop condition, whereas corrosive (chemical) wear takes place due to the interaction of the surfaces with the lubricant. The viscosity of the oil plays a vital role in this regime. At raised temperatures, a significant decrease in viscosity may result in metal-to-metal contact. However, the decrease in temperature may lead to an increase in viscosity which results in high friction. This is due to the decrease in the internal resistance between the layers of lubricating oil. Thus, due care should be taken while selecting the oils when working in this regime by taking into consideration the different operating conditions.

1.1.3 SOLID LUBRICATION

The term 'solid lubrication' seems to be contrary to liquid lubrication, but is of great importance. The automotive sector has always been of great concern for researchers due to the involvement of sliding parts. In the process of evolution, liquid lubrication has shown remarkable performance. However, the prolonged usage of liquid lubrication has also brought into picture various shortcomings such as accessibility issues, vaporization at high temperatures and regular lubrication. The majority of these shortcomings have been overcome by solid lubricants. It cannot be said that solid lubrication is the replacement for liquid lubrication; rather, it is an addition to liquid lubrication.

Solid lubricants are also used in combination with liquid lubricants. They are being added to the lubricating oils in certain proportions and have shown promising results, especially in the case of boundary lubrication regimes. Graphite and molybdenum disulphide are the most commonly used solid lubricants in oils. Other solid lubricants include hexagonal boron nitride, tungsten disulphide and polytetrafluoroethylene (PTFE) (Scharf & Prasad, 2013). The tribological materials have been defined as the materials that modify the friction and wear depending upon the desired application; e.g., clutches and brakes require high friction and low wear, whereas bearings require low friction as well as low wear. The modification in friction and wear behaviour is attained by introducing a shear-accommodating layer between the contacting surfaces. The applications where liquid lubrication is not possible, such as high-temperature applications, vacuum or in conditions where liquids cannot be introduced, solid lubricant is the most promising solution (Scharf & Prasad, 2013). It is not more than half a century when solid lubrication was one amongst various arts. However, in the present times, it has become an integral part of materials science. There is not much difference in the functioning of solid lubricants as compared to that of liquid lubricants. The former shear easily and facilitate the tribological behaviour between the sliding surfaces.

Solid lubricants basically belong to the category of solid materials that are used to mitigate friction and wear of the rubbing parts by averting direct contact between the respective surfaces. They can be associated with solid, liquid and semisolid materials. Solid lubricants are mostly available in powder forms. The basic functioning of solid lubricants is to level the peaks and valleys at the contact surfaces, thereby adhering to the substratum. Under extreme operating conditions, the solid lubricants tend to deliver efficient boundary lubrication, thus improving friction and minimizing the wear. It can be said that a category of solid lubricants being used to mitigate friction and wear of the contact surfaces by restraining their direct contact is solid lubricants. These have been classified mainly into soft metals, lamellar solids, oxides, halides, sulphates, etc. Table 1.1 shows the classification of solid lubricants (Sharma & Anand, 2016). Amongst various lubricants, graphite, molybdenum disulphide, boron nitride and fluorides are the most commonly used lubricants that have been used with various metal-based base matrices. The succeeding sections present the association of some of the commonly used lubricants with metal-based matrices such as iron, copper, aluminium and magnesium.

1.2   SUSTAINABLE MATERIALS

The unending demand for energy and upcoming environmental issues has paved the way for the development of sustainable materials and processes. Evolution of

TABLE 1.1

Classification of Solid Lubricants

Types	Examples
Soft metals	Ag, Pb, Au, In, Sn
Oxides/mixed oxides	ZnO, TiO_2, Re_2O_7, MoO_3, B_2O_3, CoO-MoO_3, Cs_2O-MoO_3, CuO-Re_2O_7, PbO-B_2O_3, NiO-MoO_3
Lamellar solids	H_3BO_3, graphite, graphite fluoride, MoS_2, WS_2, h-BN
Alkaline earth metals, sulphates, halides	$CaSO_4$, $BaSO_4$, $SrSO_4$, CaF_2, BaF_2, SrF_2
Carbon-based solids	Glassy carbon, hollow carbon, diamond-like carbon, diamond, fullerene

machinery in defence, aerospace, marine, automotive sector, etc., has created a huge demand for materials with enhanced mechanical and tribological characteristics. The development of such materials has been a big challenge for scientists and researchers. They have explored a lot in the area of sustainable materials, and as a result, several alloys and composites have been developed (Haq & Anand, 2018a). As per the requirement, several alloys and composites have also been associated with various lubricants, whether it be liquid, semisolid or solid. From the literature, it is known that in order to improve the existing properties, researchers have developed several composites by reinforcing the conventional materials (metals, polymers, alloys, etc.) with various reinforcements (Bodunrin et al., 2015).

The following subsections discuss various materials associated with other metals or lubricants and their respective properties.

1.2.1 ALUMINIUM-BASED MATERIALS

The development of lightweight materials has led to fuel saving and cost reduction, in general and especially in sectors like automotive and aerospace. The hazardous emissions were lowered with these materials, and this has resulted in a positive influence on the environment. Lightweight metals such as aluminium and magnesium have the ability to satisfy this need in the present-day industry (Omrani et al., 2016). However, the poor mechanical and tribological properties of aluminium and magnesium limit their use in the applications where structural and tribological properties are of prime importance.

The reinforcement of ceramics in these lightweight materials significantly enhances the mechanical and anti-wear properties (Singh et al., 2018a and b). The ease of machining, recyclability, availability in abundance and ease of manufacturing at large scale in the case of aluminium make it a tough candidate in comparison with magnesium. In many engineering applications such as automotive, marine, and aerospace, it is desired that the materials, apart from having better mechanical properties, should also possess better tribological properties. Aluminium is not suitable for structural applications, and it has also shown poor performance in sliding applications. However, in order to take advantage of the better strength-to-weight ratio of aluminium, it is reinforced with other metals, ceramics and other reinforcements. Depending upon the desired properties, the reinforcement, fabrication route and post-processing methods can be selected.

1.2.1.1 Aluminium Alloys

Despite being readily available, the soft and ductile nature of aluminium made it unsuitable for any structural application. The density of aluminium is around one-third of that of the commonly used metals, except for titanium and magnesium. Aluminium is alloyed with various agents such as copper, manganese and silicon for the improvement of various properties (Santos et al., 2016). Aluminium alloys are grouped into wrought alloys and casting alloys. The properties of these alloys can be tailored by different heat treatments such as solution heat treatment, precipitation or age hardening and quenching. However, some of the alloys do not respond to heat treatments and are usually used in the as-cast condition. In order to distinguish the different aluminium alloys, Aluminum Association Inc. has designated a four-digit system (Callister & Rethwisch, 2007). The first digit denotes the group, and the last two digits denote the minimum aluminium percentage. Table 1.2 presents the major aluminium alloys, their alloying elements and their specific properties.

TABLE 1.2
Aluminium Alloy Designations: Properties and Applications

Alloy Designation	Alloying Elements	Properties	Applications
1XXX	None (99% aluminium)	Very ductile / Good corrosion resistance	Food, electrical and packaging industries
2XXX	Copper	Heat-treatable / Better strength / Poor weldability / Poor corrosion resistance	Aerospace sector as airframes
3XXX	Manganese	Non-treatable / Medium strength / Good weldability, formability and corrosion resistance	Vehicle panelling and marine applications
4XXX	Silicon	Lower melting point / Good corrosion resistance	Welding of aluminium alloys
5XXX	Magnesium	Moderate-to-high strength / Non-heat-treatable alloys / Excellent corrosion resistance	Pressure vessels, road and rail tankers and marine applications
6XXX	Magnesium and silicon	Easy to machine / Good weldability / Medium strength	Automotive, marine and construction industries
7XXX	Zinc	Heat-treatable / Very high strength-to-weight ratio / Good ballistic deterrent properties	Aerospace, defence, aircraft, automotive and sports equipment

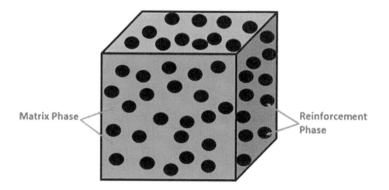

FIGURE 1.2   Schematic of a metal matrix composite with particle reinforcements.

1.2.1.2   Aluminium Composites

Composites with aluminium and its alloys as the base matrix are termed as aluminium composites. The major advantages associated with such composites in comparison with the aluminium alloys include higher strength, better stiffness, better thermal resistance, improved damping capacity, better fatigue resistance and improved resistance to wear. Various ceramic reinforcements such as SiC, Al_2O_3, B_4C, TiO_2, ZrO_2, TiB_2 and TiC have been added to aluminium and its alloys for the development of ceramic-reinforced composites (Bains et al., 2016; Bodunrin et al., 2015; Singh et al., 2019). Figure 1.2 shows a schematic diagram of a particle-reinforced metal matrix composite.

A lot of research has been focused towards the development of hybrid composites wherein two reinforcements are added to the matrix, primarily to produce a synergetic effect of the two reinforcements. The addition of ceramic reinforcements alone leads to improvement in the mechanical properties such as hardness and compression strength. The composites with at least one of the reinforcements as a solid lubricant are termed as self-lubricating composites. The solid lubricant phase is embedded in the matrix phase, and upon sliding with the counter-face, the embedded solid lubricant aids in lowering friction and wear by virtue of their film formation capability (Omrani et al., 2016). This lubricity is offered even in the absence of a liquid lubricant. Owing to the soft nature of these solid lubricant additives, these materials have poor mechanical properties. Therefore, in order to improve upon the mechanical properties, ceramic reinforcements are added as primary reinforcements and solid lubricants as secondary reinforcements in aluminium-based matrices (Moghadam et al., 2015; Haq & Anand, 2018b). The challenge is to develop a composite with optimized mechanical and tribological properties. This requires a good combination of ceramic reinforcement and the solid lubricant, apart from the ratio of these reinforcements.

1.2.2   Iron-Based Materials

Amongst various materials, iron (Fe) is the one that is available in large resources and is cheap. The automotive industry nowadays has been using iron-based materials due to their availability and mechanical properties. Fe-based materials have been used in various industrial applications as wear-resistant materials (Gupta et al., 2019). The wear resistance mainly depends on the mechanical properties of materials at room and high temperatures. Thus, Fe as a low-cost material is a promising candidate for tribological applications. Due to the properties of Fe, it has replaced many ceramics for many high-temperature applications (Erdoğan, 2019; Hernandez et al., 2019).

Iron-based materials are majorly categorized into pure iron, cast iron and steel. These classifications are made according to the presence of carbon content. The ease of manufacturing along with the extensive range of values of yield strength and modulus of elasticity has also made manufacturers to adopt iron for automotive applications (Ghassemieh, 2011). Amongst the mentioned iron-based materials, steels are used in many industrial applications, automotive applications, lead frames and other electronic materials by reinforcing them with other elements, mainly chromium and nickel, to it. Iron-based alloys developed through powder metallurgy (P/M) have also been used for structural parts in mechanical devices due to their low cost, corrosion resistance, ductility, tensile strength, etc. (Potecaşu et al., 2015). The alloying of low-carbon steel with carbon has been investigated for automotive applications. The findings have shown that a range of hardness values were obtained. Martensite, austenite and cementite structure with hardness of 650 Hv, 500 Hv and 850 Hv, respectively, were obtained (Abboud et al., 2007). Various metals, ceramics, solid lubricants, etc., have been associated with iron, and a lot of new alloys and composites have been developed. Iron-based materials have been regularly used in the engine parts such ass connecting rod, cylinder and crankshaft. Most of the bearings have been fabricated with iron-based materials.

The tribological behaviour of Fe-Mo base matrix with molybdenum (Mo) and Mo–graphite as reinforcements was investigated at room temperature and also at temperatures of 320°C and 450°C. The composites were prepared using P/M route, and the formation of graphite film at RT and oxides (Fe_2O_3, Fe_3O_4 and $FeO \cdot Cr_2O_3$) at high temperatures played a vital role in mitigating friction and wear (Ma et al., 2013). The composites exhibiting good tribological behaviour over the range of temperatures from RT to HT were developed. Molybdenum (Mo), silver (Ag) and copper oxide (CuO) were reinforced in Fe-Cr matrix using P/M. The investigation of the composites was carried out for wear and friction, and the results have shown improved tribological behaviour due to the formation of oxides such as MoO3, $CuMoO_4$ and $Cu_3Mo_2O_9$.(Cui et al., 2020). The ceramics have been combined with Fe in order to improve upon wear and frictional properties.

Kim et al. (2001) studied the influence of manganese (Mn) on the wear behaviour of Fe-20Cr matrix hard alloys from 25°C to 450°C. The results have shown $\gamma \rightarrow \varepsilon$ martensitic transformation resulting in increased wear resistance of alloys when Mn content exceeded 10 wt. %. The wear resistance of alloys with Fe as base material could be increased by adding hard ceramic particles, such as SiC, TiC and Al_2O_3 (Aghili et al., 2014; Modi et al., 2004). A base matrix comprising Fe, Al and Cr was reinforced with TiC nanoparticles and tested at 800°C. The findings have shown excellent wear resistance in the resulting composite due to the presence of TiC particles (Zhang et al., 2013).

Solid lubricants along with other metals/ceramics have also been introduced in Fe-based materials. Fe-Cr-based materials have been reinforced with 8.5 wt. % Mo, 8.0 wt. % BaF_2 and 10.0 wt. % Ag using P/M. The improvement in wear and friction was due to (1) the formation of a high-temperature lubricant, $BaMoO_4$, and (2) solid solution strengthening achieved due to the presence of molybdenum (Cui et al., 2014). Some of the commonly used iron-based matrices are Fe-C, Fe-C-Ni, Fe-Cu-C, Fe-Cu and Fe-Mo. Graphite is being reinforced in Fe-based materials as it causes a change in the microstructure, which makes changes in the properties. With an increase in graphite content from 0.5% to 2% in iron-based matrices, changes in various phases can be achieved (Vadiraj et al., 2011; Zhang et al., 2012).

Nickel and graphite were reinforced in Fe matrix via P/M. The objective was to enhance wear and friction properties. The finally developed alloy (Fe-C-Ni) was then subjected to various heat treatments such as annealing, tempering and quenching. As a result, the occurrence of various hard phases improved wear resistance (Tekeli & Gural, 2007). The incorporation of talc and graphite solid lubricants in the base matrix comprising Fe-Cu-Sn was achieved through the P/M route. The change in microstructure of the developed composites was achieved through high-temperature sintering, and the findings have shown improved tribological properties as compared to wrought bronze alloys (Mamedov, 2004). Iron has been used with graphene as reinforcement due to the high strength and easy shearing capability of graphene. Researchers have found graphene to improve tribological properties of the iron-based matrices (Šuštaršič et al., 2001). The incorporation of MoS_2 into the base matrix comprising Fe-C-Cu increased the density of the developed composites. On adding 3 wt. % MOS_2 to Fe-Cu-C, the wear and frictional properties were found to increase, and high wt. % of MoS_2 resulted in increased wear and friction (Dhanasekaran & Gnanamoorthy, 2007a and b).

Fu et al. (2011) reinforced MoS_2 in different weight percentages in Fe-Cu matrix using induction hot pressing. The results have shown formation of solid solution of Fe-Mo at 8 wt. % MoS_2 resulting in the improvement of mechanical and tribological properties. Apart from this, MoS_2 has also been associated with Fe-Cu-C-Ni and 316 stainless steel base matrices, and the results have shown enhancement in the tribological properties, especially at room temperature (Dhanasekaran & Gnanamoorthy, 2007a and b) (Mahathanabodee et al., 2014). Solid lubricants such as MnS, h-BN, WS_2 and CaF_2 have been tried with Fe matrices for room- and high-temperature applications. Solid lubricants WS_2 and CaF_2 in combination were used with steel and titanium nitrate bases. The lubrication was observed up to a temperature of 500°C, and film of $CaSO_4$ was found to improve the tribological properties (John et al., 1998). Fluorides (CaF_2) have been used with Fe-Cu-C and investigated for wear and friction at room and high temperatures. The authors have also investigated the sustainability of the developed composites at high speeds. The findings have revealed that the composite Fe-Cu-C-CaF_2 can be used for tribological applications at room and high temperatures, and energy dissipated in friction can be saved (Sharma & Anand, 2017, 2018; Anand & Sharma, 2017). h-BN has been tried with Fe, steel, etc. The mechanical properties of Fe-h-BN composites were found to increase. The friction and wear as observed in 316 steel–h-BN composites decreased at high weight percentages of h-BN (Mahathanabodee et al., 2013; da Costa Gonçalves et al., 2014).

1.2.3 COPPER-BASED MATERIALS

Copper metal matrix composites have been categorized as the most promising material for engineering-related applications, prominently the applications where high temperature and microstructural stability are required. The sustainability in copper-based materials has mainly resulted with the use of ceramic, metal and solid lubricant reinforcements (Jamwal et al., 2020).

Copper has been used widely for different tribological applications (Pleskachevsky et al., 1997; Kang, 1993; Kim et al., 1988). The majority of the copper-based materials have been developed through the P/M route. Copper has been associated with different metals to form various alloys. These alloys further are reinforced with solid lubricants, and thus, copper-based self-lubricating composites are developed. The major limitations of copper and its alloys, as observed by Ziyuan and Deqing (2000), are the low strength and less wear resistance. Due to this, several reinforcements had been tried with copper-based matrices.

Graphite is considered to be the most substantial solid lubricating component in Cu-based composites, primarily due to its high-temperature lubricating capabilities and thermal conductivity (Ma et al., 2009).

Chen et al. (2003) discovered that the addition of graphite to copper greatly affects the sintering capabilities and friction behaviour of the resulting composites. The effect of an increase in the addition of graphite content to the Cu-Fe base matrix has been investigated by the researchers. A composite was developed with Cu-10Ni-3Sn-3Pb-Y_2O_3 as base matrix and graphite as reinforcement. The processing technique used was P/M. The findings have revealed that the addition of 2 wt. % graphite possesses better self-lubricating properties. These composites with a low coefficient of friction and high strength are suitable for tribological applications under a dry friction condition (Suiyuan et al., 2011). Graphite in different weight percentages was reinforced in Cu-Ni-SiC-Y_2O_3-MoS_2 matrix using P/M, and it was observed that raising the solid lubricant content resulted in improved mechanical and tribological properties (Chen et al., 2010).

In order to increase the bonding between copper and the solid lubricant, Kato et al. (2003) coated graphite and MoS_2 particles with copper and reinforced the same in Fe-10 wt. % Sn base matrix. The results have shown enhanced frictional properties in the graphite-reinforced composite as compared to the MoS_2-reinforced composite. Cu-10 Sn, a material used in bearing manufacturing, was tested for wear and friction in vacuum and atmospheric conditions. The findings had evidently showed that the coefficient of friction increased in vacuum as compared to that in atmospheric conditions. However, lower wear loss was shown by samples tested in vacuum condition (Küçükömeroğlu et al., 2008).

Bronze was reinforced with uncoated and nickel-coated graphite using P/M. The developed composites were tested for tribological properties, and a decreased wear rate was observed in nickel-coated composites with a high weight percentage of nickel (Cui et al., 2012). The frictional behaviour of copper–graphite composites has been investigated at high speeds by Ma and Lu (2011). At such speeds, formation of lubricating film was observed, thus making these composites suitable for high-speed applications. In order to explore possibilities of improving wear and friction, Cu has been associated with carbides. SiC was reinforced in Cu using friction stir processing. The distribution of SiC particles was maintained through a net of holes drilled on the surface of pure Cu sheet. The investigation revealed enhanced wear resistance in the developed material (Akramifard et al., 2014). The addition of Al_2O_3 microparticles in Cu did not produce any favourable results; however, the addition of Al_2O_3 nanoparticles resulted in decreased friction and wear (Vencl et al., 2014). The material suitable for self-lubricating machine was developed with Cu-Ni-Sn-Pb as base matrix strengthened by incorporating the matrix with W and nano-Al_2O_3. MoS_2 along with graphite was used in different weight percentages in the base matrix (Chen et al., 2007). This paper presents a facile and effective method for preparing $Ni/NbSe_2$ composites in order to improve the wettability of $NbSe_2$ and copper matrix, which is helpful in enhancing the friction-reducing and anti-wear properties of copper-based composites. The P/M technique was used to fabricate copper-based composites with different weight fractions of $Ni/NbSe_2$, and tribological properties of composites were evaluated by using a ball-on-disc friction-and-wear tester. Results indicated that tribological properties of copper-based composites were improved by the addition of $Ni/NbSe_2$. In particular, copper-based composites containing 15 wt. % $Ni/NbSe_2$ showed the lowest friction coefficient (0.16) and wear rate (4.1×10^{-5} mm³/N/m) amongst all composites. Wear and friction of $Cu-NbSe_2$ composites improve with the improvement in wettability. Zhang et al. (2019) improved the wettability of copper and $NbSe_2$ by reinforcing $Ni/NbSe_2$ in copper-based composite, and thus, low friction and wear rates were achieved.

1.2.4 MAGNESIUM-BASED MATERIALS

The environmental consciousness amongst the customers, stringent environmental regulations, growing pollution levels and depleting energy resources have forced the automotive industry to look for lightweight and sustainable materials. According to an estimate, a 10% reduction in the weight of the vehicle can improve the fuel efficiency by around 7% (Kumar & Sasanka, 2017). In other words, around 20 kg of CO_2 emission is reduced corresponding to every 1 kg weight reduction. The weight reduction can be achieved best by using sustainable and lightweight materials such as magnesium.

Owing to its low density, better mechanical strength, good resistance to corrosion and wear and comparatively lower thermal expansion coefficient, magnesium-based composites have evolved as a new class of materials with numerous applications in industries in general and the automotive industry in particular (Avedesian & Baker, 1999). In addition to these benefits, magnesium also offers excellent castability and damping capacity. The high strength-to-weight ratio of magnesium makes it a potential candidate for automotive parts where lower weight is preferred. Further, the low melting point of magnesium makes liquid metallurgy route the most preferred route for the synthesis of magnesium-based composites. Nevertheless, the P/M route has also been used for the fabrication of magnesium-based alloys as P/M offers the advantage to add higher contents of reinforcement which is otherwise difficult by liquid metallurgy routes (Kumar & Sasanka, 2017). Most recently, friction stir processing has been used for the development of surface magnesium-based composites, wherein the reinforcement is added at the surface rather than the bulk (Sunil et al., 2016).

A limitation of magnesium-based composites is their low corrosion resistance and wear resistance in comparison with aluminium alloys, which restricts their applications (Song et al., 2004; Chen & Alpas, 2000). Purohit et al. (2018) developed a magnesium-based composite with SiC as the reinforcement by using the P/M method and have concluded that the mechanical properties such as density, hardness and tensile strength improve by around 30% due to the addition of the ceramic material; however, there was a slight increase in the porosity of the developed composites.

In a recent study, Abdo et al. (2020) have tried ceramic nanofiller (TiO_2) and carbon nanofibers as reinforcement in magnesium metal matrix. The study revealed that the ultimate compression strength as well as hardness improved with an addition of the TiO_2, whereas the addition of carbon nanofibers in Mg matrix led to degradation of the mechanical properties. Omrani et al. (2016) in their work related to tribology of magnesium self-lubricating materials have reported that when 0.5 wt. % CNTs were added to the magnesium, the COF and wear rate decrease considerably.

In a study by Zhang et al. (2008), the authors studied the friction and wear behaviour of AZ91D-0.8% Ce by adding graphite solid lubricant of different sizes and Al_2O_3 as a ceramic reinforcement. The authors observed that the wear loss has an inverse relationship with the size of the graphite particles. Moreover, a different trend and distinct wear mechanisms were observed at different loads. Lim et al. (2005) developed Mg-based composites with Al_2O_3 nanoparticles (50 nm, vol. % of 0.22, 0.66 and 1.11%). The study showed that the wear rate of the developed nanocomposites showed a decreasing trend. This behaviour was ascribed to the increased hardness and strength of the materials with the increased content of the Al_2O_3 nanoparticles.

From the above studies, it can be summarized that the friction and wear of the materials can be tailored by adding a hard ceramic reinforcement and a solid lubricant. The various parameters that influence the friction and wear include properties such as hardness, strength, contact pressure and speed of sliding. Further, the protective layer formation during the sliding plays a vital role in determining the friction and wear properties of these materials. The challenge is to optimize the various parameters and develop materials suitable for a variety of applications.

1.3 CONCLUSIONS

The growing concern for greener technologies has forced the materials scientists to explore materials that are light in weight, cost-efficient and easy to manufacture. Friction and wear being very common in engineering applications, the materials in addition to being light in weight are expected to have possess better tribological performance. The work summarized in this chapter related to lightweight and self-lubricating materials particularly with aluminium, copper, iron and magnesium as the base matrices suggests that there is still tremendous scope to explore new reinforcing agents. Further, these materials can be extensively tried in various engineering applications; however, the material development is not yet application specific. Moreover, the cost-effectiveness of these materials remains an issue particularly in the case of materials with nano-reinforcements. The application domain of these materials can be widened by working on economic aspects of these materials.

REFERENCES

Abboud, J. H., Benyounis, K. Y., Olabi, A. G., & Hashmi, M. S. J. (2007). Laser surface treatments of iron-based substrates for automotive application. *Journal of Materials Processing Technology*, *182*(1–3), 427–431.

Abdo, H. S., Khalil, K. A., El-Rayes, M. M., Marzouk, W. W., Hashem, A. F. M., & Abdel-Jaber, G. T. (2020). Ceramic nanofibers versus carbon nanofibers as a reinforcement for magnesium metal matrix to improve the mechanical properties. *Journal of King Saud University-Engineering Sciences*, *32*(5), 346–350.

Aghili, S. E., Enayati, M. H., & Karimzadeh, F. (2014). Synthesis of (Fe, Cr) $3Al$–Al_2O_3 nano-composite through mechanochemical combustion reaction induced by ball milling of Cr, Al and Fe_2O_3 powders. *Advanced Powder Technology*, *25*(1), 408–414.

Akramifard, H. R., Shamanian, M., Sabbaghian, M., & Esmailzadeh, M. (2014). Microstructure and mechanical properties of Cu/SiC metal matrix composite fabricated via friction stir processing. *Materials & Design (1980–2015)*, *54*, 838–844.

Anand, A., Haq, M. I. U., Vohra, K., Raina, A., & Wani, M. F. (2017). Role of green tribology in sustainability of mechanical systems: A state of the art survey. *Materials Today: Proceedings*, *4*(2), 3659–3665.

Anand, A., & Sharma, S. M. (2017). High temperature friction and wear characteristics of Fe–Cu–C based self-lubricating material. *Transactions of the Indian Institute of Metals*, *70*(10), 2641–2650.

Avedesian, M. M., & Baker, H. (Eds.). (1999). *ASM Specialty Handbook: Magnesium and Magnesium Alloys*. ASM International.

Aziz, R., Haq, M. I. U., & Raina, A. (2020). Effect of surface texturing on friction behaviour of 3D printed polylactic acid (PLA). *Polymer Testing*, *85*, 106434.

Baba, Z. U., Shafi, W. K., Haq, M. I. U., & Raina, A. (2019). Towards sustainable automobiles-advancements and challenges. *Progress in Industrial Ecology, an International Journal*, *13*(4), 315–331.

Bains, P. S., Sidhu, S. S., & Payal, H. S. (2016). Fabrication and machining of metal matrix composites: a review. *Materials and Manufacturing Processes*, *31*(5), 553–573.

Bodunrin, M. O., Alaneme, K. K., & Chown, L. H. (2015). Aluminium matrix hybrid composites: A review of reinforcement philosophies; mechanical, corrosion and tribological characteristics. *Journal of materials research and technology*, *4*(4), 434–445.

Callister, W. D., & Rethwisch, D. G. (2007). *Materials Science and Engineering: An Introduction* (Vol. 7, pp. 665–715). New York: John Wiley & Sons.

Chen, H., & Alpas, A. T. (2000). Sliding wear map for the magnesium alloy Mg-9Al-0.9 Zn (AZ91). *Wear*, *246*(1–2), 106–116.

Chen, S. Z., Lin, J. H. C., & Ju, C. P. (2003). Effect of graphite content on the tribological behavior of a Cu-Fe-C based friction material sliding against FC30 cast iron. *Materials Transactions*, *44*(6), 1225–1230.

Chen, S. Y., Liu, Y. J., Liang, J., & Liu, C. S. (2010). Structure and properties of Cu-based self-lubricating composite with high graphite content [J]. *Journal of Northeastern University (Natural Science)*, *31*(9), 1283–1287.

Chen, S. Y., Xing, R. R., Liu, C. S., & Zhang, S. D. (2007). Preparation of a new Cu-based self-lubricating composite. *Journal-Northeastern University Natural Science*, *28*(9), 1285.

Cui, G., Liu, Y., Gao, G., Liu, H., & Kou, Z. (2020). Microstructure and high-temperature wear performance of FeCr matrix self-lubricating composites from room temperature to 800°C. *Materials*, *13*(1), 51.

Cui, G., Lu, L., Wu, J., Liu, Y., & Gao, G. (2014). Microstructure and tribological properties of Fe–Cr matrix self-lubricating composites against Si_3N_4 at high temperature. *Journal of Alloys and Compounds*, *611*, 235–242.

Cui, G., Niu, M., Zhu, S., Yang, J., & Bi, Q. (2012). Dry-sliding tribological properties of bronze–graphite composites. *Tribology Letters*, *48*(2), 111–122.

da Costa Gonçalves, P., Furlan, K. P., Hammes, G., Binder, C., Binder, R., de Mello, J. D. B., & Klein, A. N. (2014) Self-lubricating sintered composites with hexagonal boron nitride and graphite mixtures as solid lubricants.

Dhanasekaran, S., & Gnanamoorthy, R. (2007a). Dry sliding friction and wear characteristics of Fe-C-Cu alloy containing molybdenum di sulphide. *Materials & Design*, *28*(4), 1135–1141.

Dhanasekaran, S., & Gnanamoorthy, R. (2007b). Microstructure, strength and tribological behavior of Fe-C-Cu-Ni sintered steels prepared with MoS_2 addition. *Journal of Materials Science*, *42*(12), 4659–4666.

Enomoto, Y., & Yamamoto, T. (1998). New materials in automotive tribology. *Tribology Letters*, *5*(1), 13–24.

Erdoğan, A. (2019). Investigation of high temperature dry sliding behavior of borided H13 hot work tool steel with nanoboron powder. *Surface and Coatings Technology*, *357*, 886–895.

Fu, C. Q., Sun, J. C., & Wang, Z. (2011). Tribological properties of $Fe-Cu-MoS_2$ and self-lubricating behaviors. In *Advanced Materials Research* (Vol. 268, pp. 389–394). Trans Tech Publications Ltd.

Ghassemieh, E. (2011). Materials in automotive application, state of the art and prospects. *New Trends and Developments in Automotive Industry*, *20*, 365–394.

Gupta, A., Mohan, S., Anand, A., Haq, M. I. U., Raina, A., Kumar, R., ... Kamal, M. (2019). Tribological behaviour of Fe-C-Ni self-lubricating composites with WS_2 solid lubricant. *Materials Research Express*, *6*(12), 126507.

Haq, M. I. U., & Anand, A. (2018a). Dry sliding friction and wear behavior of $AA7075-Si_3N_4$ composite. *Silicon*, *10*(5), 1819–1829.

Haq, M. I. U., & Anand, A. (2018b). Dry sliding friction and wear behaviour of hybrid $AA7075/Si_3N_4$/Gr self-lubricating composites. *Materials Research Express*, *5*(6), 066544.

Haq, M. I. U., Raina, A., Vohra, K., Kumar, R., & Anand, A. (2018). An assessment of tribological characteristics of different materials under sea water environment. *Materials Today: Proceedings*, *5*(2), 3602–3609.

Hernandez, S., Hardell, J., Courbon, C., Winkelmann, H., & Prakash, B. (2014). High temperature friction and wear mechanism map for tool steel and boron steel tribopair. *Tribology-Materials, Surfaces & Interfaces*, *8*(2), 74–84.

Holmberg, K., Andersson, P., & Erdemir, A. (2012). Global energy consumption due to friction in passenger cars. *Tribology International*, *47*, 221–234.

Holmberg, K., & Erdemir, A. (2015). Global impact of friction on energy consumption, economy and environment. *FME Trans*, *43*(3), 181–5.

Holmberg, K., Kivikytö-Reponen, P., Härkisaari, P., Valtonen, K., & Erdemir, A. (2017). Global energy consumption due to friction and wear in the mining industry. *Tribology International*, *115*, 116–139.

Holmberg, K., Matthews, A., & Ronkainen, H. (1998). Coatings tribology—contact mechanisms and surface design. *Tribology International*, *31*(1–3), 107–120.

Jamwal, A., Mittal, P., Agrawal, R., Gupta, S., Kumar, D., Sadasivuni, K. K., & Gupta, P. (2020). Towards sustainable copper matrix composites: Manufacturing routes with structural, mechanical, electrical and corrosion behaviour. *Journal of Composite Materials*, *54*, 2635–2649.

John, P. J., Prasad, S. V., Voevodin, A. A., & Zabinski, J. S. (1998). Calcium sulfate as a high temperature solid lubricant. *Wear*, *219*(2), 155–161.

Jost, H. P. (2005). Tribology micro & macroeconomics: A road to economic savings. *Tribology & Lubrication Technology*, *61*(10), 18.

Kang, S. (1993). A study of friction and wear characteristics of copper-and iron-based sintered materials. *Wear*, *162*, 1123–1128.

Kato, H., Takama, M., Washida, K., Sasaki, Y., & Miyashita, S. (2003). Mechanical and wear properties of sintered Cu-Sn composites containing copper-coated solid lubricant powders. *Journal of the Japan Society of Powder and Powder Metallurgy, 50*(11), 968–972.

Kerni, L., Raina, A., & Haq, M. I. U. (2019). Friction and wear performance of olive oil containing nanoparticles in boundary and mixed lubrication regimes. *Wear, 426*, 819–827.

Kim, J. W., Kang, B. S., Kang, S. S., & Kang, S. J. L. (1988). Effect of sintering temperature and pressure on sintered and friction properties of a Cu base friction material. *Powder Metallurgy, 20*(3), 32–34.

Kim, J. K., Kim, G. M., & Kim, S. J. (2001). The effect of manganese on the strain-induced martensitic transformation and high temperature wear resistance of Fe-20Cr-1C-1Si hardfacing alloy. *Journal of Nuclear Materials, 289*(3), 263–269.

Kovalchenko, A., Ajayi, O., Erdemir, A., Fenske, G., & Etsion, I. (2005). The effect of laser surface texturing on transitions in lubrication regimes during unidirectional sliding contact. *Tribology International, 38*(3), 219–225.

Küçükömeroğlu, T., Pürçek, G., Saray, O., & Kara, L. (2008). Investigation of friction and wear behaviours of CuSn10 alloy in vacuum. *Journal of Achievements in Materials and Manufacturing Engineering, 30*(2), 172–176.

Kumar, D. S., & Sasanka, C. T. (2017). Magnesium and its alloys. In *Lightweight and Sustainable Materials for Automotive Applications*, ed. by O. Faruk, J. Tjong, & M. Sain (pp. 329–368). CRC Press, Boca Raton. ISBN 9781498756877.

Lansdown, A. R. (2013). *Lubrication: A Practical Guide to Lubricant Selection.* Elsevier, UK.

Lim, C. Y. H., Leo, D. K., Ang, J. J. S., & Gupta, M. (2005). Wear of magnesium composites reinforced with nano-sized alumina particulates. *Wear, 259*(1–6), 620–625.

Ma, W., & Lu, J. (2011). Effect of sliding speed on surface modification and tribological behavior of copper–graphite composite. *Tribology Letters, 41*(2), 363–370.

Ma, W., Lu, L., Guo, H., Wang, J., Jia, H., Zhang, S., & Lv, J. (2013). Tribological behavior of Fe-Mo-graphite and Fe-Mo-Ni-graphite composites at elevated temperature. *Tribology, 9*, 475–480.

Ma, W., Lu, J., & Wang, B. (2009). Sliding friction and wear of Cu–graphite against 2024, AZ91D and Ti$_6$Al$_4$V at different speeds. *Wear, 266*(11–12), 1072–1081.

Mahathanabodee, S., Palathai, T., Raadnui, S., Tongsri, R., & Sombatsompop, N. (2013). Effects of hexagonal boron nitride and sintering temperature on mechanical and tribological properties of SS316L/h-BN composites. *Materials & Design, 46*, 588–597.

Mahathanabodee, S., Palathai, T., Raadnui, S., Tongsri, R., & Sombatsompop, N. (2014). Dry sliding wear behavior of SS316L composites containing h-BN and MoS$_2$ solid lubricants. *Wear, 316*(1–2), 37–48.

Mamedov, V. (2004). Microstructure and mechanical properties of PM Fe-Cu-Sn alloys containing solid lubricants. *Powder metallurgy, 47*(2), 173–179.

Modi, O. P., Prasad, B. K., Jha, A. K., Deshmukh, V. P., & Shah, A. K. (2004). Effects of material composition and microstructural features on dry sliding wear behaviour of Fe-TiC composite and a cobalt-based stellite. *Tribology Letters, 17*(2), 129–138.

Moghadam, A. D., Omrani, E., Menezes, P. L., & Rohatgi, P. K. (2015). Mechanical and tribological properties of self-lubricating metal matrix nanocomposites reinforced by carbon nanotubes (CNTs) and graphene–a review. *Composites Part B: Engineering, 77*, 402–420.

Omrani, E., Moghadam, A. D., Menezes, P. L., & Rohatgi, P. K. (2016). New emerging self-lubricating metal matrix composites for tribological applications. In *Ecotribology* (pp. 63–103). Springer, Cham.

Pleskachevsky, Y. M., Kovtun, V. A., & Kirpichenko, Y. E. (1997). Tribological coatings formed by electrocontact sintering of powder systems. *Wear, 203*, 679–684.

Potecaşu, F., Marin, M., Potecaşu, O., Tamara, R. A. D. U., & Istrate, G. (2015). Influence of alloying elements on corrosion resistance of some iron-based sintered P/M Alloys.

The Annals of "Dunarea de Jos" University of Galati. Fascicle IX, Metallurgy and Materials Science, 38(1), 15–18.

Purohit, R., Dewang, Y., Rana, R. S., Koli, D., & Dwivedi, S. (2018). Fabrication of magnesium matrix composites using powder metallurgy process and testing of properties. *Materials Today: Proceedings, 5*(2), 6009–6017.

Raina, A., & Anand, A. (2017). Tribological investigation of diamond nanoparticles for steel/steel contacts in boundary lubrication regime. *Applied nanoscience, 7*(7), 371–388.

Raina, A., & Anand, A. (2018a). Lubrication performance of synthetic oil mixed with diamond nanoparticles: effect of concentration. *Materials Today: Proceedings, 5*(9), 20588–20594.

Raina, A., & Anand, A. (2018b). Effect of nanodiamond on friction and wear behaviour of metal dichalcogenides in synthetic oil. *Applied Nanoscience, 8*(4), 581–591.

Raina, A., & Anand, A. (2018c). Influence of surface roughness and nanoparticles concentration on the friction and wear characteristics of PAO base oil. *Materials Research Express, 5*(9), 095018.

Santos, M. C., Machado, A. R., Sales, W. F., Barrozo, M. A., & Ezugwu, E. O. (2016). Machining of aluminium alloys: a review. *The International Journal of Advanced Manufacturing Technology, 86*(9–12), 3067–3080.

Scharf, T. W., & Prasad, S. V. (2013). Solid lubricants: a review. *Journal of Materials Science, 48*(2), 511–531.

Shafi, W. K., Raina, A., & Haq, M. I. U. (2018a). Friction and wear characteristics of vegetable oils using nanoparticles for sustainable lubrication. *Tribology-Materials, Surfaces & Interfaces, 12*(1), 27–43.

Shafi, W. K., Raina, A., & Haq, M. I. U. (2018b). Tribological performance of avocado oil containing copper nanoparticles in mixed and boundary lubrication regime. *Industrial Lubrication and Tribology, 70*(5), 865–871.

Shafi, W. K., Raina, A., & Haq, M. I. U. (2019). Performance evaluation of hazelnut oil with copper nanoparticles-a new entrant for sustainable lubrication. *Industrial Lubrication and Tribology, 71*(6), 749–757.

Shafi, W. K., Raina, A., Haq, M. I. U., & Khajuria, A. (2018c, Feb). Interdisciplinary aspects of tribology. *International Research Journal of Engineering and Technology (IRJET), 5*(02), e-ISSN: 2395-0056.

Shafi, W. K., Raina, A., Haq, M. I. U., & Khajuria, A. (2018d). Applications of industrial tribology. *International Research Journal of Engineering and Technology, 5*(1), 1285–1289.

Sharma, S. M., & Anand, A. (2016). Solid lubrication in iron based materials--A review. *Tribology in Industry, 38*(3).

Sharma, S. M., & Anand, A. (2017). Friction and wear behaviour of Fe-Cu-C based self lubricating material with CaF_2 as solid lubricant. *Industrial Lubrication and Tribology, 69*(5), 715–722.

Sharma, S. M., & Anand, A. (2018). Effect of speed on the tribological behavior of Fe-Cu-C based self lubricating composite. *Transactions of the Indian Institute of Metals, 71*(4), 883–891.

Singh, H., Haq, M. I. U., & Raina, A. (2019). Dry sliding friction and wear behaviour of $AA6082$-TiB_2 in situ composites. *Silicon*, 1–11.

Singh, N., Mir, I. U. H., Raina, A., Anand, A., Kumar, V., & Sharma, S. M. (2018a). Synthesis and tribological investigation of Al-SiC based nano hybrid composite. *Alexandria Engineering Journal, 57*(3), 1323–1330.

Singh, H., Raina, A., & Haq, M. I. U. (2018b). Effect of TiB_2 on mechanical and tribological properties of aluminium alloys–a review. *Materials Today: Proceedings, 5*(9), 17982–17988.

Smith, R. A. (2008). Enabling technologies for demand management: Transport. *Energy Policy, 36*(12), 4444–4448.

Song, G., Bowles, A. L., & StJohn, D. H. (2004). Corrosion resistance of aged die cast magnesium alloy AZ91D. *Materials Science and Engineering: A, 366*(1), 74–86.

Suiyuan, C., Jing, W., Yijie, L., Jing, L., & Changsheng, L. (2011). Synthesis of new Cu-based self-lubricating composites with great mechanical properties. *Journal of Composite Materials, 45*(1), 51–63.

Sunil, B. R., Reddy, G. P. K., Patle, H., & Dumpala, R. (2016). Magnesium based surface metal matrix composites by friction stir processing. *Journal of Magnesium and Alloys, 4*(1), 52–61.

Šuštaršič, B., Kosec, L., Jenko, M., & Leskovšek, V. (2001). Vacuum sintering of water-atomised HSS powders with MoS_2 additions. *Vacuum, 61*(2–4), 471–477.

Tekeli, S. & Güral, A. (2007). Dry sliding wear behaviour of heat treated iron based powder metallurgy steels with 0.3% Graphite+ 2% Ni additions. *Materials & Design, 28*(6), 1923–1927.

Tung, S. C., & McMillan, M. L. (2004). Automotive tribology overview of current advances and challenges for the future. *Tribology International, 37*(7), 517–536.

Vadiraj, A., Kamaraj, M., & Sreenivasan, V. S. (2011). Wear and friction behavior of alloyed gray cast iron with solid lubricants under boundary lubrication. *Tribology International, 44*(10), 1168–1173.

Vencl, A., Rajkovic, V., & Zivic, F. (2014). Friction and wear properties of copper-based composites reinforced with micro-and nano-sized Al_2O_3 particles. In *Proceedings of the 8th International Conference on Tribology* (pp. 357–364).

Wakuda, M., Yamauchi, Y., Kanzaki, S., & Yasuda, Y. (2003). Effect of surface texturing on friction reduction between ceramic and steel materials under lubricated sliding contact. *Wear, 254*(3–4), 356–363.

Zhang, F. X., Chu, Y. Q., & Li, C. S. (2019). Fabrication and tribological properties of copper matrix solid self-lubricant composites reinforced with $Ni/NbSe_2$ composites. *Materials, 12*(11), 1854.

Zhang, M. J., Liu, Y. B., Yang, X. H., Jian, A. N., & Luo, K. S. (2008). Effect of graphite particle size on wear property of graphite and Al_2O_3 reinforced AZ91D-0.8% Ce composites. *Transactions of Nonferrous Metals Society of China, 18*, s273–s277.

Zhang, X., Ma, J., Fu, L., Zhu, S., Li, F., Yang, J., & Liu, W. (2013). High temperature wear resistance of Fe-28Al-5Cr alloy and its composites reinforced by TiC. *Tribology International, 61*, 48–55.

Zhang, X., Ma, F., Ma, K., & Li, X. (2012). Effects of graphite content and temperature on microstructure and mechanical properties of Iron-based powder metallurgy parts. *Journal of Materials Science Research, 1*(4), 48.

Ziyuan, S., & Deqing, W. (2000). Surface dispersion hardening Cu matrix alloy. *Applied Surface Science, 167*(1–2), 107–112.

2 Tribology of Lightweight Materials

V. Anandakrishnan and S. Sathish
National Institute of Technology Tiruchirappalli

Manoj Gupta
National University of Singapore (NUS)

CONTENTS

2.1   INTRODUCTION

Tribology is the term introduced by Jost in 1966, which is derived from the Greek word *tribos* that denotes rubbing. Tribology refers to the study of surfaces that are subjected to relative motion, and majorly, it focuses on the wear, friction and lubrication. The tribology is not limited to the industrial applications, but it also blends with our daily life. Here, the tribology of lightweight materials focuses on the wear and friction behaviour of most fascinating materials in industries that are based on magnesium and aluminium. The low strength and poor wear resistance of these lightweight materials limit their applications. The enhancement of friction and wear resistance of these materials will undoubtedly unlock their applications in various industrial sectors. The present chapter summarizes the earlier attempts on the improvement of tribology of the most commonly used lightweight materials, especially those that are based on magnesium and aluminium.

2.2   TIMELINE OF TRIBOLOGY – LIGHTWEIGHT MATERIALS

In 1880 BC, the attention towards the reduction of friction was focused on the cartwheels through the practice of using a liquid lubricant (i.e. water). Figure 2.1 shows the timeline of tribology that placed a milestone in the concepts of tribology (Bhushan, 2013). Further research endeavours led to several milestones in the history of tribology, and the application of lightweight materials in the automotive,

FIGURE 2.1   Milestones in tribology.

railway and aircraft industries opened the importance of tribological research on lightweight materials. The 20th century was the time when the lightweight materials were employed in the transportation sector that triggered remarkable research progress in the tribology of lightweight materials.

2.3   STANDARD WEAR TESTING METHODS AND MEASURING METHODS

2.3.1   STANDARD SLIDING WEAR TEST METHODS

The tribological applications of lightweight materials are most targeted on the sliding contacts. Therefore, it is required to perform a standard test method to understand the sliding wear behaviour of materials. The objective of the test experimentation is to understand the fundamentals of wear behaviour, and the effects of wear parameters on it. The above understanding will certainly help in the material selection and characterization and the usage of lubrication for a specific application. During various applications, the materials are subjected to different modes of contact that necessitated different methods of wear testing. The wear testing methods are grouped depending on the point of contact during the wear tests, and either the sample (pin) or the counterpart (disc) is driven to simulate the relative motion. The relative motion among the system may be in the form of sliding, rolling, spin and impact. Figure 2.2 illustrates the scheme of contact of the specimen in different sliding wear testing methods (Bhushan, 2013).

Four-ball – In the four-ball test method, the balls are kept in an equilateral tetrahedron position, three balls are held stationary, and one ball rotates against three balls to ensure the relative motion under a specified load (Figure 2.2a). The four-ball method is often used to understand the wear behaviour of the material under the lubricated condition and the performance of lubricants. The wear loss is recorded with the measurements of variations/surface scar in the diameter of the ball.

Crossed cylinder – In the crossed cylinder test method, two cylinders are positioned perpendicular to each other, and one will be stationary and the other will rotate (Figure 2.2b). The rotating cylinder will be a solid object, whereas the stationary cylinder may be either a solid or hollow cylinder to assist the cooling.

Pin-on-disc – In the pin-on-disc test method, the pin is positioned perpendicular to the disc and kept stationary, whereas the disc will rotate against the pin (Figure 2.2c). The pin may be in the form of a cylinder or a prism. The ends of a cylinder may be flat or hemispherical and the ends of a prism may be rectangular or square. The pin-on-disc test method is the most often used test method to examine the wear performance of materials subjected to tribological applications.

Flat-on-cylinder – In the flat-on-cylinder test method, the cylinder is rotated against the stationary rectangular flats (Figure 2.2d). One or two rectangular flats may be used. In some cases, V-shaped blocks and two bearing shells are used instead of flats.

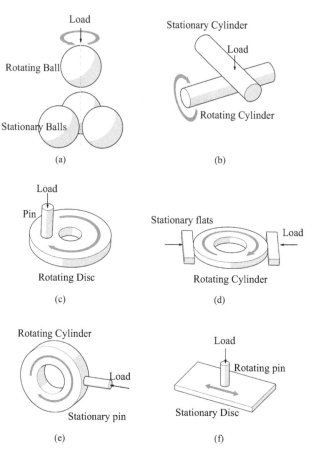

FIGURE 2.2 Illustration of the scheme of specimen contact in different sliding wear test methods. (a) Four-ball, (b) crossed cylinder, (c) pin-on-disc, (d) flat-on-cylinder, (e) pin-on-cylinder and (f) pin-on-flat.

Pin-on-cylinder – The pin-on-cylinder test method is found to be similar to the pin-on-disc method, and the only difference is that the loading of pin is perpendicular to the rotational axis (Figure 2.2e).

Pin-on-flat – In the pin-on-flat test method, a stationary pin is kept against the reciprocating flat disc (Figure 2.2f). In certain conditions, a ball or pin of a hemispherical tip may be used, and the pin may be made as a reciprocating one.

2.3.2 STANDARD WEAR MEASURING METHODS

In actual conditions, the wear of materials is observed to be relatively smaller, and the measurement of wear needs to be more precise and accurate. The common representation of wear is in the form of volume loss, weight loss and thickness loss. Though several methods such as weight loss measurement, linear wear measurement, surface profile measurement, grooving method, indentation method and radioactive method are available to measure the amount of wear (Wen and Huang, 2017), the most often used methods are discussed here.

Weight loss measurement – In the weight loss method, the specimen pin is weighed with a high-precision weighing balance and usually an electronic balance of 0.1 mg accuracy is used. The weight of the specimen pin is weighed before and after the wear test to record the amount of wear. As the balance is of high precision, the range of measurement is restricted to only smaller specimens. This method is even suitable when the specimen pin is deformed plastically, whereas the deformed/wiped metals may be accumulated at the pin edge, which results in consequences that need to be taken care.

Linear wear measurement – In the linear wear measurement method, the wear loss is measured with the support of a non-contact linear measurement device. The linear measurement device continuously records the change in specimen height from the initial stage till the end of the experiment. Then, the recorded linear wear is converted to volume loss using the standard mathematical relation. As per ASTM standards, it is advised to observe the change in shape/length of wear track in pin and disc (American Society for Testing and Materials, 2017).

2.4   FACTORS INFLUENCING THE TRIBOLOGY OF LIGHTWEIGHT MATERIALS

Generally, the properties of materials are mostly related to their metallurgical features, and when dealing with composites, it depends on the reinforcement types, morphology of reinforcement and its percentage addition. Apart from the metallurgical factors, the tribology of materials is greatly influenced by the tribological parameters and the operating conditions. The critical parameters that influence the wear performance are load, sliding velocity, sliding distance and temperature. There are not much research attempts on the effect of parameters such as atmospheric conditions, humidity and counterpart material, which shows their limited significance in the selection of wear parameters. Typically, the amount of wear follows an increasing trend with increased load, due to the effect of increased normal force. A similar trend is observed with the increase in sliding distance. As per Archard's equation, the material wear is directly proportional to the applied load and sliding distance (Eq. 2.1). This shows that the increased load and sliding distance directly increase wear as a result of temperature rise. In some instances, it is not in line with the above perception; i.e., material wear loss is reduced with increased load. Likewise, a constant trend, i.e. either an increase or decrease in wear loss, is not observed with sliding velocity too. In some instances, it follows a decreasing trend to a certain velocity and then it increases, and in some instances, it is vice versa. When considering the percentage of reinforcement, it follows a reduction in wear to a specific limit and then it shows an increased wear. Moreover, the observations in wear loss will vary according to the operating conditions; i.e., under the combination of parameters, it will differ. Also, the tribo-characteristics will differ upon the material property that

is subjected to test, and hence, it is mandated to understand the tribological characteristics of developed/advanced materials through experimentations.

$$\text{Volume loss } \alpha \text{ Normal load} * \text{Sliding distance} \qquad (2.1)$$

2.5 TRIBOLOGY OF MAGNESIUM ALLOYS AND COMPOSITES

Magnesium, the most fascinating material discovered in 1775, attracts most industries and researchers owing to its less density of 1.74 g/cm^3 (Gupta and Sharon, 2011).
The limitations in using pure magnesium are its low strength, low stiffness, poor
corrosion and wear resistance. The strength of pure magnesium has been improved
with the insertion of alloying elements and reinforcements. Often the inclusion of
metallic/ceramic elements in pure magnesium and its alloys results in grain refinement, secondary-phase formation and retainment of insoluble metallic particles such
as titanium. This leads to an improvement in material properties such as hardness,
tensile strength, ductility, corrosion and wear resistance. Several research attempts
involving the addition of metallic elements, namely aluminium, titanium, copper,
nickel and molybdenum, have successfully improved the properties of magnesium. The hardness and ductility have a significant influence on the wear behaviour of magnesium-based materials. The increase in material hardness improves the
load-bearing capacity and resistance to penetration and deformation. Likewise, the
increase in ductility indicates the improved ultimate strength of the material, which
directly resists the material fracture. A sudden increase in the ductility of magnesium is seen at 225°C due to the non-basal slip plane activation. It is noted that pure
magnesium will reach the flash temperature of 200°C–500°C at a sliding speed of
3 m/s. The most prominent method in property enhancement is with the addition of
reinforcement to accomplish the composite strengthening. In composite strengthening, the dispersed reinforcement acts as a load-bearing member, as it takes up the
load acting on the contact surface. Thus, the reduced pressure on the contact sliding
surface leads to increased wear resistance.

The addition of the metallic elements aluminium and titanium in the pure magnesium significantly improved the wear resistance owing to the improved ductility and
load-bearing capacity of the Mg-3Al-5.6Ti composite (Sathish et al., 2019b, 2020).
Similarly, the addition of nano-boron carbide in the Mg-3Al-5.6Ti composite further
improved the wear resistance of the magnesium composite (Sathish et al., 2019a).
Some of the most used magnesium alloys are AM50, AM60, AZ63, AZ81, AZ91,
HK31, HZ32, QE22, QH21, WE54, WE43, ZK51, ZK61, ZE41, ZC63 and Elektron
21. The magnesium alloy AZ91 was made through three different conditions, one by
casting and the next by powder metallurgy (P/M) with 1 and 10h of milling (Shanthi
et al., 2007). The three modes of processing induced significant changes in the grain
size, hardness and ductility. The grain size of the cast sample and of the P/M sample
with milling for 1 and 10h is reported as 28.9, 11.4 and 0.6 μm. Likewise, the hardness of the cast composite is reported as 75.4 RH with 12.6% ductility and the hardness of the P/M 10-h-milled composite is 90.2 RH with 2.81% ductility. The above
variations in the material property due to the modifications of processing conditions
proved the significance of microstructural characteristics. The reduction in the grain

size, increase in hardness and reduction in ductility significantly contribute to the wear performance of materials. The wear rate is found to be higher for the P/M 10-h-milled composite and lower for the cast composite. In the worn surface analysis of the cast composite, the material displacement along the sides of the grooves is observed owing to its higher ductility. On the contrary, the composite with lower ductility exhibits a ribbon strip, small fragments and chips due to the material removal through the abrasion mechanism.

The AZ91 magnesium composite with the addition of 100-μm boron carbide particles of 0, 1, 3, 5, 10 and 15 wt. % was made under the cast and extruded conditions to analyse tribological behaviour (Mohammadi et al., 2019). The cast condition showed a higher rate of wear compared with the extruded condition at all weight percentages of boron carbide. The magnesium alloy for the extruded condition exhibited a nearly 18% reduction in wear compared to the as-cast condition, whereas for the Mg-1 wt. % B_4C composite, the extruded condition exhibited a maximum of 37% reduction in wear compared to the as-cast condition. The wear resistance was found improved with the addition of B_4C: the addition of 1 wt. % B_4C contributes a reduction in wear of 31% and 53% for the cast and extruded composites, respectively. For the Mg-1 wt. % B_4C composite, a 46% reduction for friction coefficient was observed for the extruded condition compared to the as-cast condition, whereas for both the as-cast and extruded conditions, the addition of B_4C resulted in an increased friction coefficient.

The wear behaviour of AZ91 magnesium alloy exhibited an improvement of 13% wear reduction due to friction stirring, when compared with that of the base alloy (Ahmadkhaniha et al., 2016). The reason for the increase in wear resistance is the improved hardness by the attainment of refined grains. Further, the addition of alumina particles improved the wear resistance of the magnesium composite to 78% compared to the friction-processed alloy. Also, the coefficient of friction decreased by 3% due to friction stir processing, and further addition of alumina by friction stirring led to a reduction of 46% in friction coefficient.

Similarly, the sliding wear behaviour of equal-channel angular-pressed AZ91 magnesium alloy exhibited a better wear resistance compared to the as-received alloy (Saleh et al., 2020). At low loads, higher wear resistance was observed in the ECAP-processed alloy compared with higher load conditions, i.e. 67% of wear reduction at low loads and 14% of wear reduction at high loads. In the dry sliding of ZE41 magnesium alloy, the increase in the amount of load exhibited an increased wear loss in all the conditions of sliding velocity, i.e. from 0.1 to 1 m/s (López et al., 2011). The increase in the applied load on the pin affects the wear resistance of the magnesium-based materials as it directly increases the pressure acting on the contact surface. This leads to the increase in temperature on the sliding surface that eventually activates the slip planes, resulting in more plastic deformation. The wear rate is observed to be higher at lower velocity and higher loads, which shows the impact of load on the wear rate, whereas the coefficient of friction highly relies on the sliding velocity, that too at a lower velocity.

In general, the increase in sliding velocity results in the formation of a mechanically mixed layer (MML) owing to the mechanical and chemical actions on the sliding surface. Further sliding disintegrates the formed MMLs into oxide particles which act as second-phase particles that improved the wear resistance. The evidence of the formation of an MML can be visualized through the microscopy analysis in the sectional view of the worn surface sample. Such formation of MMLs is observed in the AZ91 magnesium composite reinforced with in situ TiB_2 particles. The MML thickness of 8.7 μm was observed at a low sliding velocity of 0.25 m/s, and the layer thickness was found to be increased to 26.2 μm with an increased sliding velocity of 0.75 m/s (Xiao et al., 2018). This suggests that upon the frictional heat generation, the thickness of the MML will vary, and it has a useful contribution in the wear resistance. Also, the increase in load decreased the frictional coefficient nearly 8%, from 12.5 to 37.5 N.

Further, the morphology, quantity of reinforcement and different treatments also significantly contribute to the wear behaviour of magnesium-based materials. The magnesium alloy AZ91 reinforced with different micron-sized silicon carbide particles at different volume percentages was produced under extruded and solid solution-treated conditions (Deng et al., 2012). The wear studies on the developed magnesium composites displayed how the particle size, quantity of reinforcement and the treatment affect the wear rate of the magnesium composite. The composite in the extruded condition exhibited a lower wear rate compared to the solution-treated composite at all volume percentages and particle sizes. The reason is that in the solution-treated composite, the silicon carbide particles were aggregated along the grain boundaries, whereas in the extruded condition, they were distributed in a homogenous way. This variation in the distribution of particles impacts on the load-bearing component in a uniform nature, and also it induces a transformation in the wear mechanism. Thus, the extruded magnesium composite resulted in a higher wear resistance than the solution-treated composite. When coming to the particle size variation, it showed two different trends for the extruded and solution-treated composites. For the extruded condition, an increase in particle size first showed a reduction in wear rate, and then, it again increased, whereas the trend in change of wear behaviour was found to be inverse due to solution treatment. This difference in the wear rate trend is obviously due to the effect of transition in wear mechanism influenced by the particle size variation. The increase in the quantity of silicon carbide shows an increased wear rate, which is in contrast to the typical statute; i.e., increased reinforcement shows a lesser wear rate. The reason for the contrary is that the increased reinforcement percentage led to the changes in the dominance of wear mechanisms, specifically dominated by abrasion and delamination.

Another significant way adopted to enhance the wear property of magnesium and its alloys is the addition of solid lubricants such as graphite, graphene, molybdenum disulphide and tungsten disulphide. The mechanism behind the adequate protection of solid lubricant during the sliding process can be sequenced as follows: (1) disintegration of solid lubricants, (2) distribution of solid lubricants and (3) uniformly splashed layers of solid lubricants. During sliding, these particles act as an effective medium in resisting the transfer of load, which in turn controlled the wear with enhanced friction coefficient. The addition of graphite at different percentages along with the 8% of alumina particles in the AZ91 magnesium alloy displayed a significant improvement in the wear resistance and frictional coefficient. With the self-lubricating property of Ti_2AlC, the tribology of the magnesium composite is

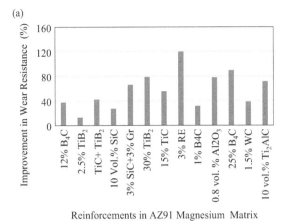

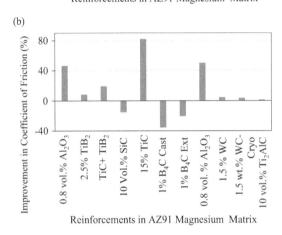

FIGURE 2.3 Tribological behaviour of AZ91 magnesium alloy reinforced with different reinforcements. (a) Improvement in wear resistance and (b) improvement in the coefficient of friction.

enhanced; i.e., a 71% reduction in wear and a 1% reduction in friction coefficient were observed. The improved wear resistance and coefficient of friction of typical AZ91 magnesium alloy with different reinforced conditions are shown in Figure 2.3a and b, respectively.

2.6 TRIBOLOGY OF ALUMINIUM ALLOYS AND COMPOSITES

The extraction of aluminium took a long time, and notably in 1845, its density was reported as 2.7 g/cm³ (Runge and Runge, 2018). Further, the interest in aluminium metal and the requirements in the property enhancement led to several developments in pure aluminium. With the alloying of silicon (4xxx and 6xxx series), copper (2xxx series), magnesium (5xxx, 6xxx and 7xxx series), manganese (3xxx series), zinc (7xxx series), lithium and nickel, a number of series from 1xxx to 9xxx (ASM International, 1990) were established. Though ample research outcomes were attained in aluminium alloys, there are gaps to fulfil the need of newer and even higher performance aluminium alloys. Initially, in the 1950s, the cast iron engine blocks were replaced with the aluminium engine blocks with cast iron liner owing to its lightweight and better thermal conductivity (ASM International, 1990). Its higher production cost unveiled the aluminium-silicon alloy with a considerable amount of wear resistance without any liners. Later, a system of engine block, piston and cylinder bore was made with aluminium alloys. Then, the components such as pistons, transmission components, crankshafts and bearings were made with aluminium alloys. Likewise, the engine block for the aircraft designed by the Wright brothers was made of aluminium alloy with the composition of aluminium and copper of 92% and 8%, respectively. Most noticeable series of aluminium alloys are 2xxx, 3xxx, 5xxx, 6xxx and 7xxx, as alloys grouped in them are used in specific industries, particularly in aircraft, automotive, defence, structural and architecture applications. Based on the addition of alloying elements, and methods using which the alloys were made, the aluminium alloys were strengthened through grain size refinement and second-phase particle formation.

The strengthening of aluminium alloys through the inclusion of reinforcements is referred to as composite strengthening that shows a greater shift in property enhancement (Embury et al., 1989). Further, the exhaustive need in the strengthening of aluminium alloys in the light of tribological aspects focused the research on the development of aluminium composites. Aluminium-based composites were made with the inclusion of particulates, fibres and whiskers, which supports the enhancement of metallurgical, mechanical and tribological properties. The inclusion of reinforcements was categorized as hardener and softener. The primary role of the hardener is to improve the strength of the developed composite, whereas the role of softener is to serve as a solid lubricant. Some of the most commonly used hardener particles are Al_2O_3, AlN, B_4C, SiC, Si_3N_4, SiO_2, TiC, TiN, TiO_2, TiB_2, ZrO_2, ZrB_2 and WC. Likewise, graphite (Gr), graphene, molybdenum disulphide (MoS_2) and tungsten disulphide (WS_2) are the most used ones as a solid lubricant.

Some of the attempts in strengthening 6xxx aluminium alloy with various reinforcements in order to improve the tribological properties are as follows. AA6061 was strengthened with the addition of silicon carbide particulates of up to 6 wt. % in the steps of 2% increment. The experimented sliding wear tests were done with a load of up to 60 N (increment of 10 N) and sliding distance of up to 6,000 m (increment of 1,000 m) at a constant sliding velocity of 2.62 m/s. The experimental results display the increased wear rate with an increase in load and sliding distance, whereas the increase in the percentage of reinforcement resulted in decreased wear rates. The addition of silicon carbide particulates of up to 6 wt. % showed a 67% reduction in wear rate compared with cast 6061 alloys. Similarly, AA6061 was strengthened with boron carbide particulates at 10 wt. % and its sliding wear performance was investigated in dry condition. The strengthened cast composite exhibited an 18% reduction in wear when compared with the cast alloy. The addition of 15 wt. % $MoSi_2$ under different milling conditions (i.e. wet blend method and rotating cube method) for 4 and 10 h strengthened the aluminium 6061 alloys through powder metallurgy. The wear experimentations revealed a significant reduction in wear loss in the produced

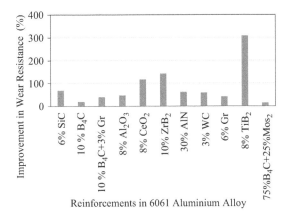

FIGURE 2.4 Improvement in wear resistance of 6061 aluminium alloy with different reinforcements.

composites in all conditions. Much higher wear resistance was observed in the composites produced through the wet blend method under 10 h of milling time. An attempt to strengthen AA6061 alloy was made with 10 wt. % ZrB_2 through the in situ cast method, which exhibited a significant reduction in wear rate. Likewise, the 6061 aluminium alloy was strengthened with the addition of TiC particulates from 3 to 7 wt. %, AlN particulates up to 30%, Al_2O_3 particulates up to 8%, CeO_2 particulates up to 8% and TiB_2 up to 12 vol. %. The strengthened composite exhibited a significant reduction in wear rate compared with base alloy. Likewise, the AA6063 was strengthened by the in situ titanium diboride particles up to 10 wt. %, and the AA6082 strengthened by silicon carbide particles exhibited an enhanced wear resistance compared with the base alloy. Figure 2.4 shows the typically improved wear resistance of AA6061 alloy when reinforced with different reinforcements.

Likewise, the AA7075 strengthened with TiC particles up to 10 wt. %, TiO_2 particles up to 10 wt. %, Si_3N_4 particles up to 10 wt. %, SiC of 10% and Al_2O_3 of 8 wt. % exhibited an enhanced wear resistance compared to the base alloy. In the same way, the AA7050 strengthened with TiC particles up to 10 wt. %, AA5252 strengthened with nano-sized SiC particles up to 7 wt. % and micron-sized SiC particles with 10 wt. %, AA5052 with in situ ZrB_2 particles up to 9 wt. % and AA5059 with SiC particles up to 15 wt. % exhibited an enhanced wear resistance. The possible reason for the increase in wear resistance is the increased strength of the developed composite. The properties of the reinforcements are considerably higher (specifically higher hardness and heat resistance), and the dispersed particles among the matrix act as a barrier during the load transfer. Hence, the composites exhibit higher hardness and strength that directly support the wear resistance. Besides, the main aspect of difference in coefficient of thermal expansion between the matrix aluminium and reinforcements supports the wear resistance. Also, it is well understood that the temperature rise in the tribopairs is unavoidable while sliding. Further, the variation in the thermal expansion coefficient of matrix and reinforcement induced a stress at the interface of matrix and reinforcements due to the formation of dislocation densities at the time of solidification. This induced stress and dislocation densities improved the wear resistance of materials as it resists the plastic deformation.

Another mode of enhancing the wear resistance is the inclusion of solid lubricant in the matrix aluminium alloy. Typical examples of inclusion of solid lubricants in aluminium alloys are as follows: AA6063 composite with the inclusion of 15 wt. % MoS_2 and graphite, AA6063 composite with the inclusion of MoS_2 up to 10 wt. % and AA7050 with 0.3 wt. % graphene, which exhibited the appreciable resistance to wear. Contrary to the above, the addition of graphite particles in AA6061 alloy led to a decrease in wear resistance, and the reason for this unusual behaviour is the increased porosity. The addition of hardener and softener also effectively improves the wear resistance. One typical example, AA6061-10 wt. % B_4C-3 wt. % graphite composite, shows an 18% improvement in wear resistance for B_4C-reinforced composite, and further addition of 3 wt. % graphite shows a 17% increase in wear resistance. This shows a 50% increase in the wear resistance with the addition of graphite. The effect of strengthening through the inclusion of reinforcement shows variations in friction coefficient. The addition of reinforcements didn't show a similar trend in friction coefficient; four different trends were noted from the earlier research works:

   i. continually increasing
  ii. continually decreasing
 iii. increasing to a specific percentage and then decreasing
  iv. decreasing to a specific percentage and then increasing

Figure 2.5 illustrates a typical comparison of the coefficient of friction of aluminium composites containing different reinforcements. The composite strengthening with nano-inclusion/formation exhibited a lesser friction coefficient compared to microparticles. The hard particle reinforcements used in the comparison of the coefficient of friction are in the range of nanometre to 2 μm level. These variations in the coefficient of friction are attributed to the change in particle size, the amount of particle

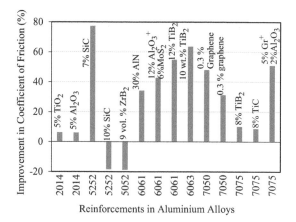

FIGURE 2.5 Variations in the coefficient of friction of different aluminium alloys with various reinforcements.

reinforcement and the load applied. When the particle size is found to be higher, the contact area of hard particles is increased, which results in the increased friction coefficient. It is also observed that there is an increase in friction coefficient even with the nano- and micron-level reinforcements. The reasons for the above observation may be the increased quantity of reinforcements, higher hardness of reinforcements and detachment of particles due to the low flexural strength at the time of sliding. The inclusion of solid lubricants has significantly reduced the friction coefficient, and the reason is the formation of lubrication film that restricts the contact of tribo-surface.

2.7 DOMINANT WEAR MECHANISMS EXHIBITED BY LIGHTWEIGHT MATERIALS

The principles behind the material wear can be understood through the study of wear mechanism. The occurrence of wear may be with an individual or combined action of mechanical, thermal and chemical reactions. The common forms of wear mechanism with the actions mentioned above are abrasion, adhesion, corrosion, oxidation, delamination and fatigue. Through the microscopic analysis on the worn surface and wear debris, the induced wear mechanisms may be identified. Some of the most visible traces in the morphology of worn surface and debris of lightweight materials (in aluminium and magnesium) are listed in Table 2.1. Figure 2.6a and b illustrates the morphology of worn surface and wear debris of the $(Mg-3Al-5.6Ti)2.5B_4C$ composite. Figure 2.6a shows a higher prevalence of deep scratches, grooves and craters, along with traces of white patches. The higher scratches, grooves and white patches put forth the dominance of the abrasion and oxidation mechanisms. Figure 2.6b shows the deformed plastic layers and yet to worn wear debris with perpendicular cracks on either side that leads to crater wear. The above mentioned worn surface characteristics evident the adhesion and delamination mechanisms, respectively.

Further, the wear debris (see Figure 2.7a) corresponding to the worn surface exhibits the ribbon fragments with segregated deformed layers on one side

TABLE 2.1
Morphology of Worn Surface and Wear Debris Observed in Lightweight Materials

Wear Mechanism	Worn Surface Topography	Debris Morphology
Abrasive	Grooves, scratches and ploughs.	Small fragments and ribbon strips.
Delamination	Sheet-like particles, craters/channels and series of fine perpendicular cracks.	Sheet-like flakes and laminates.
Oxidative	White layers.	Fine powder.
Adhesive	Rows of furrows, smearing and plastic deformation, series of detached fragments and extensive material transfer.	Irregular and blocky-shaped and wedge-shaped fragments (plastic shearing of the successive layer) and thin plate debris.
Thermal softening and melting	The extruded layer at the sample edge.	Large sheets/flakes appearing smooth and featureless and irregular lumps.

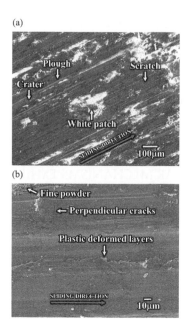

FIGURE 2.6 Worn surface of the (Mg-3Al-5.6Ti)2.5B$_4$C composite. (a) 9.81 N load, 1 m/s sliding velocity, 3000 m sliding distance. (b) 9.81 N load, 3 m/s sliding velocity, 2000 m sliding distance.

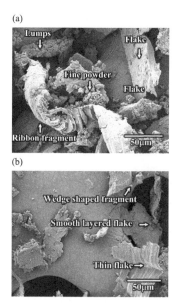

FIGURE 2.7 Wear debris of the (Mg-3Al-5.6Ti)2.5B$_4$C composite. (a) 9.81 N load, 1 m/s sliding velocity, 3000 m sliding distance. (b) 9.81 N load, 3 m/s sliding velocity, 2000 m sliding distance.

TABLE 2.2

Dominant Wear Mechanisms Observed in Aluminium Alloys and Aluminium Matrix Composites

		Mechanisms Observed				
S. No.	Materials	Abrasion	Adhesion	Delamination	Oxidation	Melting and Softening
1	AA5252	√	√			
2	AA5252+SiC	√	√	√		
3	AA5052	√			√	
4	AA5052+ZrB$_2$	√		√		
5	AA6061	√	√	√		
6	AA6061+AlN	√	√	√	√	
7	AA6061+TiC	√	√	√		
8	AA6061+ZrB$_2$	√		√		
9	AA6061+TiB$_2$	√	√	√		
10	AA6061+WC	√		√		
11	AA6063+TiB$_2$	√	√	√		
12	AA6082+SiC	√		√		
13	AA7075	√		√		
14	AA7075+Si$_3$N$_4$	√				
15	AA7075+TiC	√		√	√	
16	AA7075+TiO$_2$	√			√	
17	AA7075+Al$_2$O$_3$+Gr	√		√	√	
18	AA7050+graphene	√	√			

(see Figure 2.6a), which confirms the occurrence of abrasion mechanism and plastic deformation. Also, flake debris with the presence of a crack, fine debris collections and lumps confirm the delamination, oxidation and thermal softening mechanisms. The wear debris (see Figure 2.7b) corresponding to the worn surface exhibit ribbon fragments and large sheets/flakes with smooth surface (see Figure 2.6b), which confirms the occurrence of abrasion mechanism and thermal softening. Also, the presence of wedge-shaped debris exhibits the adhesion mechanism. Some of the typical wear mechanisms observed in the aluminium alloys and aluminium matrix composites are listed in Table 2.2, and those observed in magnesium alloys and magnesium matrix composites are listed in Table 2.3.

2.8 SUMMARY

The enhancements in the lightweight materials have made them inevitable choices for multiple industrial and domestic applications. The understanding of the tribological behaviour of a material is most vital, and it is dependent on the processing conditions, reinforcement morphology, reinforcement quantity and the tribological parameters. It is well proven that the alloying and the composite strengthening can

TABLE 2.3

Dominant Wear Mechanisms Observed in Magnesium Alloys and Magnesium Composites

	Materials	Abrasion	Adhesion	Delamination	Oxidation	Melting and Softening
				Mechanisms Observed		
1	Pure Mg	√	√	√		√
2	Mg-3Al-5.6Ti	√	√	√	√	√
3	(Mg-3Al-5.6Ti)2.5B$_4$C	√	√	√	√	√
4	AZ91+0.8 vol. % Al$_2$O$_3$	√			√	
5	AZ91+12 wt. % B$_4$C	√	√			
6	AZ91+8% Al$_2$O$_3$+20% Gr	√			√	
7	AZ91+ 2.5 wt. % TiB$_2$	√		√	√	
8	AZ91+ TiC+ TiB$_2$	√		√	√	√
9	AZ91+10 vol. % SiC	√		√	√	√
10	AZ91+3 wt. % SiC+3 wt. % Gr	√	√	√		
11	AZ91+30 wt. % TiB$_2$	√			√	
12	AZ91+15 wt. % TiC	√	√		√	
13	AZ91+3 wt. % RE		√		√	
14	AZ91+10 vol. % Ti$_2$AlC	√			√	
15	AM60B	√		√	√	
16	ZE41A	√	√	√	√	
17	AZ31B	√	√	√	√	√
18	AZX915+12 wt. % TiC			√	√	

greatly enhance the wear resistance of lightweight materials. The wear mechanism is the one that stimulates the transition in the wear behaviour of light materials, and it needs to be identified for a better understanding of the tribology of lightweight materials. The wear of a material is expressed in different forms such as weight loss in grams, wear loss in mm^3/m, wear rate in mm^3/m and wear rate in mg/m. Also, there is no specific rule in concluding the levels of operating parameters. Henceforth, it is very difficult to arrive a conclusion from the reported findings of the wear behaviour. Hence, it is concluded that it is mandatory to identify the tribological behaviour of the newly developed lightweight materials in order to ease their passage into real-time applications.

REFERENCES

Ahmadkhaniha, D., Heydarzadeh Sohi, M., Salehi, A. and Tahavvori, R. (2016), "Formations of AZ91/Al$_2$O$_3$ nano-composite layer by friction stir processing", *Journal of Magnesium and Alloys*, Vol. 4 No. 4, pp. 314–318. Available at Doi: 10.1016/j.jma.2016.11.002.

American Society for Testing and Materials. (2017), "ASTM G99-17: Standard test method for wear testing with a pin-on-disk apparatus", *Annual Book of ASTM Standards*, pp. 1–6. Available at Doi: 10.1520/G0099-17.

ASM International. (1990), *Properties and Selection : Nonferrous Alloys and Special-Purpose Materials*, Available at Doi: 10.31399/asm.hb.v02.9781627081627.

Bhushan, B. (2013), *Introduction to Tribology*. John Wiley & Sons, Ltd. Doi: 10.1002/9781118403259.

Deng, K.K., Wang, X.J., Wu, Y.W., Hu, X.S., Wu, K. and Gan, W.M. (2012), "Effect of particle size on microstructure and mechanical properties of SiCp/AZ91 magnesium matrix composite", *Materials Science and Engineering A*, Vol. 543, pp. 158–163. Available at Doi: 10.1016/j.msea.2012.02.064.

Embury, J.D., Lloyd, D.J. and Ramachandran, T.R. (1989), "Strengthening mechanisms in aluminum alloys", *Treatise on Materials Science & Technology*, Vol. 31, pp. 579–601. Available at Doi: 10.1016/B978-0-12-341831-9.50027-9.

Gupta, M. and Sharon, N.M.L. (2010), *Magnesium, Magnesium Alloys, and Magnesium Composites*. John Wiley & Sons, Ltd. Available at Doi: 10.1002/9780470905098.

López, A.J., Rodrigo, P., Torres, B. and Rams, J. (2011), "Dry sliding wear behaviour of ZE41A magnesium alloy", *Wear*, Vol. 271 No. 11–12, pp. 2836–2844. Available at Doi: 10.1016/j.wear.2011.05.043.

Mohammadi, H., Emamy, M. and Hamnabard, Z. (2019), "The microstructure, mechanical and wear properties of AZ91-x%B4C metal matrix composites in as-cast and extruded conditions", *Materials Research Express*, Vol. 6 No. 12, Available at Doi: 10.1088/2053-1591/ab5405.

Runge, J.M. and Runge, J.M. (2018), *A Brief History of Aluminum and Its Alloys, The Metallurgy of Anodizing Aluminum*, Available at Doi: 10.1007/978-3-319-72177-4_1.

Saleh, B., Jiang, J., Xu, Q., Fathi, R., Ma, A., Li, Y. and Wang, L. (2020), "Statistical analysis of dry sliding wear process parameters for AZ91 alloy processed by RD-ECAP using response surface methodology", *Metals and Materials International*, The Korean Institute of Metals and Materials, No. 0123456789, Available at Doi: 10.1007/s12540-020-00624-w.

Sathish, S., Anandakrishnan, V. and Manoj, G. (2019a), "Optimization of wear parameters of Mg-(5.6 Ti+ 3Al)-2.5 B4C composite", *Industrial Lubrication and Tribology*, No. August, Available at Doi: 10.1108/ILT-08-2019-0326.

Sathish, S., Anandakrishnan, V. and Manoj, G. (2020), "Optimization of tribological behavior of magnesium metal-metal composite using pattern search and simulated annealing techniques", *Materials Today: Proceedings*, Vol. 21, pp. 492–496.

Sathish, S., Anandakrishnan, V., Sankaranarayanan, S. and Gupta, M. (2019b), "Optimization of wear parameters of magnesium metal-metal composite using Taguchi and GA technique", *Jurnal Tribologi*, Vol. 23 No. November, pp. 76–89.

Shanthi, M., Lim, C.Y.H. and Lu, L. (2007), "Effects of grain size on the wear of recycled AZ91 Mg", *Tribology International*, Vol. 40 No. 2 SPEC. ISS., pp. 335–338. Available at Doi: 10.1016/j.triboint.2005.11.025

Wen, S. and Huang, P. (2017), *Principles of Tribology*, Available at Doi: 0.1002/9781119214908.

Xiao, P., Gao, Y., Xu, F., Yang, C., Li, Y., Liu, Z. and Zheng, Q. (2018), "Tribological behavior of in-situ nanosized TiB2 particles reinforced AZ91 matrix composite", *Tribology International*, Vol. 128, pp. 130–139. Available at Doi: 10.1016/j.triboint.2018.07.003.

3 Self-Lubricating Iron-Based Metal Matrix Composites

Shuhaib Mushtaq
Islamic University of Science and Technology Awantipora

M.F. Wani
National Institute of Technology Srinagar

Carsten Gachot
Institute for Engineering Design and Logistics Engineering
Vienna University of Technology (TUW) Vienna, Austria

Mohd Nadeem Bhat
Indian Institute of Technology Kanpur

CONTENTS

3.1   INTRODUCTION

The demand for developing efficient and low-cost materials has become a challenging job for designers and engineers. It is a well-known fact that designers and engineers of various industries, such as automotive, aerospace and marine industries, have replaced the classical materials with advanced materials which possess better mechanical and tribological properties. The backbone of human civilization and industrial growth is metals. Metals and alloys are being extensively used in applications ranging from a simple blade to complex space structures. In addition, new materials such as ceramics and plastics have also been developed. The literature reveals that researchers throughout the globe are working on developing composites by reinforcing metals/alloys, ceramics and polymers with stronger and stiffer elements to improve their properties for enhanced performance in existing and futuristic technologies. In this way, the properties of low-strength metals and alloys are significantly improved to a large extent. The two main constituents in a composite are base matrix and reinforcement. The reinforcement when uniformly distributed in matrix component provides improved strength and stiffness. In composites, matrix materials can be metals, ceramics or polymers. Metals, ceramics and polymers are used as matrix phase and are reinforced with appropriate solid lubricant for fabrication of self-lubricating composites.

The basic objective in the fabrication of self-lubricating metal matrix composites (SLMMCs) is to combine or reinforce a metal with appropriate solid lubricant in order to develop a new material with desired or enhanced tribological properties (Omrani et al. 2016). The powder metallurgy (P/M) and stir casting techniques are used for the development of SLMMCs (Gupta et al. 2013). The fabrication process is selected by the cost aspect; i.e., the P/M is the most valuable route (Rosso 2006). In large, the importance of P/M route is that, less temperature is used against the other routes, and the matrix and reinforcement interface is low. The P/M route provides less interaction between the reinforcement and matrix phase and gives homogenous distribution in the metal matrix, thus finding the great future in a bearing system having numerous advantages with the capability of permanent usage in tribological applications (Evans and Senior 1982). The ability of SLMMCs in tailoring the properties of materials is widely employed in the development of various mechanical elements. Figure 3.1 shows the plot of a bulk SLMMC that contains the solid lubricant to improve the friction and wear due to tough phases which support the load. Solid

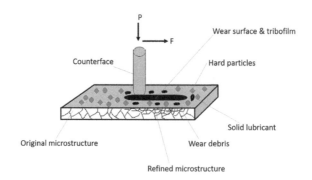

FIGURE 3.1 A schematic of bulk SLMMC, formation of solid lubricant-based tribofilm and the transfer film between the two tribopairs.

lubricants impart the self-lubrication to the metal matrix composite by the formation of mechanical mixed layer or tribofilm.

Since from decades, SLMMCs have been used by the industries to lower friction and wear in a number of tribological applications. During the sliding process, wear particles of reinforced solid lubricant develop a lubricating film at the interface. This lubricating film guides in lowering the friction coefficient and wear rate (Erdemir 2000; Evans and Senior 1982; Rohatgi, Ray and Liu 1992). Graphite, molybdenum disulphide (MoS_2) and boron nitride (BN) are the regularly used inorganic solid lubricants with lamellar structures. The crystal lattice of these materials has a layered structure consisting of hexagonal rings forming thin parallel planes. Each atom inside the plane is strongly bound to other atoms (covalent bonding), which increases the capacity of the film to bear the load between the contact surfaces. The planes are connected by weak van der Waals forces to each other. These less properties like lubrication and strength of the materials are determined by weak bonding between the planes.

Iron-based self-lubricating materials for bearing applications have improved dramatically over copper-based ones over the last few decades, due to their higher strength, low cost, and iron powder availability (Teisanu and Gheorghe 2011). In order to improve metallurgical and tribological properties, iron (Fe) is mixed with other metals, such as C, Cu, P, Ni and Sn. Pre-alloyed ferrous powders have contributed to high-strength martensitic microstructures when worn with additives such as Cu, Ni and C (Causton and Fulmer 1992). The addition of manganese and carbon to Fe-based materials has shown promising results for mechanical properties and dimensional changes (Šalak and Selecká 2002). It is well known that the addition of copper will increase the strength of iron because it melts and dissolves rapidly in iron (Annamalai, Upadhyaya, and Agrawal 2013; Narasimhan and Semel 2007). Tin is the main element in antifriction alloys because it facilitates liquid-phase sintering and helps in improving fatigue strength and tribological properties (Teisanu and Gheorghe 2011).

Various alloy systems based on iron were developed, such as Fe-C, Fe-Cu, Fe-Cu-C and Fe-Cu-Sn (Donnet and Erdemir 2004; Nowosielski and Pilarczyk 2005). They contain a large amount (15%–40%) of solid lubricant particles such as MoS_2 and h-BN, low-melting metals such as Pb, Sn, Ag and graphite, and polymers such as PTFE (Dangsheng 2001; Moustafa 2002). h-BN is very similar to graphite in properties. It has high thermal conductivity, low thermal expansion, excellent resistance to thermal shock and good machinability and is non-abrasive and lubricious (Chu and Lin 2000). Fe-SiC-h-BN composites fabricated by the P/M technique were studied with different amounts of h-BN, and with the increased addition of solid lubricant, researchers and authors found better wear properties. However, due to the dispersion of lubricant around iron particles, the mechanical properties were adversely affected (Da Costa Gonçalves et al. 2014). Boron addition in Fe matrix helps in decreasing the wear rate and improving the density and hardness due to the formation of liquid-phase sintering (Mamedov 2004). The tribological properties of h-BN were also studied with Fe-1.5 Mo-1.0Si-0.8 C alloy. Five weight percentage of h-BN was added to the ferrous matrix. It was observed that the mechanical properties were healed up and the maximum wt.

% of h-BN forms the lubricious layer over the interface thus results in decrease in coefficient of friction (COF) and wear properties (Gülsoy et al. 2007).

Graphite is used in most of the ferrous-based composites, as it helps to create different phases that contribute to changes in the characteristics of friction and wear. Fe-C-Ni, one such composite, has been prepared by powder metallurgy. Iron–copper–tin (Fe-Cu-Sn) can be alloyed with graphite, and graphite results in weak mechanical and tribological properties due to incomplete diffusion (Mamedov 2004).

3.2   DEVELOPMENT OF IRON-BASED ALLOY SYSTEMS

Copper readily dissolves in iron and improves the strength, as it melts and absorbs into iron easily. Huge pores occur in the composition of iron–copper (Fe-2 wt. % Cu) since copper melts at 1,083°C and spreads between iron particles and results in the compact's swelling. At room temperature, the solubility of copper in iron is around 0.4 wt. %, which results in the precipitation of excess copper (Annamalai, Upadhyaya, and Agrawal 2013). This is the cause for the formation after sintering of homogeneous clusters of copper in Fe matrix. In addition to the constant copper content of Fe-Cu products, Fe-Cu with varying Cu wt. % and fixed cobalt (Co) content was also checked by the researchers (Barbosa et al. 2008).

To boost mechanical strength, hardness and toughness, iron is alloyed with carbon (graphite) and other elements (Nowosielski and Pilarczyk 2005). In a high-energy mill (shaker type), a mixture of 93.3 Fe and C of 6.67 powders by weight percentages was mechanically alloyed in a controlled atmosphere for varying time, resulting in the creation of an amorphous matrix with a nanocrystalline phase. Owing to the change in the size of the microstructure and chemical composition of the crystallite, the alloying period increases, rendering the alloy more amorphous. The better mechanical and tribological properties can be obtained after sintering.

3.2.1   HEXAGONAL BORON NITRIDE (H-BN) AS A SOLID
LUBRICANT IN IRON BASE ALLOY

Hexagonal boron nitride (h-BN) has very similar properties to graphite. It has better thermal properties, less thermal expansion, improved resistance to thermal shock and low dielectric constant (Bernard and Miele 2014). In the case of Fe-h-BN composites, ferritic grain size reduces with maximum h-BN content without change in mechanical properties (Sueyoshi, Tagami, and Rochman 2001).

Liersch, Danninger and Ratzi (2007) investigated the influence of h-BN on sintered Fe and Fe-C materials. The smooth grains of h-BN have more effect on mechanical properties than coarse grains revealed by results. Kelin et al. (2013) studied the development of Fe-h-BN composite by the double pressing/double sintering (DPDS) method. They varied the vol. % of h-BN from 0% to 10% in their study. It was concluded that due to the dispersion between the iron particles during mixing, the addition of h-BN particles affects the mechanical properties of the composite. The use of the DPDS approach led to a marked reduction in the porosity of the metallic matrix and thus to improved continuity. Tensile strength and a small improvement in elongation resulted in increased continuity of the metallic matrix, thus mitigating the effects of applying strong lubricant.

Mahathanabodee et al. (2013) concluded in their study that h-BN was changed into a boride liquid that deshapes the lubricating film, resulting in weak friction properties. Mamedov (2004) investigated the effect of boron addition in Fe matrix, and they observed that the addition of boron to Fe helps in decreasing the wear rate and improves density and hardness due to the formation of liquid-phase sintering.

Gulsoy et al. (2007) studied the tribological properties of h-BN with Fe-1.5 Mo-1.0Si-0.8 C alloy. Five weight percentage of h-BN was added to the ferrous matrix. It was observed that the mechanical properties were healed up and the more wt. % of h-BN forms the lubricious layer over the interface thus results in lowering the friction coefficient and wear rate. Da Costa Gonçalves et al. (2014) investigated the maximum addition of solid lubricant (h-BN), and they observed a change in scuffing resistance. However, due to the formation of lubricious film around iron particles, the mechanical properties were adversely affected.

3.2.2 POWDER METALLURGY

Mixing, compaction and sintering are the three main steps in the formation of composites by P/M (Rohatgi, Ray, and Liu 1992). The main challenges in the development of SLMMCs for various applications such as high-temperature applications are high porosity and less bonding strength. In order to overcome porosity and low bonding of alloying elements, different alloying elements such as copper and tin are. Figure 3.2 exhibits a typical SEM image of Fe-Cu-Sn-05 wt. % h-BN composite fabricated by P/M (Mushtaq and Wani 2017).

During sintering at 1,000°C, h-BN decomposes in Fe-Cu-Sn matrix and leads to the formation of Ferrite. (Mushtaq and Wani 2017). The incorporation of h-BN into the metal matrix withstands high sintering temperatures, but leads to a decrease in strength (Mahathanabodee et al. 2013). The high sintering temperature can be controlled by using solid lubricants exhibiting higher oxidation resistance (Mahathanabodee et al. 2012, 2013).

The solid lubricant and metal matrix reinforcement should be taken into consideration for porosity and bonding strength. Sintering temperature plays a vital role

FIGURE 3.2   SEM image of mixed powdered and sintered Fe-Cu-Sn-5 wt. % h-BN composite.

in the porosity of composites; i.e., inadequate sintering temperature is not feasible for adequate bonding of composite components, decreased porosity and improved bonding strength effectively (Shi et al. 2014). In order to promote porosity, the new technique glow discharge plasma is adequate which helps in quick and full heating process (Cardenas et al. 2016). Finally, after processing the extrusion improves density and hardness of Cu-Sn-h-BN composites (Furlan et al. 2015).

3.3 TRIBOLOGICAL BEHAVIOUR OF IRON-BASED SLMMCS CONTAINING GRAPHITE

3.3.1 MICROSTRUCTURE EXAMINATION

The Fe-Cu-Sn sintered specimens (0–3 wt. % of graphite) and their optical micrographs are shown in Figure 3.3a–d. These specimens are etched with natal solution. The microstructural change can be easily observed in the micrographs with the change in graphite content. The interaction between Cu and Sn speeds up the inter-diffusion between Cu and Fe during the sintering process. A Fe-based solid solution of Fe-Cu-Sn with bright bronze inclusions alloyed by Fe is distinguished by the matrix of the alloys studied. The transitive character of liquid phases that occur during sintering, results in microstructure production for all alloys. At a temperature of 232°C, a liquid Sn process exists. In the bronze liquid phase, the higher sintering temperature facilitates the oxide reduction and dissolution process of solid particles of Fe and Cu.

3.3.2 FRICTION AND WEAR

The varying graphite wt. % (0–3) in Fe-Cu-Sn (disc) alloy and tribopair with EN8 steel ball under dry conditions. The variation in friction and wear parameters is discussed in the following.

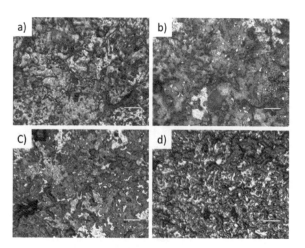

FIGURE 3.3 Microstructure of the sintered Fe-Cu-Sn containing (a) 0, (b) 1, (c) 2 and (d) 3 wt. % graphite as self-lubricating composite.

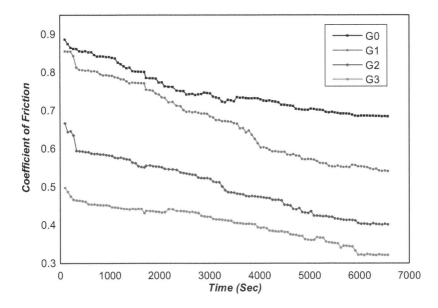

FIGURE 3.4 Coefficient of friction with time at 5 N load and 50°C temperature.

3.3.2.1 Coefficient of Friction (COF)

The tribological tests were performed in order to study the effect of graphite content on friction coefficient by varying sliding distance and load. These tests were performed two to three times in order to eliminate any error in the experimentation. Figures 3.4 and 3.5 show the results of COF obtained from these experiments. Figure 3.4 demonstrates the COF variance with (time) sliding distance for alloys dependent on Fe-Cu-Sn with different graphite wt. % at 5 N normal load and 0.157 m/s sliding velocity. COF is observed as it illustrates the reducing trend for all compositions with an improved sliding distance. Composition G0 represents the highest COF value, as opposed to the G3 composition, which indicates the lowest COF value. For all compositions at 5 N standard load, Figure 3.5 shows the difference in the sliding velocity of average COF. COF is observed to decrease in all compositions, but at a sliding velocity of 0.261 m/s, there is a sudden increase in COF in the G0 composition. This may be due to the extreme adhesion at the maximum sliding speed induced by the frictional heat.

3.3.2.2 Wear Behaviour

Figure 3.6 shows the sliding distance vs wear rate for Fe-Cu-Sn alloys with distinct graphite wt. % at 5 N normal load and 0.157 m/s sliding velocity. With the rise in sliding distance, the wear rate declines steadily. The order of 10^{-4} mm^3/N m is the specific wear rate. The composition of the G3 has the minimum wear rate value obtained.

3.4 APPLICATIONS, CHALLENGES AND FUTURE DIRECTIONS

SLMMCs, especially considering global green manufacturing initiatives and improved feasibility in engineering design, remain a significant class of engineering

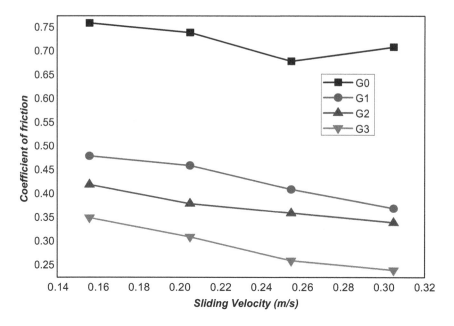

FIGURE 3.5   Coefficient of friction with sliding velocity at 5 N load.

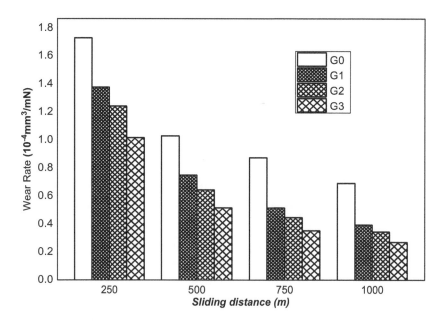

FIGURE 3.6   Wear rate vs sliding distance at 5 N load and 0.157 m/s sliding velocity.

materials. Such materials are used in tribological contacts with low friction, for example in the automobile sector, spaceships, marine applications and different industries. Various solid lubricants (e.g. MoS_2, Gr, h-BN, CNTs, G and WS_2) are used for the development of various metal matrixes. The development of SLMMCs by the P/M technique, from the mixing of Fe-Cu-Sn alloys with Gr, has been reviewed here. Thus for SLMMCs, there is an important body of scientific literature and knowledge base which could have a significant effect on the understanding of lubrication mechanisms of SLMMCs and also on the optimized SLMMCs' properties.

The Lubrication in SLMMC's is attributed to the formation of tribofilm, which ultimately helps in decreasing the friction between different tribopairs. Riahi and Alpas (2001) in their work felt the presence of graphite as the main component in the formation of tribofilm. The description of formation of tribofilms for SLMMCs, which is the main component of composites, should be made accurate.

In general, the thick mechanical layer of the solid lubricant affects greatly the lubrication mechanism of SLMMCs. The volume of the solid lubricant should be enough to make a successful lubricious film (Zhang et al. 2012). The flow recirculation or debris out of contact creates the flaws in the wear track. This implies that the mechanisms (Mogonye et al. 2016) observed are more significant. The solid lubricant particles and observation of the tribofilm mechanism is the main design factor for better SLMMCs. The plastic deformation or wear causes the solid lubricant flow over the surface, and it depends upon the fabrication route of the composite. The matrix and morphology of the lubricant are the main parts of composite performance.

New processing routes on the manufacturing side provide SLMMCs with new possibilities. A new perspective is provided by additive manufacturing in particular. The additive manufacturing techniques in producing the bulk components from metals and alloys can be better used for aligning the properties of SLMMCs with good control on lubricant content and morphology. The removal of decomposition of solid lubricants by cold spraying is unavoidable for the production process. The enhanced feedstock development and accurate control of powder delivery system are challenges for cold spray techniques. This will give a new way of using cold spray to manufacture SLMMCs.

ACKNOWLEDGEMENTS

The authors would like to express their gratitude towards the Director, National Institute of Technology (NIT), Hazratbal, Srinagar, India; and all the staff of the tribology laboratory of NIT Srinagar (India) and Rtec Instruments (USA) for helping in carrying out the fabrication and tribological experimentation.

REFERENCES

Annamalai, R., A. Upadhyaya, and D. Agrawal. 2013. "An investigation on microwave sintering of Fe, Fe-Cu and Fe-Cu-C alloys." *Bulletin of Materials Science* 36(3): 447–56.
Barbosa, A. P. et al. 2008. "Processing and mechanical properties of PM processed Fe-Cu-Co alloys." *Materials Science Forum* 591–593: 247–51.

Bernard, S., and P. Miele. 2014. "Nanostructured and architectured boron nitride from boron, nitrogen and hydrogen-containing molecular and polymeric precursors." *Materials Today* 17(9): 443–50.

Cardenas, A., Y. Pineda, A. Sarmiento Santos, and E. Vera. 2016. "Effect of glow discharge sintering in the properties of a composite material fabricated by powder metallurgy." *Journal of Physics: Conference Series* 687(1): 012025.

Causton, R. J., and J. J. Fulmer. 1992. "Sinter hardening low-alloy steels." *Advances in Powder Metallurgy* 5: 17–52.

Chu, H. Y., and J. F. Lin. 2000. "Experimental analysis of the tribological behavior of electroless nickel-coated graphite particles in aluminum matrix composites under reciprocating motion." *Wear* 239(1): 126–42.

Da Costa Gonçalves, P. et al. 2014. "Self-Lubricating sintered composites with hexagonal boron nitride and graphite mixtures as solid lubricants." *Advances in Powder Metallurgy and Particulate Materials -2014, Proceedings of the 2014 World Congress on Powder Metallurgy and Particulate Materials, PM* 2014: 910–17.

Dangsheng, X. 2001. "Lubrication behavior of Ni-Cr-based alloys containing MoS_2 at high temperature." *Wear* 250–251(PART 2): 1094–99.

Donnet, C., and A. Erdemir. 2004. "Historical developments and new trends in tribological and solid lubricant coatings." *Surface and Coatings Technology* 180–181: 76–84.

Erdemir, A. 2000. "Solid lubricants and self-lubricating films." In *Modern Tribology Handbook: Volume One: Principles of Tribology*, CRC Press, Columbus, OH, 787–825.

Evans, D. C., and G. S. Senior. 1982. "Self-lubricating materials for plain-bearings." *Tribology International* 15(8): 243.

Furlan, K. P. et al. 2015. "Influence of alloying elements on the sintering thermodynamics, microstructure and properties of Fe-MoS_2 composites." *Journal of Alloys and Compounds* 652: 450–58.

Gülsoy, H. Ö., M. K. Bilici, Y. Bozkurt, and S. Salman. 2007. "Enhancing the wear properties of iron based powder metallurgy alloys by boron additions." *Materials and Design* 28(7): 2255–59.

Gupta, P., D. Kumar, O. Parkash, and A. K. Jha. 2013. "Structural and mechanical behaviour of 5% Al_2O_3-reinforced Fe metal matrix composites (MMCs) produced by powder metallurgy (P/M) route." *Bulletin of Materials Science* 36(5): 859–68.

Kelin, A.N., Hammes, G., Binder, C. and Furlen, K. 2013. Fe-H BN Composites Produced by Double Pressing and Double Sintering. In *Proceedings of EURO PM2009,* Federal University of Santa Catarina (UFSC), Materials Laboratory Brazil.

Kováčik, J., Š. Emmer, J. Bielek, and L. Keleši. 2008. "Effect of composition on friction coefficient of Cu-graphite composites." *Wear* 265(3–4): 417–21.

Liersch, A., H. Danninger, and R. Ratzi. 2007. "The role of admixed hexagonal boron nitride in sintered steels 2. effect on sliding wear and machinability." *Powder Metallurgy Progress* 7(2): 1–11.

Mahathanabodee, S. et al. 2012. "Effect of H-BN content on the sintering of SS316L/h-BN composites." *Advanced Materials Research* 410: 216–19.

Mahathanabodee, S. et al. 2013. "Effects of hexagonal boron nitride and sintering temperature on mechanical and tribological properties of SS316L/h-BN composites." *Materials and Design* 46: 588–97.

Mogonye, J. E. et al. 2016. "Solid/self-lubrication mechanisms of an additively manufactured Ni-Ti-C metal matrix composite." *Tribology Letters* 64(3): 1–12.

Moustafa, S. F., S. A. El-Badry, A. M. Sanad, and B. Kieback. 2002. "Friction and wear of copper-graphite composites made with Cu-coated and uncoated graphite powders." *Wear* 253(7–8): 699–710.

Mushtaq, S., and M. F. Wani. 2017. "Self-lubricating tribological characterization of lead free Fe-Cu based plain bearing material." *Jurnal Tribologi* 12(July): 18–37.

Narasimhan, K. S., and F. J. Semel. 2007. "Sintering of powder premixes - a brief overview." In *Advances in Powder Metallurgy and Particulate Materials -2007, Proceedings of the 2007 International Conference on Powder Metallurgy and Particulate Materials, PowderMet* 2007, 51–524.

Nowosielski, R., and W. Pilarczyk. 2005. "Structure and properties of Fe-6.67% C alloy obtained by mechanical alloying." *Journal of Materials Processing Technology* 162–163(SPEC. ISS.): 373–78.

Omrani, E., A. D. Moghadam, P. L. Menezes, and P. K. Rohatgi. 2016. "New emerging self-lubricating metal matrix composites for tribological applications." In J.P. Davim (ed.), *Ecotribology, Materials forming, Machining and Tribology*, 63–103, Springer, Cham.

Riahi, A. R., and A. T. Alpas. 2001. "The role of tribo-layers on the sliding wear behavior of graphitic aluminum matrix composites." *Wear* 250–251(PART 2): 1396–1407.

Rohatgi, P. K., S. Ray, and Y. Liu. 1992. "Tribological properties of metal matrix-graphite particle composites." *International Materials Reviews* 37(1): 129–52.

Rosso, M. 2006. "Ceramic and metal matrix composites: Routes and properties." *Journal of Materials Processing Technology* 175(1–3): 364–75.

Šalak, A., and M. Selecká. 2002. "New aspects for sinter boriding of PM steels." *Powder Metallurgy Progress* 2(3): 161–69.

Shi, X. et al. 2014. "Tribological behavior of Ni_3Al matrix self-lubricating composites containing WS_2, Ag and hBN tested from room temperature to 800°C." *Materials and Design* 55: 75–84.

Sueyoshi, H., K. Tagami, and N. T. Rochman. 2001. "Damping capacity of graphite-dispersed composite steel." *Materials Transactions* 42(6): 965–69.

Teisanu, C., and S. Gheorghe. 2011. "Development of new PM iron-based materials for self-lubricating bearings." *Advances in Tribology* 2011: 11. doi: 10.1155/2011/248037

4 Metal Matrix Nanocomposites

Physical, Mechanical and Tribological Properties

Mahendra Kumar Soni and Ovais Gulzar
Islamic University of Science and Technology

Mir Irfan Ul Haq
Shri Mata Vaishno Devi University

M.F. Wani
NIT Srinagar

CONTENTS

4.1 INTRODUCTION

Nanotechnology offers a new paradigm in various engineering applications including aerospace, automotive and energy sectors (Gulzar, Qayoum, & Gupta, 2018). The use of nano-based technologies not only helps to achieve higher efficiencies but also helps to reduce weight and address sustainability issues (Anand et al., 2020; Gulzar, Qayoum, & Gupta, 2019). To state a few, nanoparticles have been successfully employed for lubricating purposes, nano-based cutting fluids, nanofluids for thermal applications, nanocomposites for aerospace and automotive applications to yield better and sustainable performance (Shafi, Raina, & Ul Haq, 2018; Singh et al., 2018).

TABLE 4.1

Types of Solid Lubricants

Solid Lubricant Type	Examples
Lamellar solids	MoS_2, graphite, h-BN
Metals	Silver, gold, zinc
Polymers	PTFE, nylon
Oxides	PbO, ZnO, FeO, Fe_2O_3
Halides and sulphides	WS_2, TaS_2, ZnS, CaF_2, BaF_2

The global demand for lightweight materials and the growing concern for sustainability have led to the development of a new class of materials known as metal matrix composites (MMCs). MMCs have found use in almost all engineering applications, particularly in aerospace and automotive applications (Baba, Shafi, Haq, & Raina, 2019). The widespread involvement of sliding surfaces in a variety of engineering applications demands materials that are having superior mechanical and tribological performance (Ul Haq & Anand, 2018). Further, the environmental concerns associated with the use of liquid lubricants and the technical difficulties while using liquid lubricants in extreme environments such as high temperature and vacuum led to the development of self-lubricating materials. Self-lubricating materials are a special class of materials with at least one constituent as a solid lubricant. Table 4.1 provides a brief classification of the various solid lubricants used in the development of the self-lubricating materials (Dorri Moghadam, Omrani, Menezes, & Rohatgi, 2015).

Generally, MMCs with hard ceramic particles or fibres as additives exhibit superior mechanical properties as compared to unreinforced alloys. Due to the soft nature of the solid lubricants, the addition of the solid lubricants alone in the matrix leads to the degradation of the mechanical strength. Therefore, to retain the mechanical strength of the resulting self-lubricating materials and impart better tribological properties, the hybrid materials are being developed, wherein ceramic materials are being added primarily to improve the mechanical strength and the solid lubricant phase is added to improve the tribological performance (Irfan Ul Haq, Raina, Anand, Sharma, & Kumar, 2020). The materials with either the ceramic phase or the solid lubricant in the nano-size range are termed metal matrix nanocomposites.

4.2   NANOCOMPOSITES

Nanocomposites refer to those composite materials wherein the reinforcement is of the order of <100 nm. Nanocomposites typically refer to the multiphase solid materials in which one of the constituents has dimensions of less than 100 nm. Previous studies have revealed that the nanocomposites exhibit better properties when compared to the unreinforced or the micro-sized counterparts (Ceschini et al., 2017; Moghadam et al., 2014). This improvement is primarily due to the nano-sized reinforcement. In the case of MMCs with reinforcement of micrometre range, the relatively bigger size of the reinforcement actually leads to inhomogeneity, which further

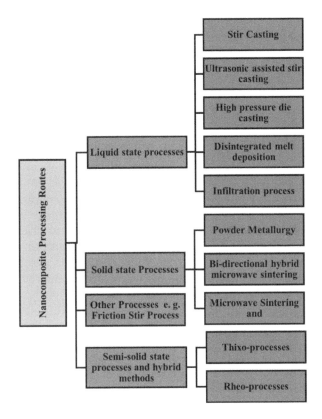

FIGURE 4.1   Different processing routes for the development of nanocomposites.

leads to uneven stress distribution and ultimately results in failure of the component even at relatively lower loads. Further, the uniform dispersion of the finer reinforcement in the case of nanocomposites leads to improvement in strength and ductility.

The development of the nanomaterials remains a subject of research, and the development of low-cost techniques to synthesize nanomaterials is also being researched extensively. The production of nanocomposites can be broadly divided into three major routes (Ceschini et al., 2017) as presented in Figure 4.1.

4.2.1   LIQUID-STATE PROCESSES

The processing of low-melting matrix-based nanocomposites, such as aluminium and magnesium-based, the liquid-state route apart from being cost-effective offers many technical advantages particularly for the production of large-sized parts. The major liquid-state processes include stir casting, ultrasonic-assisted stir casting, infiltration process, disintegrated melt deposition and high-pressure die casting. While conventional stir casting remains the most widely used and an economical processing route for development of these composites however poor wettability of the nanoparticles in the molten melt, increased porosity, non-homogeneous distribution and clustering of nanoparticles needs to be addressed (Hashim, Looney, & Hashmi, 1999). To over-come these disadvantages, advanced techniques such as ultrasonic-assisted stir cast-ing, infiltration process, disintegrated melt deposition and high-pressure die casting have been developed. The higher melting point of iron and copper matrices has lim-ited the use of liquid-based methods in the development of nanocomposites based on Fe and Cu.

4.2.2 SOLID-STATE PROCESSES

These routes are capable of incorporating higher volume fractions of reinforcements as compared to the liquid-state methods and can produce near-net-shape products. The major solid-state methods include powder metallurgy, microwave sintering and bidirectional hybrid microwave sintering. While methods such as powder metallurgy have been extensively used in the production of parts for industrial applications, the solid-state processes in general suffer from some disadvantages such as high cost of powders and difficulty in the development of large-sized parts. However, the issues of poor wettability and non-homogeneity in the case of liquid-state processes can be solved to a great extent by the use of solid-state routes.

4.2.3 SEMISOLID-STATE PROCESSES AND HYBRID METHODS

These methods employ the mixing of the matrix and the reinforcement phase in a semisolid phase, and a major advantage is the lower operating temperature range. While two main types of semisolid state include thixo-processes and rheo-processes (compocasting), the literature suggests that very little effort has been made to utilize these methods in the production of nanocomposites.

In order to achieve better properties and to exploit the potential of the afore-mentioned production routes, a combination of the processing routes has also been employed for the production of nanocomposites. Further, the conventional tech-nique to produce surface composites known as friction stir processing has also been employed for the development of bulk nanocomposites.

4.3 PHYSICAL AND MECHANICAL PROPERTIES

In this section, the effect of various nanoparticle reinforcements on the various phys-ical and mechanical properties will be discussed. Table 4.2 summarizes the work related to the physical and mechanical properties of these composites such as density, tensile strength, microstructure and hardness. The density of the developed compos-ites depends on the fabrication route apart from the relative density of the reinforcing agent. The density of the nanocomposites improves considerably as the small-sized nanoparticles fill the voids (Vaziri, Shokuhfar, & Afghahi, 2019). The microstruc-tural changes occurring due to the reinforcement may have a direct impact on various properties such as hardness and wear resistance. With regard to microstructure, grain refinement has been observed by many researchers. This grain refinement alters the mechanical properties. The nanoparticle addition leads to strengthening the com-posites through the Hall–Petch strengthening mechanism (Sree Manu, Arun Kumar,

TABLE 4.2

Brief Summary of Literature concerning the Physical and Mechanical Properties of Nanocomposites

Base Matrix	Reinforcement	Fabrication Route	Nanoparticle Content	Major Findings	Ref.
AA2124	Graphene	Ball milling and hot extrusion	3, 5 wt. %	A 133% increase in the hardness, 34% decrease in the wear rate and 25% decrease in the COF for 3 wt. % composites	El-Ghazaly, Anis, and Salem (2017)
Al	WS$_2$	Powder metallurgy	0–16 wt. %	A 40% reduction in COF as compared to pure Al	Vaziri, Shokuhfar, and Afghahi (2019)
AA2024	Multi-walled carbon nanotube	Powder metallurgy	5 vol. %	A decrease in thermal expansion Around 2.5 times better strength and a reduction of around 20% in the coefficient of thermal expansion	Shin, Ko, and Bae (2016)
AA2219	Nano-SiC	Stir casting	0.5, 1, 1.5, 2 and 2.5 wt. %	A 66% enhancement in the hardness for 2.5% nano-SiC filler Around 20% improvement in the tensile and flexural properties	Faisal and Kumar (2018)
ZA27	Nano-graphite	Ball milling and hot pressing	0.5%, 1%, 1.5%, 3%, 5% and 7%	Density of the developed composites decreases while the porosity increases	Dalmis, Cuvalci, Canakci, and Guler (2016)
Al	TiB$_2$ and Al$_2$O$_3$	Ultrasonic-assisted stir casting	1 wt. % TiB$_2$ and 1 wt. % Al$_2$O$_3$	Higher strength, but low ductility	Moghadam, Omrani, Menezes, and Rohatgi (2016)
Al-1100	MWCNT	Surface impregnation by solid-state process	-	An improvement in microhardness by around 2.86 times	Sahoo, Narsimhachary, and Paul (2019)
Mg	Nano-Al$_2$O$_3$	Melt deposition technique and hot extrusion	1.1 wt. %	• Reduced porosity • Uniform distribution achieved	Hassan and Gupta (2005)
Mg	CNT	Disintegrated melt deposition (DMD) technique	1.3 wt. %	Improved fatigue strength after the addition of CNTs	Goh, Wei, Lee, and Gupta (2008)
ZK60A	Nano-Al$_2$O$_3$	DMD	1.5 vol. %	• A decrease in yield strength • An improvement in ultimate tensile strength • Enhanced tensile and compressive failure strains	Paramsothy, Chan, Kwok, and Gupta (2011)

Rajan, Riyas Mohammed, & Pai, 2017). Hardness is an important property for the development of nanocomposites as the wear resistance is a function of hardness of the material as per Archard's wear law. Depending on the nature of the reinforcement used and the content of the reinforcement, the hardness of the nanocomposites varies. The processing route of the nanocomposites also has a significant impact on the properties such as microstructure, density and hardness. With regard to the hardness of the nanocomposites, the nanofillers lead to significant improvement in hardness mainly due to the following factors (Gupta & Wong, 2015):

a. Harder particles acting as barriers to dislocation
b. Mismatch in the elastic moduli of the matrix and reinforcement
c. A decrease in the grain size

The tensile strength of the nanocomposites in general is improved by the addition of the nanofillers, while the ductility shows a decreasing trend. The processing route, type of the nanoparticle and the content of the reinforcement also affect the strength. Orowan strengthening, Hall–Petch effect, effective transfer of load from the matrix to the reinforcements and the work hardening are the main reasons behind this improvement (Gupta & Wong, 2015). A better bonding between the matrix and the filler is an important parameter (Malaki et al., 2019) in the resulting mechanical properties as the interface (metal–reinforcement) otherwise acts a potential site for crack initiation and propagation. Moreover, the porosity levels also increase due to void formation at the interface. Previous literature suggests that the mechanical properties such as fatigue and creep behaviour of the nanocomposites have not been studied much.

4.4 FRICTION AND WEAR PROPERTIES

The involvement of friction and wear in the major engineering applications, the developed nanocomposites are expected to possess improved tribological properties apart from having better mechanical strength. The type of reinforcement greatly affects the tribological properties of the resulting composite. The addition of hard ceramic reinforcements generally leads to improvement in the wear resistance; however, the coefficient of friction (COF) can't be altered much by the ceramic reinforcements. Therefore, the need for the addition of a solid lubricating agent in the metal matrix arises. However, the addition of softer solid lubricating agents leads to degradation of properties; therefore, the hybrid composite system of composites is a good option to arrive at optimized mechanical and tribological properties.

The wear resistance of the developed composites is improved generally by the addition of ceramic reinforcement which is primarily linked to the improvement in the hardness of the developed composites. Further, the wear process being a phenomenon of material removal, the improvement in the mechanical strength and the resistance by the hard particles to the crack propagation also aid in improving the wear resistance. The behaviour of these hard particles as load-bearing agents also improves the wear behaviour. The friction and wear of the nanocomposites in general depend on a variety of factors, such as the following:

a. Composition of the nanocomposite
b. Processing route
c. Type of heat treatment
d. Mechanical strength, particularly hardness and compression strength
e. Contact pressure and the sliding speed

The tribological issues associated with the micron-sized composites such as counter-face abrasion, higher material loss due to property variation, and better matrix and filler are obviated in the case of nanocomposites. Table 4.3 provides a brief overview of some of the studies carried out in the area of tribological behaviour of nanocomposites.

4.5 APPLICATIONS AND CHALLENGES

The improved properties of the nanocomposites make them appropriate for a range of engineering applications; however, the equipment needed in the sectors related to automobiles, aviation, aerospace and defence can be potential application areas owing to the need for materials with higher strength-to-weight ratio. Figure 4.2 depicts the main application areas of nanocomposites. Nevertheless, the high cost associated with the nanoparticles is still a barrier for the large-scale use of these nanocomposites. The complexity involved in the synthesis and fabrication and the difficulty in storage and handling of the nanoparticles are also a challenge. Furthermore, the issue of wettability resulting in weaker matrix–nanoparticle interface particularly in processes such as casting also limits the widespread use of these nanocomposites for industrial applications.

4.6 CONCLUSIONS

With the advancements and the cost-efficient techniques being developed to synthesize nanomaterials, metal matrix nanocomposites have evolved as potential materials for various applications. The complex mechanisms involved in these nanocomposites demands a thorough investigation. Nevertheless, the capability of the nano-sized reinforcements to alter the properties even when added in small concentrations makes

FIGURE 4.2 Application areas of nanocomposites.

TABLE 4.3

Brief Summary of Literature concerning the Friction and Wear Properties of Nanocomposites

Base Matrix	Reinforcement	Fabrication Route	Nanoparticle Content	Major Findings	Ref.
Ni-Ti-C MMC	TiC–graphite	Mechanical alloying and spark plasma sintering	TiC: 3%–10% Graphite: 5%–20%	• Microhardness improved • Tribological properties improved	Patil et al. (2020)
Cu matrix composites	Nano-boron carbide	Powder metallurgy	0–1.5 wt. %	The addition of 1.5 wt. % increased the micro-Vickers hardness from 32 to 93 Reduction in wear loss as compared to pure Cu matrix Maximum wear resistance was observed in the case of 1.5 wt. % nano-boron carbide	Lakshmanan, Dharmaselvan, Paramasivam, Kirubanandan, and Vignesh (2019)
Al6061	Al_2O_3	Ultrasonic stir casting	0–3.0 wt. %	Heat treatment increased the hardness and strength of composite leading to the increase in wear resistance	Sahu et al. (2015)
Al6061	SiC	Ultrasonic stir casting	1.2%	• A 73% increase in the hardness of the nanocomposite • Superior wear resistance as compared to Al alloy	Manivannan et al. (2017)
Al6061	Al_2O_3	Mechanical alloying	1–5	Wear rate reduction by grain refinement	Hosseini, Karimzadeh, Abbasi, and Enayati (2012)
Al7075	Si_3N_2/ZrO_2	Stir casting	2, 4 and 6 wt. % of ZrO_2 8% of Si_3N_4	• 4 wt. % of ZrO_2 resulted in enhanced UTS	Rathaur, Katiyar, and Patel (2019)

(Continued)

TABLE 4.3 (*Continued*)

Brief Summary of Literature concerning the Friction and Wear Properties of Nanocomposites

Base Matrix	Reinforcement	Fabrication Route	Nanoparticle Content	Major Findings	Ref.
Al2009	GNPs	Conventional powder metallurgy and multi-pass friction stir processing (FSP)	1 wt. %	FSP led to an improvement in the distribution of graphene Both tensile strength and yield strength improved	Zhang, Liu, Xiao, Ni, and Ma (2018)
Al7055	Graphene	Spark plasma sintering	1.0–5.0 wt. %	1 wt. % graphene addition led to an improvement in mechanical properties such as hardness and compressive strength • Stronger interface formed between matrix and graphene	Tian et al. (2016)
AZ31	ZrO$_2$	Friction stir processing		A decrease in grain size An improvement in mechanical properties	Mazaheri, Jalilvand, Heidarpour, and Jahani (2020)
Al	SiC–nAl$_2$O$_3$–WS$_2$	SiC (10 wt. %) n-Al$_2$O$_3$ (2 wt. %)	WS$_2$ (0, 5 and 9 wt. %)	Hardness and density of the composites increased Impact of sliding speed on wear highest Impact of WS$_2$ content on COF highest	Singh et al. (2018)

them interesting materials. Further, the issues associated with micro-sized composites also make them relevant. The mechanical and tribological properties of these composites depend upon a variety of parameters, and hence, to develop a material for a particular application, proper selection of the parameters is needed. Further, to increase the application arena of these composites, cost-effective techniques need to be developed for their fabrication.

REFERENCES

Anand, R., Raina, A., Ul Haq, M. I., Mir, M. J., Gulzar, O., & Wani, M. (2020). Synergism of TiO_2 and graphene as nano-additives in bio-based cutting fluid-an experimental investigation. *Tribology Transactions*, 1–21. Doi: 10.1080/10402004.2020.1842953.

Baba, Z. U., Shafi, W. K., Haq, M. I. U., & Raina, A. (2019). Towards sustainable automobiles-advancements and challenges. *Progress in Industrial Ecology*, *13*(4), 315–331. Doi: 10.1504/PIE.2019.102840.

Ceschini, L., Dahle, A., Gupta, M., Jarfors, A. E. W., Jayalakshmi, S., Morri, A., … Singh, R. A. (2017). *Aluminum and Magnesium Metal Matrix Nanocomposites*. Doi: 10.1007/978-981-10-2681-2.

Dalmis, R., Cuvalci, H., Canakci, A., & Guler, O. (2016). Investigation of graphite nano particle addition on the physical and mechanical properties of ZA27 composites. *Advanced Composites Letters*, *25*(2), 096369351602500. Doi: 10.1177/096369351602500202.

Dorri Moghadam, A., Omrani, E., Menezes, P. L., & Rohatgi, P. K. (2015). Mechanical and tribological properties of self-lubricating metal matrix nanocomposites reinforced by carbon nanotubes (CNTs) and graphene - A review. *Composites Part B: Engineering*, *77*, 402–420. Doi: 10.1016/j.compositesb.2015.03.014.

El-Ghazaly, A., Anis, G., & Salem, H. G. (2017). Effect of graphene addition on the mechanical and tribological behavior of nanostructured AA2124 self-lubricating metal matrix composite. *Composites Part A: Applied Science and Manufacturing*, *95*, 325–336. Doi: 10.1016/j.compositesa.2017.02.006.

Faisal, N., & Kumar, K. (2018). Mechanical and tribological behaviour of nano scaled silicon carbide reinforced aluminium composites. *Journal of Experimental Nanoscience*, *13* (sup1), S1–S13. Doi: 10.1080/17458080.2018.1431846

Goh, C. S., Wei, J., Lee, L. C., & Gupta, M. (2008). Ductility improvement and fatigue studies in Mg-CNT nanocomposites. *Composites Science and Technology*, *68*(6), 1432–1439. Doi: 10.1016/j.compscitech.2007.10.057.

Gulzar, O., Qayoum, A., & Gupta, R. (2018). Photo-thermal characteristics of hybrid nanofluids based on Therminol-55 oil for concentrating solar collectors. *Applied Nanoscience*, *352*, 436–444. Doi: 10.1007/s13204-018-0738-4

Gulzar, O., Qayoum, A., & Gupta, R. (2019). Experimental study on stability and rheological behaviour of hybrid Al_2O_3-TiO_2 Therminol-55 nano fl uids for concentrating solar collectors. *Powder Technology*, *352*, 436–444. Doi: 10.1016/j.powtec.2019.04.060.

Gupta, M., & Wong, W. L. E. (2015, July 1). Magnesium-based nanocomposites: Lightweight materials of the future. *Materials Characterization*, *105*, 30–46. Doi: 10.1016/j.matchar.2015.04.015.

Hashim, J., Looney, L., & Hashmi, M. S. J. (1999). Metal matrix composites: Production by the stir casting method. *Journal of Materials Processing Technology*, *92–93*, 1–7. Doi: 10.1016/S0924-0136(99)00118-1.

Hassan, S. F., & Gupta, M. (2005). Enhancing physical and mechanical properties of Mg using nanosized Al_2O_3 particulates as reinforcement. *Metallurgical and Materials Transactions A*, *36*(8), 2253–2258.

Hosseini, N., Karimzadeh, F., Abbasi, M. H., & Enayati, M. H. (2012). A comparative study on the wear properties of coarse-grained Al6061 alloy and nanostructured Al6061-Al$_2$O$_3$ composites. *Tribology International, 54*, 58–67. Doi: 10.1016/j.triboint.2012.04.020.

Irfan Ul Haq, M., Raina, A., Anand, A., Sharma, S. M., & Kumar, R. (2020). Elucidating the effect of MoS$_2$ on the mechanical and tribological behavior of AA7075/Si$_3$N$_4$ composite. *Journal of Materials Engineering and Performance, 29*(11), 7445–7455. Doi: 10.1007/s11665-020-05197-8.

Lakshmanan, P., Dharmaselvan, S., Paramasivam, S., Kirubanandan, L. K., & Vignesh, R. (2019). Tribological properties of B4C nano particulates reinforced copper matrix nanocomposites. *Materials Today: Proceedings*, 16, 584–591. Doi: 10.1016/j.matpr.2019.05.132.

Malaki, M., Xu, W., Kasar, A., Menezes, P., Dieringa, H., Varma, R., & Gupta, M. (2019). Advanced Metal Matrix Nanocomposites. *Metals, 9*(3), 330. Doi: 10.3390/met9030330.

Manivannan, I., Ranganathan, S., Gopalakannan, S., Suresh, S., Nagakarthigan, K., & Jubendradass, R. (2017). Tribological and surface behavior of silicon carbide reinforced aluminum matrix nanocomposite. *Surfaces and Interfaces, 8*, 127–136. Doi: 10.1016/j.surfin.2017.05.007.

Mazaheri, Y., Jalilvand, M. M., Heidarpour, A., & Jahani, A. R. (2020). Tribological behavior of AZ31/ZrO$_2$ surface nanocomposites developed by friction stir processing. *Tribology International, 143*, 106062. Doi: 10.1016/j.triboint.2019.106062.

Moghadam, A. D., Omrani, E., Menezes, P. L., & Rohatgi, P. K. (2016). Effect of in-situ processing parameters on the mechanical and tribological properties of self-lubricating hybrid aluminum nanocomposites. *Tribology Letters, 62*(2), 25.

Moghadam, A. D., Schultz, B. F., Ferguson, J. B., Omrani, E., Rohatgi, P. K., & Gupta, N. (2014, April 5). Functional metal matrix composites: Self-lubricating, self-healing, and nanocomposites-an outlook. *JOM, 66*, 872–881. Doi: 10.1007/s11837-014-0948-5.

Paramsothy, M., Chan, J., Kwok, R., & Gupta, M. (2011). The effective reinforcement of magnesium alloy ZK60A using Al$_2$O$_3$ nanoparticles. *Journal of Nanoparticle Research, 13*(10), 4855–4866. Doi: 10.1007/s11051-011-0464-2.

Patil, A., Walunj, G., Torgerson, T. B., Koricherla, M. V., Khan, M. U. F., Scharf, T. W., … Borkar, T. (2020). Tribological behavior of in situ processed Ni-Ti-C nanocomposites. *Tribology Transactions*. Doi: 10.1080/10402004.2020.1800880.

Rathaur, A. S., Katiyar, J. K., & Patel, V. K. (2019). Experimental analysis of mechanical and structural properties of hybrid aluminium (7075) matrix composite using stir casting method. *IOP Conference Series: Materials Science and Engineering, 653*(1), 012033. Doi: 10.1088/1757-899X/653/1/012033.

Sahoo, B., Narsimhachary, D., & Paul, J. (2019). Surface mechanical and self-lubricating properties of MWCNT impregnated aluminium surfaces. *Surface Engineering, 35*(11), 970–981. Doi: 10.1080/02670844.2019.1584959.

Sahu, K., Rana, R. S., Purohit, R., Koli, D. K., Rajpurohit, S. S., & Singh, M. (2015). Wear behavior and micro-structural study of Al/Al$_2$O$_3$ nano-composites before and after heat treatment. *Materials Today: Proceedings, 2*(4–5), 1892–1900. Doi: 10.1016/j.matpr.2015.07.143.

Shafi, W. K., Raina, A., & Ul Haq, M. I. (2018). Friction and wear characteristics of vegetable oils using nanoparticles for sustainable lubrication. *Tribology - Materials, Surfaces and Interfaces, 12*(1), 27–43. Doi: 10.1080/17515831.2018.1435343.

Shin, S. E., Ko, Y. J., & Bae, D. H. (2016). Mechanical and thermal properties of nanocarbon-reinforced aluminum matrix composites at elevated temperatures. *Composites Part B: Engineering*, 106, 66–73. Doi: 10.1016/j.compositesb.2016.09.017.

Singh, N., Mir, I. U. H., Raina, A., Anand, A., Kumar, V., & Sharma, S. M. (2018). Synthesis and tribological investigation of Al-SiC based nano hybrid composite. *Alexandria Engineering Journal, 57*(3), 1323–1330. Doi: 10.1016/j.aej.2017.05.008

Sree Manu, K. M., Arun Kumar, S., Rajan, T. P. D., Riyas Mohammed, M., & Pai, B. C. (2017). Effect of alumina nanoparticle on strengthening of Al-Si alloy through dendrite refinement, interfacial bonding and dislocation bowing. *Journal of Alloys and Compounds*, 712, 394–405. Doi: 10.1016/j.jallcom.2017.04.104.

Tian, W., Li, S., Wang, B., Chen, X., Liu, J., & Yu, M. (2016). Graphene-reinforced aluminum matrix composites prepared by spark plasma sintering. *International Journal of Minerals, Metallurgy, and Materials*, 23(6), 723–729. Doi: 10.1007/s12613-016-1286-0.

Ul Haq, M. I., & Anand, A. (2018). Dry sliding friction and wear behavior of AA7075-Si$_3$N$_4$ composite. *Silicon*, 10(5), 1819–1829. Doi: 10.1007/s12633-017-9675-1.

Vaziri, H. S., Shokuhfar, A., & Afghahi, S. S. S. (2019). Investigation of mechanical and tribological properties of aluminum reinforced with tungsten disulfide (WS2) nanoparticles. *Materials Research Express*, 6(4), 045018. Doi: 10.1088/2053-1591/aafa00.

Zhang, Z. W., Liu, Z. Y., Xiao, B. L., Ni, D. R., & Ma, Z. Y. (2018). High efficiency dispersal and strengthening of graphene reinforced aluminum alloy composites fabricated by powder metallurgy combined with friction stir processing. *Carbon*, 135, 215–223. Doi: 10.1016/j.carbon.2018.04.029.

5 Tribological Properties of Green Hybrid Metal Matrix Composites Reinforced with Synthetic and Industrial– Agricultural Wastes

Pawandeep Singh, R.K. Mishra,
Balbir Singh, and Vivudh Gupta
Shri Mata Vaishno Devi University

CONTENTS

5.1 INTRODUCTION

Materials having good strength-to-weight ratio arise in modern engineering designs where efficient machinery and lower fuel consumption are the essential necessities to be fulfilled. Also, materials used in the development of modern infrastructures, machinery and equipment require a better combination of properties to compete with the service demands (Alaneme, Bodunrin, & Awe, 2018). Composites are mainly classified, on the basis of matrix nature, into three types, namely ceramic matrix composites (CMCs), polymer matrix composites (PMCs) and metal matrix composites (MMCs) (Menezes, Nosonovsky, Ingole, Kailas, & Lovell, 2013). MMCs are progressively being used in many engineering applications such as automotive, electronics and aerospace industries owing to their high specific strength, enhanced resistance to wear and corrosion, low coefficient of thermal expansion and high stiffness (Kumar & Murugan, 2012). Some mechanical properties of monolithic alloys were decreased when subjected to elevated temperature, and it is entail to improve the mechanical properties of such composites over monolithic alloys (Anbuchezhiyan, Mohan, Sathianarayanan, & Muthuramalingam, 2017). The elucidation of materials is primarily dependent upon selection of optimal ceramic reinforcement materials. Ceramic reinforcement has been incorporated into the metal matrix for the purpose of strengthening metal matrices, in order to create a composite material with superior mechanical properties (Pagounis & Lindroos, 1998). A composite material system consists of a combination of two or more different materials with different chemical composition and form. Typically, a composite has two constituents, which are matrix (metal) and reinforcement. They are homogeneous at the macrostructure level and heterogeneous at the microstructure level. MMCs are formed by a combination of materials which have better mechanical properties than monolithic alloy (Kumar, Lal, & Kumar, 2013). Ceramic particles are incorporated into different monolithic light metal alloys, such as zinc (Zn), aluminium (Al), copper (Cu) and magnesium (Mg), and alloys of stainless steel for additional weight saving (Anbuchezhiyan, Muthuramalingam, & Mohan, 2018). MMCs are mainly reinforced with particulates instead of continuous fibres, as they have the competitive advantages of low cost, ease of manufacturing and nominal isotropic properties (Chawla & Shen, 2001). Hard ceramic particles such as B_4C, Al_2O_3, MgO, SiC, WC and SiO_2 and fly ash, groundnut shell ash (GSA), coconut shell ash (CSA), melon ash, rice husk ash, bean pod ash, etc., are used to develop MMCs with better mechanical and tribological properties (Aigbodion, Atuanya, Edokpia, Odera, & Offor, 2016; Paknia, Pramanik, Dixit, & Chattopadhyaya, 2016; Pramanik, 2016; Rao, Sahoo, Ganguly, Dash, & Narasaiah, 2017). Nowadays, hybrid MMCs are produced by addition of two different types of reinforcements, exclusively by combining hard ceramic particles and industrial waste/natural minerals, which are finding wider application due to their enhanced mechanical and tribological properties at a lower cost of production (Shaikh, Arif, & Siddiqui, 2018).

5.1.1 Processing Routes for the Production of MMCs

Traditionally, several processing routes are adopted for the production of MMCs, such as stir casting, compocasting, spray deposition, powder metallurgy and mechanical alloying (Tong, 1998). Figure 5.1 shows various processing routes for MMC production.

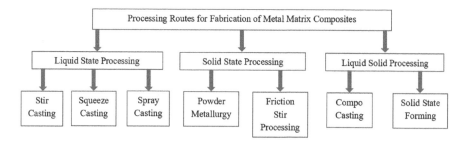

FIGURE 5.1 Different processing routes for fabrication.

5.2   TRIBOLOGICAL PROPERTIES OF NATURAL AND SYNTHETIC REINFORCED HYBRID MMCS

Faisal and Prabagaran (2018) investigated mechanical and wear properties of B_4C, graphite and fly ash-reinforced aluminium hybrid composites. Al7075/B_4C/ fly ash composites showed better hardness than Al7075/B_4C/graphite composites. Decreases in wear and COF were also observed with the addition of B_4C and fly ash. Subramaniam, Arunachalam, Nallasivam, and Pramanik (2019) studied the tribo-mechanical behaviour of sugarcane bagasse ash (SBA) and SiC-reinforced Al-Si10-Mg hybrid composites. Hardness, tensile strength and impact strength of the composites were increased with the addition of SiC. Lower wear rate and COF were also observed in hybrid composites. Reinforcement particles helped to improve the mechanical and tribological properties of hybrid composites.

Singh and Chauhan (2019) investigated dry sliding wear and frictional behaviour of red mud and SiC-reinforced Al2024 hybrid composites using Taguchi's approach. With the addition of red mud particles, wear resistance was improved. ANOVA revealed that COF is controlled by the magnitude of applied load along with red mud weight percentage, the size of the particle and sliding distance. An increase in the amount of red mud aids in reducing the plastic deformation of wear surface by supporting the development of mechanically mixed layer (MML) on the wear surface which causes the reduction in frictional contact between the sliding surfaces. Kurapati and Kommineni (2017) studied the wear parameters effect on dry sliding wear of AA2024/fly ash/SiC-reinforced hybrid composites developed through the stir casting method. Results showed a decrease in wear rate with an increase in the weight addition of fly ash and SiC particles in the matrix. Also, when the applied load increased along with constant sliding distance, the wear rate of the hybrid MMCs was also increased. Also, wear resistance in the case of hybrid composites was better than AA2024 alloy. Alaneme and Sanusi (2015) studied the mechanical and wear behaviour of hybrid aluminium matrix composites reinforced with RHA, graphite and alumina particles prepared using a two-step stir casting method. Hardness of the composites was found to be decreased with increased weight ratio of RHA and graphite particles. Wear resistance of graphite-containing composites was higher as compared to composites with RHA only. A decrease in wear resistance was observed with an increase in graphite content from 0.5 to 1.5 wt.%.

Manoj Kumar and Shanmuga Prakash (2017) investigated wear characteristics of graphite and fly ash particulate-reinforced Al6063 hybrid composites. Dry sliding wear tests were performed on hybrid composites using a pin-on-disc wear tester. It was observed that wear resistance of the hybrid composites increased with the addition of reinforcement particulates. Senthil Kumar, Thiagarajan, and Chandrasekaran (2016) investigated mechanical and wear characteristics of silicate and fly ash-reinforced LM24 hybrid composites fabricated using the vortex method. They concluded that hardness of the composites was increased with an increase in silicate content and was higher than that of the base alloy. Wear loss of base alloy was higher than that of the hybrid composites. With the increase in silicate content, wear loss of the composites was also decreased. Reddy, Kishore, Theja, and Rao (2020) evaluated the wear properties of Al 7075/B_4C/fly ash composites manufactured using stir casting. It was perceived that wear resistance of hybrid composites was better in the case of hybrid composites. Manikandan and Arjunan (2020) found that Al7075 hybrid composites containing B_4C and cow dung ash (CDA) particles showed better wear resistance than the composites reinforced with single reinforcement.

Alaneme and Bamike (2018) performed mechanical and wear characterization on AA6063 hybrid composites reinforced with quarry dust (QD) and SiC. A marginal decrease in hardness and wear resistance of composites with an increase in QD was reported. QD-based results revealed that it can be used as a substitute for SiC for the development of aluminium-based hybrid composites. Raju, Panigrahi, Ganguly, and Rao (2017) investigated mechanical and tribological behaviour of aluminium hybrid composites reinforced with CSA and graphite particles. The addition of graphite particles improved the mechanical properties. Results showed that hybrid composites provide better wear resistance when compared to non-hybrid composites. This was attributed to the combined presence of graphite and alkaline metal oxides in CSA. Almomani, Hayajneh, and Draidi (2016) studied tribological behaviour of zamak alloys reinforced with Al_2O_3 and fly ash produced by the compocasting technique. Hardness and wear tests were conducted on composite specimens. Composites hardness was increased with an increase in weight fractions of the reinforcements. Moreover, wear resistance was also increased. Fly ash particles' effect on the wear resistance was statistically significant than the effect of Al_2O_3 particles. Zhu and Yan (2017) studied wear behaviour of A356 hybrid composites fabricated through the squeeze casting method by employing fly ash and mullite as reinforcement. Compared to the base alloy, composites showed better wear resistance. Alaneme et al. (2014) studied the wear properties of hybrid Al-Mg-Si composites reinforced with RHA and SiC particles. COF and wear resistance of hybrid composites were compared, and it was found that RHA can be used as a substitute for SiC for high-wear-resistance applications. Mahendra and Radhakrishna (2010) studied mechanical and wear behaviour of Al-4.5% Cu alloy where fly ash and SiC particles were used as reinforcements. Hybrid composites are prepared using the stir casting method. Mechanical properties increased with an increase in particulates percentage. Also, dry sliding wear resistance improved with the rise in particulates percentage. Prasat and Subramanian (2013) performed an investigation on tribological behaviour AlSi10Mg hybrid composites reinforced with fly ash and graphite and synthesized using the stir casting method. The authors found that hybrid composites showed greater hardness and tensile strength. COF and wear rate reduced with the incorporation of reinforcement particles. Improvement in tribological properties was the result of the load-carrying ability of fly ash and development of lubricating film by the graphite among the sliding interfaces.

Subramaniam, Natarajan, Kaliyaperumal, and Chelladurai (2019) performed wear studies on B_4C/CSA-reinforced Al7075 hybrid composites. It was reported that an increase in the percentage of reinforcement particles in the matrix resulted in improved wear resistance of hybrid composites. The addition of hard reinforcement improved the wear resistance. Satheesh and Pugazhvadivu (2019) investigated mechanical properties and wear characteristics of Al6061/SiC/CSA hybrid composites. Results showed a 46% increase in hardness. However, 10% CSA addition decreased the hardness by 5% when compared to hybrid composites. Due to the cushioning effect of

CSA particles, wear resistance was improved in the case of hybrid composites. Prasat, Subramanian, Radhika, and Anandavel (2011) investigated dry sliding wear performance of AlSi10Mg/fly ash/graphite hybrid composites using Taguchi's technique. Wear loss was reduced when both fly ash content and sliding speed were increased. At low loads, the mechanism responsible for wear was abrasive wear mechanism. When load was increased, the wear mechanism changed to mixed abrasion and delamination.

5.3 CONCLUSION

1. Different types of hybrid composites were fabricated using industrial/agricultural waste and synthetic reinforcement particles through the most commonly used liquid metallurgy route.
2. Hybrid composites showed improved mechanical properties with incorporation of reinforcement particles as compared to base alloy.
3. Wear loss and COF were also reduced as a result of hard ceramic reinforcements in the matrix.
4. Industrial/agricultural waste and synthetic reinforcements provide strengthening to composites due to the load-bearing capacity of reinforcement particles and can be used in the manufacturing of components that are required for tribological applications.

REFERENCES

Aigbodion, V., Atuanya, C., Edokpia, R., Odera, R., & Offor, O. (2016). Experimental study on the wear behaviour of Al–Cu–Mg/bean pod ash nano-particles composites. *Transactions of the Indian Institute of Metals, 69*(4), 971–977.

Alaneme, K. K., Adewale, T. M., & Olubambi, P. A. (2014). Corrosion and wear behaviour of Al–Mg–Si alloy matrix hybrid composites reinforced with rice husk ash and silicon carbide. *Journal of Materials Research and Technology, 3*(1), 9–16.

Alaneme, K. K., & Bamike, B. J. (2018). Characterization of mechanical and wear properties of aluminium based composites reinforced with quarry dust and silicon carbide. *Ain Shams Engineering Journal, 9*(4), 2815–2821.

Alaneme, K. K., Bodunrin, M. O., & Awe, A. A. (2018). Microstructure, mechanical and fracture properties of groundnut shell ash and silicon carbide dispersion strengthened aluminium matrix composites. *Journal of King Saud University-Engineering Sciences, 30*(1), 96–103.

Alaneme, K. K., & Sanusi, K. O. (2015). Microstructural characteristics, mechanical and wear behaviour of aluminium matrix hybrid composites reinforced with alumina, rice husk ash and graphite. *Engineering Science and Technology, an International Journal, 18*(3), 416–422.

Almomani, M., Hayajneh, M. T., & Draidi, M. (2016). Tribological investigation of Zamak alloys reinforced with alumina (Al_2O_3) and fly ash. *Particulate Science and Technology, 34*(3), 317–323.

Anbuchezhiyan, G., Mohan, B., Sathianarayanan, D., & Muthuramalingam, T. (2017). Synthesis and characterization of hollow glass microspheres reinforced magnesium alloy matrix syntactic foam. *Journal of Alloys and Compounds, 719*, 125–132.

Anbuchezhiyan, G., Muthuramalingam, T., & Mohan, B. (2018). Effect of process parameters on mechanical properties of hollow glass microsphere reinforced magnesium alloy syntactic foams under vacuum die casting. *Archives of Civil and Mechanical Engineering, 18*(4), 1645–1650.

Chawla, N., & Shen, Y. L. (2001). Mechanical behavior of particle reinforced metal matrix composites. *Advanced Engineering Materials, 3*(6), 357–370.

Faisal, M., & Prabagaran, S. (2018). Investigation on mechanical and wear properties of aluminium based metal matrix composite grand fly ash reinforced with B4C. *Advanced Manufacturing and Materials Science: Selected Extended Papers of ICAMMS 2018*, 379.

Kumar, A., Lal, S., & Kumar, S. (2013). Fabrication and characterization of A359/Al$_2$O$_3$ metal matrix composite using electromagnetic stir casting method. *Journal of Materials Research and Technology, 2*(3), 250–254.

Kumar, B. A., & Murugan, N. (2012). Metallurgical and mechanical characterization of stir cast AA6061-T6–AlNp composite. *Materials & Design, 40*, 52–58.

Kurapati, V. B., & Kommineni, R. (2017). Effect of wear parameters on dry sliding behavior of Fly Ash/SiC particles reinforced AA 2024 hybrid composites. *Materials Research Express, 4*(9), 096512.

Mahendra, K., & Radhakrishna, K. (2010). Characterization of stir cast Al-Cu-fly ash+ SiC) hybrid metal matrix composites. *Journal of Composite Materials, 44*(8), 989–1005.

Manikandan, R., & Arjunan, T. (2020). Studies on micro structural characteristics, mechanical and tribological behaviours of boron carbide and cow dung ash reinforced aluminium (Al 7075) hybrid metal matrix composite. *Composites Part B: Engineering, 183*, 107668.

Manoj Kumar, M., & Shanmuga Prakash, R. (2017). Wear characteristics of hybrid Al 6063 matrix composites reinforced with graphite and fly ash particulates. In *Paper Presented at the Applied Mechanics and Materials* (Vol. 854, pp. 1–9). Trans Tech Publications Ltd.

Menezes, P. L., Nosonovsky, M., Ingole, S. P., Kailas, S. V., & Lovell, M. R. (2013). *Tribology for Scientists and Engineers*. New York: Springer.

Pagounis, E., & Lindroos, V. (1998). Processing and properties of particulate reinforced steel matrix composites. *Materials Science and Engineering: A, 246*(1–2), 221–234.

Paknia, A., Pramanik, A., Dixit, A., & Chattopadhyaya, S. (2016). Effect of size, content and shape of reinforcements on the behavior of metal matrix composites (MMCs) under tension. *Journal of Materials Engineering and Performance, 25*(10), 4444–4459.

Pramanik, A. (2016). Effects of reinforcement on wear resistance of aluminum matrix composites. *Transactions of Nonferrous Metals Society of China, 26*(2), 348–358.

Prasat, S. V., & Subramanian, R. (2013). Tribological properties of AlSi10Mg/fly ash/graphite hybrid metal matrix composites. *Industrial Lubrication and Tribology, 65*(6), 399–408.

Prasat, S. V, Subramanian, R., Radhika, N., & Anandavel, B. (2011). Dry sliding wear and friction studies on AlSi10Mg–fly ash–graphite hybrid metal matrix composites using Taguchi method. *Tribology-Materials, Surfaces & Interfaces, 5*(2), 72–81.

Raju, R. S. S., Panigrahi, M., Ganguly, R., & Rao, G. S. (2017). Investigation of tribological behavior of a novel hybrid composite prepared with Al-coconut shell ash mixed with graphite. *Metallurgical and Materials Transactions A, 48*(8), 3892–3903.

Rao, R. G., Sahoo, K., Ganguly, R., Dash, R., & Narasaiah, N. (2017). Effect of flyash treatment on the properties of Al-6061 alloy reinforced with SiC-Al$_2$ O$_3$-C mixture. *Transactions of the Indian Institute of Metals, 70*(10), 2707–2717.

Reddy, T. P., Kishore, S. J., Theja, P. C., & Rao, P. P. (2020). Development and wear behavior investigation on aluminum-7075/B 4 C/fly ash metal matrix composites. *Advanced Composites and Hybrid Materials*, 1–11.

Satheesh, M., & Pugazhvadivu, M. (2019). Investigation on physical and mechanical properties of Al6061-Silicon Carbide (SiC)/Coconut shell ash (CSA) hybrid composites. *Physica B: Condensed Matter, 572*, 70–75.

Senthil Kumar, B., Thiagarajan, M., & Chandrasekaran, K. (2016). Investigation of mechanical and wear properties of LM24/silicate/fly ash hybrid composite using vortex technique. *Advances in Materials Science and Engineering, 2016*.

Shaikh, M. B. N., Arif, S., & Siddiqui, M. A. (2018). Fabrication and characterization of aluminium hybrid composites reinforced with fly ash and silicon carbide through powder metallurgy. *Materials Research Express, 5*(4), 046506.

Singh, J., & Chauhan, A. (2019). Investigations on dry sliding frictional and wear characteristics of SiC and red mud reinforced Al2024 matrix hybrid composites using Taguchi's approach. *Proceedings of the Institution of Mechanical Engineers, Part L: Journal of Materials: Design and Applications, 233*(9), 1923–1938.

Subramaniam, S., Arunachalam, B., Nallasivam, K., & Pramanik, A. (2019). Investigations on tribo-mechanical behaviour of Al-Si10-Mg/sugarcane bagasse ash/SiC hybrid composites. *China Foundry, 16*(4), 277–284.

Subramaniam, B., Natarajan, B., Kaliyaperumal, B., & Chelladurai, S. J. S. (2019). Wear behaviour of aluminium 7075—boron carbide-coconut shell fly ash reinforced hybrid metal matrix composites. *Materials Research Express, 6*(10), 1065d1063.

Tong, X. (1998). Fabrication of in situ TiC reinforced aluminum matrix composites Part I: Microstructural characterization. *Journal of Materials Science, 33*(22), 5365–5374.

Zhu, J., & Yan, H. (2017). Fabrication of an A356/fly-ash-mullite interpenetrating composite and its wear properties. *Ceramics International, 43*(15), 12996–13003.

6 Mechanical and Tribological Properties of Natural Fiber Reinforced Polymer Composites

A. Vivek Anand
MLR Institute of Technology

R. Arvind Singh and S. Jayalakshmi
Wenzhou University

K. Srinivas Rao
MLR Institute of Technology

Xizhang Chen
Wenzhou University

CONTENTS

6.1   INTRODUCTION

Synthetic fibre-reinforced polymer (FRP) composites are widely known for their high specific strength (i.e. strength/density) and are widely used in aerospace and automobile sectors. A critical concern of these composites is that they are not environment-friendly, as they are not biodegradable [1,2]. Due to this reason, research and development of natural composites is being actively pursued [3]. However, natural fibre-reinforced polymer composites (NFRPCs) have high water absorption characteristics and relatively weak mechanical properties. In order to enhance the behaviour of natural composites, combinations of natural and artificial/synthetic fibres as reinforcements are widely researched [4]. It has been reported that when natural fibres are alkali treated and when incorporated in conjunction with synthetic fibres, composites show improved mechanical characteristics [5,6].

Mechanical properties of NFRPCs depend upon the resin/fibre material, fibre volume fraction and water absorption characteristics. Further, the existence of void content in NFRPCs significantly reduces their mechanical properties. The void content of NFRPCs increases with an increase in the length of fibre and loading conditions on the fibre [7]. While tensile and flexural strengths of NFRPCs can be enhanced by increasing the length of fibre, which occurs due to better adhesive bonding of fibre and resin, there exists a certain critical fibre length beyond which the strength will decrease due to curling effect of long fibres [8].

Water absorption by a fibre directly affects the bonding between fibre and resin, which degrades mechanical properties of the composite material. Water absorption of natural fibre leads to the movement of moisture from higher concentration area to lower concentration area by three different mechanisms. Based on the operating mechanism, the behaviour of the composite, i.e. whether or not it follows Fickian behaviour, can be determined [9,10]. The first mechanism initiates between the microgaps present between the polymeric chains of the natural fibres. The second mechanism arises at the gaps present between the reinforcement and resin interface. The third mechanism predominantly occurs in the natural fibres where microcracks initiate in the moisture-absorbed fibre and the cracks grow in the resin [11,12]. Moisture absorption characteristics in natural fibres can be controlled by using proper coupling agents or treating the surface of the fibres. Fibre surface treatments can enhance interlocking between fibres and resin that will in turn improve mechanical properties of NFRPCs [13]. Chemical treatments can be easily undertaken for natural fibres as they contain hydroxyl groups in their cellulose and lignin constituents [13]. The use of coupling agents or fibre surface treatments can tackle the moisture absorption issue in natural fibres [14,15].

Apart from water absorption, tensile, compression, impact and flexural strength properties, the tribological properties (i.e. wear and friction) of natural composites are important as they have good tribological application potential in clutches, bearings and brake pads [16]. A detailed overview of the applications of NFRPCs in the automotive sector can be found in ref. [17]. Reports on tribological studies of natural composites reveal that reinforcement having combinations of natural fibres and artificial/synthetic fibres increases the wear resistance of composites [18,19]. In the present work, fibre-reinforced composites with natural fibre (coir and sisal) and synthetic fibre (E-glass) combination as reinforcements in novolac resin were fabricated. Coir is a coconut fibre, and sisal fibre is extracted from the leaves of *Agave sisalana* plant. These fibres have low density and low price (as they are abundantly available). Cardanol-based novolac resin derived from cashew nutshell liquid was used as the matrix, as it has good antimicrobial property [20]. The fabricated composites were investigated for their water absorption, tensile strength and tribological properties.

6.2 EXPERIMENTAL METHODS

6.2.1 FIBRE PREPARATION

Combinations of short fibres of sisal, coir and E-glass were used as reinforcement, and cardanol-based novolac was used as resin. Before reinforcing in the resin, the natural fibres were heated in sodium hydroxide (6% NaOH) solution at 30°C in a closed container for 60 h. The excess sodium hydroxide stacked on the surface of the fibres was removed by soaking the treated fibres in distilled water, followed by sun-drying for 5 h. Next, cardanol was gently coated on all natural fibres, and the coated fibres were cured at ambient temperature for 24 h. This step was undertaken in order to close the voids/holes on the natural fibres [4]. The cured fibres were then chopped into 20 mm length and mixed in equal proportion with synthetic fibres (E-glass) for all the fibre combinations.

6.2.2 COMPOSITE MOULDING

The hand layup technique was used to fabricate the composite laminates. A steel mould (350 × 150 mm) was designed to prepare the laminates with a thickness of 4 mm. The volume fraction of fibre (V_f) of the NFRPC laminates for all cases was kept constant at 0.4. A thin polythene paper coated with wax was placed on the surface of the mould. Chopped fibres were placed randomly over the wax-coated paper layer by layer. Subsequently, the novolac resin was poured gently over each layer. The entrapped air voids/bubbles between the layers were removed using a hand-roller. While curing the composites, in order to ensure the removal of air voids/bubbles, a hydraulic pressure of 10 bar was applied on the mould via a universal testing machine (UTM). Post-curing was done by placing the samples in an oven at 70°C for 3 h. Samples from the fabricated composite laminates were cut using a diamond saw cutter as per the ASTM standards for each test. Photographs of the fabricated NFRPCs are shown in Figure 6.1.

Nomenclature to denote the type of composite is given in Table 6.1.

6.2.3 WATER ABSORPTION TEST

Water absorption test was performed on the NFRPC laminate samples in accordance with ASTM570 standard. Prior to the test, the initial weight of all the samples was measured using a microbalance. The samples were immersed in deionized water at

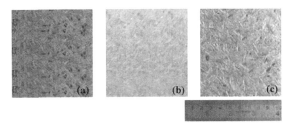

FIGURE 6.1 Photographs of the fabricated NFRPC laminates with reinforcements: (a) coir/glass, (b) sisal/glass and (c) coir/sisal.

TABLE 6.1

Nomenclature of the Fabricated Composite Laminates

Coir/Glass	Sisal/Glass	Coir/Sisal
CG	SG	CS

23°C and their weight was measured periodically until the sample gained the maximum weight (saturation point) [17]. Water absorption of the samples was estimated by taking the difference between the initial weight (i.e. before immersion) and the final weight of the samples (i.e. after immersion), using the following equation:

$$\text{Water Absorption } (\%) = \frac{\text{Weight of the Wet Specimen} - \text{Initial Weight of the Specimen}}{\text{Intial Weight of the Specimen}} \times 100$$

6.2.4 MECHANICAL AND TRIBOLOGICAL TESTS

Tensile tests and tribological tests were conducted on the natural composite laminates. Tensile test was conducted as per ASTM standard D3039 using servo UTM with a cross-head speed of 2 mm/min. For the tension tests, the composite samples were cut to the dimension $120 \times 25 \times 4$ mm. Tribological tests were conducted with sample dimension 4×8 mm using a pin-on-disc equipment, under constant applied normal loads of 5 and 10 N. Test samples were fixed to steel pins by using a super adhesive and were slid against AISI 52100 bearing steel discs at the constant speed of 1 m/s. Each experiment was conducted for 20 min. Wear in terms of height loss (microns) and friction force (N) were continuously recorded during the test duration. All tests were repeated for three times, and the mean values are reported. Figure 6.2 shows a schematic of a composite laminate, and the directions in which the tensile test and sliding wear test were conducted. Tension tests were performed along the longitudinal direction, and tribology tests were carried out along the transverse direction.

6.3 RESULTS AND DISCUSSION

6.3.1 WATER ABSORPTION

Water absorption (%) of the NFRPCs with respect to exposure time (t) is shown in Figure 6.3. The amount of absorbed water by the NFRPC laminate specimens increases linearly with time, initially. After a certain time period, the water absorption reaches a saturation point (maximum weight gained), which proves the samples are following a Fickian diffusion pattern [9,19]. The samples reach their saturation point at 60 h of immersion time.

Natural fibres are made of cellulose, lignin, hemicelluloses, pectin and wax. These features define the properties of fibres, and their content varies according to the type of fibre. Cellulose is a semicrystalline polysaccharide that contains polar hydroxyl groups, which makes natural fibres hydrophilic [21]. Cellulose and hemicellulose determine moisture absorption [21]. Lignin is a cross-linked biopolymer

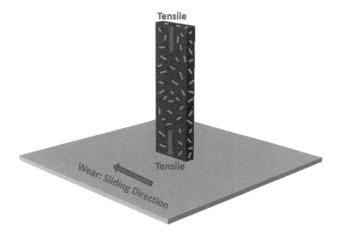

FIGURE 6.2 A schematic of a composite laminate showing the directions in which the tensile tests and sliding wear tests were conducted.

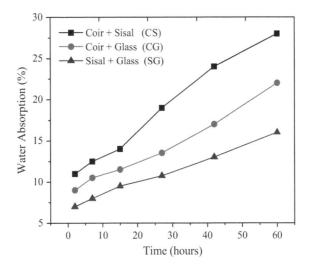

FIGURE 6.3 Variation in water absorption of the three NFRPC laminates with time.

that gives rigidity to fibres [21]. Lignin has the least water absorption property [22]. Comparing the constituents of sisal fibre and coir fibre, (1) sisal has higher cellulose content than coir (sisal: 78 wt. %; coir: 37 wt. %), (2) sisal has lower lignin content than coir (sisal: 8 wt. %; coir: 42 wt. %), and (3) sisal has 10 wt. % hemicellulose [23]. Therefore, raw sisal fibre has relatively higher water absorption property than raw coir ('raw' refers to the untreated state) [23]. Another feature of natural fibres that contributes towards moisture absorption from environment is the presence of hollow cavity called 'lumen' [21]. When natural fibres are treated with NaOH solution (reaction: Fibre—OH + NaOH = Fibre—O−NA+ + H_2O [21]), lignin and hemicellulose get removed from their surfaces [21,22]. As lignin content is more in coir than

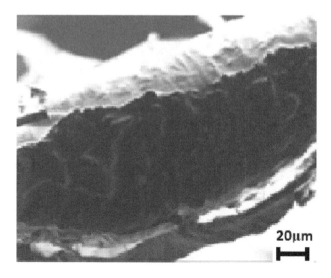

FIGURE 6.4   SEM image of SG sample, showing novolac resin over sisal fibre.

TABLE 6.2
Tensile Properties of the Three NFRPC Laminates

Sample	Tensile Strength (N/mm²)	Elongation (%)
CS	297	3.1
CG	381	4.8
SG	498	5.6

sisal, loss of lignin during the alkali treatment makes the coir fibre to absorb more water than sisal. Due to this reason, it is observed that coir-containing composites, namely CS and CG, show higher water absorption than SG. Amongst CS and CG samples, CG has less water absorption as glass fibres are hydrophobic and do not absorb water. SG shows least water absorption as novolac resin covers the surface of sisal fibre as a protective coating (Figure 6.4) and does not permit water to enter into the fibre [4]. Therefore, the amount of water absorption (%) in the composites was observed to be in the order CS > CG > SG.

6.3.2   TENSILE STRENGTH

Uniaxial tensile test results of the composites are listed in Table 6.2. Tensile strength of composites having a combination of fibres as reinforcement, as in the present case, depends on three factors: (1) combination of the strengths of the individual fibres, (2) moisture absorption and (3) interaction of fibres with resin.

The effects of the above-mentioned three factors are elaborated here.

  i. The tensile strengths of the individual fibres are as follows: sisal, 530–640 MPa; coir, 175 MPa; and E-glass, 3,400 MPa [23]. The combined strength of the fibre combinations is in the order SG > CG > CS. Hence, it is observed that the experimentally determined tensile strength of the composite laminates follows the same order.

ii. The main aim of treating the natural fibres with NaOH alkali treatment was to increase the bonding at the interface between the fibres and resin, so as to enhance the strength properties of the NFRPCs. The alkali treatment is known to increase the surface roughness of fibres resulting in increased surface area [13,24]. As a consequence, better interlocking is achieved between fibres and resin, which would promote the interfacial strength [13,24]. When the interfacial strength is enhanced, load transfer from resin to fibre occurs more efficiently, thereby increasing the overall strength of composites. However, the alkali treatment has a side effect of increasing water absorption in coir fibres, as it removes the high lignin content [22]. Increased water absorption is deleterious to strength properties as it reduces the interfacial strength between fibre and resin [21]. Therefore, composites having coir fibre show lower tensile strength (i.e. CG and CS; Table 6.2) and that the trend in the strength of the composites is inverse to that of their water absorption trend. It is also observed that the

iii. elongation (%) of the composites follows the same trend as that of their tensile strength.

iv. Novolac resin interacts with sisal fibres and coir fibres in different ways. As mentioned earlier, novolac covers the surface of sisal fibre and this will increase the interfacial bonding [25], consequently increasing the load-bearing capacity of the SG composite. The bonding of novolac with sisal fibres is promoted by the removal of wax on the sisal fibres during NaOH alkali treatment; otherwise, in the raw condition (i.e. untreated condition) the presence of wax on the fibres would resist easy spreading of novolac resin over the sisal fibres [21]. In contrast, novolac turns coir fibres brittle, which makes them more vulnerable to crack initiation and premature failure [26]. Due to this reason, the coir-containing composites (i.e. CG and CS; Table 6.2) show lower strength properties under tensile loads.

Taking all the three factors together, it is seen that the tensile strength of the composites is in the order SG > CG > CS.

6.3.3 WEAR AND FRICTION

Wear of the composite laminates in terms of height loss with respect to the sliding time at 5 and 10 N loads is shown in Figure 6.5.

During the initial running-in stage, the composite samples show high wear. Natural composite surfaces have uneven surface topography, and they experience non-conformal contact with the counter-face disc. Uneven surfaces have higher roughness that increases contact pressure due to smaller contact areas (i.e. real contact area < geometrical contact area). Contact pressure induces high wear in the samples initially, as can be seen from Figure 6.5. When the time of sliding increases, the wear of the samples reaches steady-state condition [19]. The steady-state wear of all

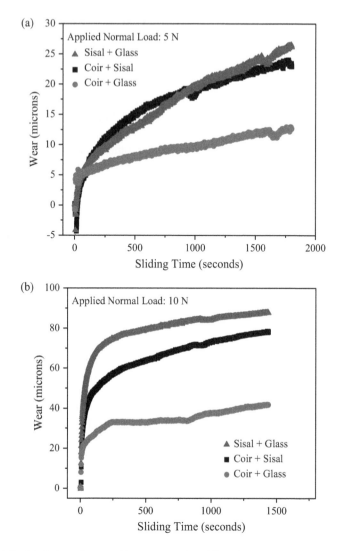

FIGURE 6.5 Variation of wear with respect to the sliding time of the three natural composite laminates at (a) constant load 5 N and (b) constant load 10 N.

the samples increases with the increase in applied normal load; i.e., wear increases when the load is increased from 5 to 10 N. This occurrence is in accordance with Archard's law, which states that wear is directly proportional to the applied load [27].

Wear indicates the undesirable loss of material under tribo-contact conditions. Wear of NFRPCs is strongly related to the type/nature of fibres: (1) coir has good wear resistance and durability than other natural fibres [22], and therefore, coir-containing composites tend to show lower wear; and (2) glass is brittle in nature and breaks easily into smaller particles under sliding, i.e. under shear force, and therefore, glass-containing composites show higher wear. Amongst the glass-containing composites (i.e. CS and CG), CG has lower wear as coir is more wear resistant when compared to sisal. Hence, it is observed that wear, i.e. material loss of the composites, is in the order SG > CS > CG.

Friction indicates the resistance to motion under tribo-contact conditions. Friction force of the composite samples is shown in Figure 6.6.

The trend of the friction force of the composites follows their wear trend, wherein the smaller glass particles generated during wear counter-abrade the surfaces of the sliding samples and increase the friction force. Nevertheless, coir fibres can withstand the abrasion by glass but not sisal fibres. Thus, it is observed that the friction force of the composites is in the order SG > CS > CG. Friction force increases

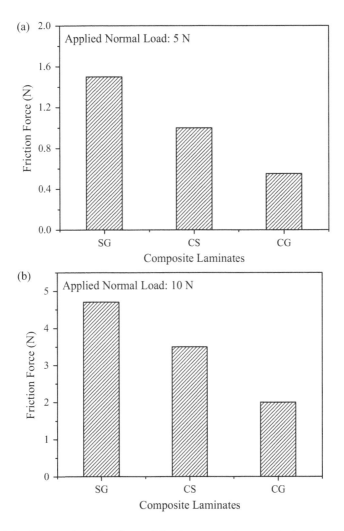

FIGURE 6.6 Measured friction force of the three natural composite laminates at (a) constant load 5 N and (b) constant load 10 N.

with the applied normal load, as the intensity of counter-abrasion increases due to increased wear.

To note, no direct correlation between the tribological and tensile test results could be identified in this work, possibly due to the difference in the direction of the tests.

6.4   CONCLUSIONS

Natural fibre-reinforced composites containing coir/sisal, coir/glass and sisal/glass fibres in novolac resin were fabricated. Their water absorption properties, tensile strength and tribological behaviour were investigated. The conclusions drawn from the studies are as follows:

- Water absorption characteristic of the NFRPCs was in the order CS > CG > SG. Their water absorption property was dependent on two factors, namely (1) loss of lignin in coir fibres due to alkali treatment, which increases water absorption in the fibres; and (2) good coating of novolac resin over the sisal fibres, which reduces water absorption in the fibres. SG composite showed the least water absorption, due to the protection by novolac resin against water absorption, and as glass fibres do not absorb water.
- Tensile strength of the NFRPCs was in the order SG > CG > CS. The strength of the composites was governed by three factors, namely (1) combined strength of the individual fibres; (2) water absorption characteristics, wherein higher water absorption leads to lower strength and, as a consequence, the strength of the composites showed an inverse trend to their water absorption trend; and (3) interaction of the novolac resin with the fibres. The resin covered the surface of sisal fibres and enhanced the interfacial strength, whereas it makes coir fibres brittle. SG composite showed the highest strength due to the high strength combination of sisal and glass, low water absorption and higher interfacial strength because of good coating of novolac resin over sisal fibres.
- Tribological properties of the natural composites were influenced mainly by two factors, namely (1) good wear resistance of coir fibres and (2) brittleness of glass that increased both wear and friction. Both the wear and friction of the NFRPCs followed the order SG > CS > CG. The CG composite showed the best tribological properties due to the high wear-resistant property of the coir fibres.
- Cardanol-based novolac resin can be a promising matrix in NFRPCs. The combination of natural fibres with artificial/synthetic fibres is effective in improving NFRPC performance, for which right selection of fibres and resin is essential.

REFERENCES

1. D. Saravana Bavan & G.C. Mohan Kumar. 2010. Potential use of natural fibre composite materials in India. *Journal of Reinforced Plastics & Composites*, 29(24), pp. 3600–3613. Available at https://journals.sagepub.com/doi/abs/10.1177/0731684410381151.

2. N. Venkateshwaran, A. ElayaPerumal & M.S. Jagatheeshwaran. 2011. Effect of fiber length and fiber content on mechanical properties of banana fiber/epoxy composite. *Journal of Reinforced Plastics & Composites*, 30(19), pp. 1621–1627. Available at https://journals.sagepub.com/doi/abs/10.1177/0731684411426810.

3. R. Kumar, M.I. Ul Haq, A. Raina & A. Anand. 2017. Industrial applications of natural fibre-reinforced polymer composites – challenges and opportunities. *Industrial Journal of Sustainable Engineering*, 12(3), pp. 1–9. Available at https://www.tandfonline.com/doi/full/10.1080/19397038.2018.1538267.

4. R. Kannan, A. Vivek Anand, V. Hariprasad, R. Arvind Singh, S. Jayalakshmi & V. Arumugam. 2017. Effect of cashew nut shell oil (Cardanol) on water absorption and mechanical characteristics of sisal fiber. *RIPE Conference (Mar 23–24)*, MIT Chennai, India. Available at https://www.researchgate.net/publication/316042001_Effect_of_Cashew_Nut_Shell_Oil_Cardanol_on_Water_Absorption_and_Mechanical_Characteristics_of_Sisal_FiberFibres.Journal.

5. S. Nunna, P.R. Chandra, S. Shrivastava & A.K. Jalan. 2012. A review on mechanical behavior of natural fibre based hybrid composites. *Journal of Reinforced Plastics and Composites*, 31(1), pp. 759–769. Available at https://journals.sagepub.com/doi/10.1177/0731684412444325.

6. L.Y. Mwaikambo & M.P. Ansel. 2002. Chemical modification of hemp, sisal, jute and kapok fibres by alkalization. *Journal of Applied Polymer Science*, 84, pp. 2222–2234. Available at https://scinapse.io/papers/2000814017.

7. M. Mehdikhani, L. Gorbatikh, I. Verpoest & S.V. Lomov. 2018. Voids in fiber-reinforced polymer composites: A review on their formation, characteristics, and effects on mechanical performance. *Journal of Composite Materials*, pp. 1–91. Available at https://journals.sagepub.com/doi/full/10.1177/0021998318772152.

8. C. Pavithran, P.S. Mukherjee & M. Brahmakumar. 1991. Coir-glass intermingled fibre hybrid composites. *Journal of Reinforced Plastics and Composites*, 10(1), pp. 91–101. Available at https://journals.sagepub.com/doi/10.1177/073168449101000106.

9. E. Muñoz & J.A. García-Manrique. 2015. Water absorption behaviour and its effect on the mechanical properties of flax fibre reinforced bioepoxy composites. *International Journal of Polymer Science*, 2015, pp. 1–10. Available at Doi: 10.1155/2015/390275.Journal.

10. H.M., Akil, L.W. Cheng, Z.M. Ishak, A.A. Bakar & M.A. Rahman. 2009. Water absorption study on pultruded jute fibre reinforced unsaturated polyester composites, *Composite Science and Technology*, 69(11–12), pp. 1942–1948. Available at Doi: 10.1016/j.compscitech.2009.04.014.

11. H.M. Akilac, C. Santulli, F. Sarasini, J. Tirillò & T. Valente. 2014. Environmental effects on the mechanical behaviour of pultruded jute/glass fibre-reinforced polyester hybrid composites. *Composite Science and Technology*, 94(9), pp. 62–70. Available at Doi: 10.1016/j.compscitech.2014.01.017.

12. C.-H. Shen & G.S. Springer. 1976. Moisture absorption and desorption of composite materials. *Journal of Composite Material*, 10, pp. 2–20. Available at https://journals.sagepub.com/doi/abs/10.1177/002199837601000101.

13. I. Siva, J.T.W. Jappes & B. Suresha., 2012. Investigation on mechanical and tribological behavior of naturally woven coconut sheath-reinforced polymer composites. *Polymer Composites*, pp. 723–732, Available at https://onlinelibrary.wiley.com/doi/epdf/10.1002/pc.22197.

14. M.S. Anbupalani, C.D. Venkatachalam & R. Rathanasamy., 2020. Influence of coupling agent on altering the reinforcing efficiency of natural fibre-incorporated polymers – A review. *Journal of Reinforced Plastics and Composites*, pp. 1–25. Available at Doi: 10.1177/0731684420918937.

15. A. Espert, F. Vilaplana & S. Karlsson. 2004. Comparison of water absorption in natural cellulosic fibres from wood and one-year crops in polypropylene composites and its influence

on their mechanical properties. *Composites Part A: Applied Science and Manufacturing*, 35(11), pp. 1267–1276. Available at Doi: 10.1016/j.compositesa.2004.04.004.

16. Z. Nusrat & A. Srivastava. 2019. Study on tribology of natural fiber reinforced polymer composites: A review. *SVOA Materials Science and Technology*, 1(1), pp. 4–10. Available at https://sciencevolks.com/materials-science/pdf/SVOA-MST-01-002.pdf.

17. L. Mohammed, M.N.M. Ansari, G. Pua, M. Jawaid, & M.S. Islam. 2015. A review on natural fiber reinforced polymer composite and its applications. *International Journal of Polymer Science*, 2015, Article ID 243947, pp. 1–15. Available at Doi: 10.1155/2015/243947.

18. T. Singh, B. Gangil, B. Singh, S.K. Verma, D. Biswas & G. Fekete. 2019. Natural-synthetic fiber reinforced homogeneous and functionally graded vinylester composites: Effect of bagasse-Kevlar hybridization on wear behavior. *Journal of Materials Research and Technology*, 8(6), pp. 5961–5971. Available at Doi: 10.1016/j.jmrt.2019.09.071.

19. N.A. Nordin, F.M. Yussof, S. Kasolang, Z. Salleh & M.A. Ahmad. 2013. Wear rate of natural fibre: Long Kenaf composite. *Procedia Engineering*, 68, pp. 145–151. Available at Doi: 10.1016/j.proeng.2013.12.160.

20. K. Krishnadevi, S. Devaraju, S. Sriharshitha, M. Alagar & Y.K. Priya. 2020. Environmentally sustainable rice husk ash reinforced cardanol based polybenzoxazine bio-composites for insulation applications. *Polymer Bulletin*, 77, pp. 2501–2520. Available at https://link.springer.com/article/10.1007/s00289-019-02854-4.

21. A. Devadas, U. Nirmal & J. Hossen., 2018. Investigation into mechanical & tribological performance of kenaf fibre particle reinforced composite. *Cogent Engineering*, 5(1), p. 1479210. Available at Doi: 10.1080/23311916.2018.1479210.

22. M. Mittal & R. Chaudhary. 2018. Experimental study on the water absorption and surface characteristics of alkali treated pineapple leaf fibre and coconut husk fibre. *International Journal of Applied Engineering Research*, 13(15), pp. 12237–12243. Available at https://www.ripublication.com/ijaer18/ijaerv13n15_75.pdf.

23. E. Omrani, P.L. Menezes & P.K. Rohatgi. 2016. State of the art on tribological behavior of polymer resin composites reinforced with natural fibers in the green materials world. *Engineering Science and Technology, An International Journal*, 19(2), pp. 717–736. Available at Doi: 10.1016/j.jestch.2015.10.007.

24. P.A. Udaya Kumar, R.B. Suresha & R. Hemanth. 2018. Mechanical and tribological behavior of vinyl ester hybrid composites. *Tribology in Industry*, 40(2), pp. 283–299. Available at Doi: 10.24874/ti.2018.40.02.12.

25. M Das, V.S. Prasad & D. Chakrabarty. 2009. Thermogravimetric and weathering study of novolac resin composites reinforced with mercerized bamboo fiber. *Polymer Composites*, 30(10), pp. 1408–1416. Available at Doi: 10.1002/pc.20705.

26. S. Jayavani, H. Deka, T.O. Varghese & S.K. Nayak. 2015. Recent development and future trends in coir fiber-reinforced green polymer composites: Review and evaluation. *Polymer Composites*, 37(11), pp. 3296–3309. Available at Doi: 10.1002/pc.23529.

27. J.F. Archard. 1953. Contact and rubbing of flat surfaces. *Journal of Applied Physics*, 24, pp. 981–988. Available at Doi: 10.1063/1.1721448.

7 Solid Lubricant Coatings
Effective Lubricating Coatings for Tribological Applications

Summèra Banday
University of Kashmir

M. Junaid Mir
Islamic University of Science and Technology

M.F. Wani
National Institute of Technology, Srinagar

CONTENTS

7.1 INTRODUCTION

The vitality misfortunes because of energy losses occurred due to wear and friction in the industrialized nations speak to a yearly expense evaluated at somewhere in the range of 5% and 7% of their gross domestic product, and almost 33% of the world's energy resource in current use show friction and wear in one form or another [1]. Consistently, countless components in industries must be disposed of because of exorbitant wear. It has also been measured that 10% of oil utilization in the United States is utilized just to reduce friction [2]. In an automobile, a decrease of friction among parts of motors could yearly diminish about 5% worldwide utilization of fuel joined by a yearly decline of 250 million tons of CO_2 discharges. However, from the perspective of both energy saving and ecological security, it is an impressive concern to reduce friction and wear through tribological standards (the reasonable utilization of materials, coatings, lubricants, surface medicines and unique structural designs).

Surface engineering, by applying surface coatings and treatments, is one of the best and adaptable solutions for solving tribological issues. Coatings modify tribological systems by reducing the coefficient of friction, inducing compressive stresses, enhancing the surface hardness and changing the surface roughness and chemistry [3]. In this way, they improve surface wear resistance and increased the life of a significant components. During the last several decades, various coatings and deposition techniques have effectively formed, and they are applied to reduce friction and/or to avoid surfaces from wear-out in various mechanical systems. The increased use of coatings for tribological applications is mostly dependent on certain reasons: (1) various researchers observe that surface is a most significant part in several engineering components, and maximum failures occur due to the properties of surface area. (2) Advanced and better performance is essential for mechanical parts and tools, which can't be acknowledged just by improving structures or choosing different materials. The utilization of coatings can enhance the properties of surface areas to reduce friction, corrosion resistance, wear resistance and various other functional attributes; simultaneously, the substrate properties do not change, which are responsible for the toughness and strength. (3) In some unique circumstances, the systems can't typically work without using advanced tribological coatings, for instance bearing systems and devices in space mechanisms working at close vacuum conditions, or the engineering components of aeroturbines that work under erosive or corrosive conditions. (4) Development of innovative techniques to produce some new deposition methods, which gives the chance of preparing coatings with greater performance that was not achieved till date.

However, with the advancement of surface engineering, there are several issues. It is difficult to prepare coatings with the entirety of the ideal properties, such as high hardness, low shear strength, high toughness and high bonding strength, in the light of the fact that some of them are in conflict with one another. For instance, high hardness will sacrifice bonding strength and toughness. Therefore, we have to choose the most appropriate one from a large number of coatings for a particular application. The utilization of coatings is a complex situation, and there is no broad guideline to help the selection of coatings for different tribological applications. The present determination technique is generally founded on two procedures. One is to attempt a coating, and on the off chance that it works it will be utilized; so some progressively reasonable coatings will be neglected. The other is experimentation (trial and error), which is costly and time-consuming.

Self-lubricating coatings or solid lubricant films have been utilized to decrease friction in sliding or rotating machine components, for example those working in ultra-high vacuum, in space or in other applications where re-lubrication of tribo-pairs is not possible or any other conventional lubricating material can't be used due to the lack of accessibility [4]. Up to now, various coating methods have been used, for example ion beam-assisted deposition (IBAD), physical vapour deposition, plasma spraying, chemical vapour deposition and their various combinations. Such coating methods provide a wide range of means and adaptability for developing self-lubricating coatings or solid lubricants on substrates of different shapes, sizes and materials [5]. Exceptionally different types of self-lubricating coatings have been created, including carbon-based coatings [6], soft metal coatings [7], fluoride [8] and sulphate coatings [9], TMD coatings [10] and polymer [11] and oxide coatings [12].

It is notable that tribological performances are not inherent or intrinsic ability of materials, but rather are emphatically method and system dependent. Also, tribological performances of various coatings and materials are emphatically affected both by working test parameters, for example counterbody materials, sliding speed and contact pressure, and by the environmental components, for example humidity, temperature, vacuum degree and atmosphere. It was also observed that self-lubricating coatings give successful lubrication within a narrow (constrained) temperature range and their output declines rapidly outside that range. In such cases, the key challenge is to design such self-lubricating coatings that could be utilized over broad working conditions. This chapter is generally aimed at reviewing tribological performances of self-lubricating/solid lubricant coatings created before.

7.2 SELF-LUBRICATING COATINGS

These coatings have been separated into three categories: transition metal dichalcogenide lubricant coatings, adaptive tribological coatings and hybridized tribological coatings.

7.2.1 TRANSITION METAL DICHALCOGENIDE LUBRICANT COATINGS

MX_2 (X = S, Se, Te; and M = W, Mo, Nb, Ta) transition metal dichalcogenides (TMD) are one of the solid lubricant materials that are nowadays used for various industrial applications [13,14]. TMD have hexagonal layered structures and hexagonal sheets are held by weak van der Waals forces, whereas metal and chalcogens are bonded by strong covalent bond. However, during sliding the hexagonal layers slip easily due to the lamellar structure of TMD. Among these TMD, molybdenum disulphide (MoS_2) is an outstanding solid lubricant for dry and vacuum environments and is extensively used in very high-precision space applications, such as gears, satellite bearings and gimbals working under different temperature ranges. Extensive literature studies are available on the tribological behaviour of MoS_2 coating in ambient condition, and in this respect, MoS_2 coating has attained a maximum attention [15–22]. However, these investigations clearly show the formation of film and metal oxides on the sliding surfaces [15–17]. The tribological properties of MoS_2 coating are mainly due to its lamellar or layer lattice structure with a weak van der Waals force existing among the lamellae [23–25].

Yukhno et al. [26] observed that with the addition of materials that have a lamellar structure, such as graphite, MoS_2 or blends of graphite, MoS_2 in polymer coatings, leads to a slight improved friction coefficient and yet at the same time shows the lubricating property when temperature declines from 293 to 77 K. Colbert et al. [27] examined tribological characterization of MoS_2 coatings at various temperatures and revealed that MoS_2 coatings reduce the coefficient of friction to 0.02 within the temperature range of −100°C to 100°C, and also 50% of wear rate is reduced at −100°C. Ostrovskaya et al. [28] investigated that with a decrease in temperature from 293 to 77 K, polycarbamide coating with the addition of MoS_2 shows an increase in friction coefficient from 0.02 to 0.04. In the case of cryogenic liquids, frictional properties in low-pressure gaseous conditions or in vacuum are not exactly the same as those at cryogenic temperature. In the contact zone, cryogenic liquids create cooling of tribopairs as well as heat removal. It has also been observed that solid lubricant coatings result in a decrease in the coefficient of friction in liquid H_2 and N_2, compared to that in open air at room temperature. The reduction in temperature prompted embrittlement as well as enhanced hardness of an individual element in a solid lubricant coating. The impact of such elements decides tribological properties of those self-lubricating coatings with a decrease in temperature. Kohli and Prakash [29] also analysed the impact of increasing temperature up to 350°C on frictional properties of burnished MoS_2 coatings. The coefficient of friction experienced consecutive decline, relatively steady and increasing patterns successively. Gamulya et al. [30] also studied the consequence of temperature on the coefficient of friction of MoS_2 coating with the addition of various additives. During tribological tests at 293–120 K, sprayed MoS_2 coatings with the addition of polytetrafluoroethylene (PTFE), graphite or both as an additive and urea–formaldehyde pitch (UFAR) as the binder material show contact friction. Brainard et al. [31] showed that disulphides are most stable in vacuum, whereas diselenides are intermediate and ditellurides are least stable.

Hu et al. [32] examined tribological conduct of MoS_2 coatings developed on laser-finished surface. It has been observed that deposition of MoS_2 coating over laser-textured surface is a viable method to improve tribological properties of counterbody. Results show that a denser coating has improved adhesion strength and tribological properties with textured steel substrate. A friction coefficient of less than 0.1 was observed in hot-pressed coatings. Also, hot-pressed coatings have 15 times more wear life than the coatings developed by burnishing. Moreover, hot-pressed coatings show a significant increase in wear life with an increase in texture density. Donnet et al. [33] studied the tribological behaviour of RF magnetron-sputtered MoS_2 coatings in various environments deposited on bearing steel substrate in an ultra-high-vacuum condition with a sphere-on-plane reciprocating tribometer. It was observed that ultra-high-vacuum condition as well as dry nitrogen condition shows a low friction coefficient of MoS_2. A low friction force in ultra-high vacuum is attributed to the super lubricating property of MoS_2 as a result of basal plane direction of MoS_2 grains, though in dry nitrogen environment, similar behaviour has been seen with no tribochemical reaction occurring between MoS_2 and nitrogen. Vierneusel et al. [34] investigated the humidity resistance of MoS_2 coatings developed by unbalanced magnetron sputtering. In these conditions, which are free from oxygen or water vapour (i.e. vacuum), MoS_2 is ideal as a solid lubricant. Humid air reduces film performance because of oxidation and enhanced wear rate and friction coefficient. The main aim of this study was further advancement of sputtered MoS_2 coatings to enhance their working properties in different environmental circumstances by improving their resistance to humidity. The coating development method is supported by the use of an experimental Box–Behnken design with change in parameters of deposition, such as temperature, distance between substrate and target, cathode voltage and argon gas pressure. Unlike traditional one-factor-at-a-time (OFAT) analysis, this methodology enables a purpose of relations among deposition method parameters and tribological mechanical performances of MoS_2 film. Results showed that hardness-to-modulus ratio and residual stress state in the films are essential for their tribological properties in vacuum and humid air conditions. After in-depth investigation of relationships among the deposition parameters and film performance, certain particular micro-structural investigations are obtained which displays a considerably basal orientation of the lattice has not only positive effects on resistance to wear but also shows anisotropic film performances, which causes fissile coating fracture when point load or strong shock occurs.

7.2.2 ADAPTIVE TRIBOLOGICAL COATINGS

Adaptive tribological coatings have been developed to offer incredible self-lubricating performance not just by using only one coating material but also by a combination of chemical or physical reactions that are developed by an operating condition and working environment [4,35]. It was also observed that the addition of metal in MoS_2 deposited by the magnetron sputtering deposition method reduces the sensitivity of coating to moisture [36].

Martins et al. [37] investigated tribological property of MoS_2/Ti low-friction coating and its capability to work in heavy-loaded rolling sliding contacts for gears. Several tests, such as ball cratering, Rockwell indentations and reciprocating pin-on-disc wear, were conducted. The average friction coefficient between a gear teeth is investigated and is compared with an uncoated steel gears. MoS_2/Ti coating endorses a decline in friction coefficient among gear teeth and causes an enormous wear volume than uncoated gears. The high-quality gears and low roughness also enhance its scuffing load capacity, even though in a lesser quantity. Stoyanov et al. [38] investigated scaling impacts of Ti–MoS_2 coating on the tribological performance at micro- and macro-levels. The tribological performance of Ti–MoS_2 coating (9 at.% Ti) was investigated using an in situ tribometer and nano-indentation instrument equipped with macroscopic length scales for microsliding experiments. Experiments and estimations were led in controlled situations both at low and high moistness (e.g. ~4% RH and ~35% RH). Tribological tests were conducted using spherical diamond tip and sapphire tip with a radius of 50 µm and 3.175 mm, respectively. For the two scales, the Hertzian contact pressures vary from 0.41 to 1.2 GPa. Friction was observed to be higher at the micro-level than at the macro-level. The main cause of higher friction coefficient is the three stages that have been encountered during the sliding phase, which were recognized with respect to different tip shapes, contact areas, contact pressures and environmental conditions. Solid lubrication behaviour with film shear transfer, interfacial sliding and micro-ploughing was observed during the first two stages. The dominant VAM ploughed for the third stage, and the normalized transfer film thickness fell below the range of 0.001–0.1. Comparisons between the two scales show that microscopic contacts on Ti–MoS_2 during dry sliding demonstrate a minor variance from macroscopic behaviour with greater micro-ploughing and friction limiting. For humid sliding, microscopic contacts significantly show deviation from macroscopic behaviour with ploughing behaviour and lack of transfer film. Banday and Wani [39,40] studied the adhesion, hardness and tribological performance of Ti/MoS_2 coating deposited by the pulse laser deposition method at the nano-level. Scratch result displays that coating starts peeling off at a load of 1327.75 µN. Tribological tests at the nano-level display that the wear rate reduces with increasing wear cycles, due to the self-lubricating effect of Ti/MoS$_2$ coating. Ti/MoS$_2$ coatings also show higher hardness than pure MoS$_2$ coating. Also, a smooth wear track was observed with no debris and cracks, which indicates that plastic flow of coating takes place.

Jing et al. [41] investigated the effect on MoS$_2$-based coatings with the addition of Ti or TiN deposited by unbalanced plasma plating. Results showed that MoS$_2$ composite coatings possess higher wear resistance and are very less sensitive to humid air as compared to MoS$_2$ coatings. Such coatings show tremendous performance for industrial drilling and turning applications. It has also been observed that the addition of Ti atom in MoS$_2$ coating clearly refined the crystallites of MoS$_2$-based lubricants and increased the density of the deposited coatings, which in turn enhances the wear life and the load-bearing capacity. Researchers have concluded that MoS$_2$-based composite coatings are used in place of cutting fluids, and TiN–MoS$_2$/TiN and TiN–MoS$_2$/Ti coatings display better tribological performance even when stored in humid air. Under various humidity environments, the impacts of Cr and Ti on the tribological behaviour of composite MoS$_2$ coatings were studied by Ding et al. [42]. The unbalanced magnetron sputtering technique was used for the development of MoS$_2$–Cr and MoS$_2$–Ti composite coatings on high-speed steel substrate. X-ray (EDX), X-ray diffraction (XRD) and nanoindenter were used to study the characterization and mechanical properties of deposited coatings. Ball-on-disc tribometer was utilized to study the tribological performance of deposited coatings in various humid conditions, and alumina ball was used as a counterbody. It was observed that 16.6 at.% Cr or 20.2 at.% Ti shows higher hardness of 7.5 and 8.4 GPa, respectively, than the other added contents of Cr or Ti, whereas small added content of Cr or Ti shows improved tribological performances in humid air. Tribological tests revealed that 10 at.% of Cr or Ti in MoS$_2$ coatings shows better tribological performance in terms of low wear rate and friction. The main cause of the outstanding tribological performance of MoS$_2$ coating with the addition of Cr or Ti is the formation of stable transferred layer on a surface of alumina ball, which acts as a lubricant and prevents metal-to-metal contact. Thus, MoS$_2$–Cr and MoS$_2$–Ti composite coatings can be widely used for various industrial applications under dry to high-humidity environment because of their excellent tribological properties. Kao [43] studied the tribological performance and high-speed drilling application of MoS$_2$–Cr coatings. Kao also observed that appropriate percentage of Cr enhances the properties of MoS$_2$ coating not only increase the coating adhesion strength with a substrate, mechanical and tribological performance but also increases its machining performance. In this study, ball- and cylinder-on-disc tribometers were used to analyse the tribological performance with various counterbodies (ball and cylinder). Additionally, the machining property of coated micro-drills was tested by high-speed through-hole drilling on PCBs. The results show that sliding of MoS$_2$–Cr8% coating against Si$_3$N$_4$ ball creating point contact has optimal tribological properties, whereas in the case of line contact, MoS$_2$–Cr8% coating or MoS$_2$–Cr8% coating shows optimal tribological properties when bronze cylinder or 6061 Al cylinder, respectively, was used as a counterbody. Finally, it was observed that MoS$_2$–Cr5% coating on micro-drill enhances its tool service life by a factor of two compared to uncoated micro-drills when tested for high-speed PCB drilling under dry conditions.

Baker et al. [44] have examined tribological performance of MoS_2 composite nanocoatings, which adjust its tribological properties with an environmental temperature and humidity. The coating design involves development of nanocrystalline hard oxide particles for enhancing resistance to wear and embeds these hard oxide particles into an amorphous matrix for increasing its toughness, and addition of nanocrystalline and/or amorphous solid lubricants for a decline in friction under various environmental conditions. Tribological tests were carried out in humid air and dry nitrogen conditions and at 500°C during heating in air. The results show that both graphitic carbon and MoS_2 coatings worked together to reduce friction under various humid environmental conditions, whereas Au coating shows low friction at an elevated temperature. Lower friction coefficient values of 0.02–0.03 were observed in dry nitrogen, whereas higher friction coefficients of 0.1–0.15 and 0.1 were found in humid air and in air at 500°C, respectively. Thus, such 'chameleon' type of coatings shows adoptable tribological behaviour with respect to different environmental conditions relevant to aerospace systems. Lince et al. [45] researched the electrical, structural and frictional performance of co-sputtered Au–MoS_2 composite nanocoatings and surface changes at nano- and macro-scale levels during the sliding test. The outcomes were compared to explain the nanostructures for the coatings. Analysis of XRD, TEM and electrical resistance shows that the composite nanocoatings with an isolated Au crystallite diameter of 2–30 nm (for Au concentrations above 16 at.%) remain in the amorphous MoS_2 matrix. Electrical conductivity imaging carried out by AFM matches these tests, displays these coatings have metallic and semi-conductive domains, constant with an unreacted Au and MoS_2. The Auger electron examination with profiling of the ion sputter depth shows a nanometre-thick MoS_2 layer that does not contain Au on top of the rest of the Au- and MoS_2-containing coatings. The nanometre-thin MoS_2 layer was analysed in a lattice-resolution AFM after multiple cycles and was observed to be crystalline and parallel to the coating surface with a basal plane (0 0 0 1). The electrical results were revealed to understand the resistance of coating, with a marginal contribution of contact resistance. Pin-on-disc tribometer was used to analyse friction performance of Au–MoS_2 coatings. Researchers analysed that both the optimum Au concentration range and the contact stress (8.5 MPa) in this study were between the values that have already been studied, and evaluated 750 and ~0.1 MPa contact stresses. The friction and electrical resistance records generally show the potential significance of such coatings for various applications such as switches, slip rings and potentiometer elements. Stoyanov et al. [46] studied tribological performances of Au and Au–MoS_2 coatings at the micro-scale level. Nanoindentation instrument was used to study the tribological performances of Au and Au–MoS_2 coatings. Due to the positive impact of MoS_2 on mechanical and tribological performances, MoS_2 was chosen as an additive for this study. Scratch tests were carried out at the micro-scale level using diamond indenter of 50 μm at varying normal loads of 0.2 and 5.0 mN. Tribological study of Au and Au–MoS_2 coatings had related to different modes of velocity and adhesion levels. It was observed that Au coating with doping of 20 mol% MoS_2 shows limiting friction and reduced adhesion strength and indeed enhances wear resistance significantly. Such coating has applications in microswitches and microcomponents because of its better electrical conductivity, less friction and high wear resistance.

Stoyanov et al. [47] also investigated the tribological behaviour of Au–MoS$_2$ nanocoatings and Au/MoS$_2$ coatings at the micro-level. Nanoindentation instrument was used to conduct scratch tests and microwear tests on pure Au, Au/MoS$_2$ and co-sputtered Au–MoS$_2$ in dry air. SEM, micro-Raman spectrometer and AFM were used to analyse the tribofilm developed on a wear track during microwear tests. The result shows that Au–MoS$_2$ coatings have better resistance to wear than pure Au films, which is mainly due to the decrease in surface adhesion forces and the increase in hardness values. Similar behaviour has been observed for Au/MoS$_2$ samples. In the case of Au, the wear rate decreases due to the top MoS$_2$ layer, which acted as a sacrificial layer in order to prevent the Au. Chien et al. [48] also studied tribological behaviour of MoS$_2$/Au coatings to improve the property of MoS$_2$ coating by depositing a nanometre layer of Au (~80 nm) over MoS$_2$ surface. It has been observed that Au film enhances the property of MoS$_2$/Au as compared to the coatings without the addition of Au. During sliding tests, it was seen that after 100 cycles, the coefficient of friction rapidly decreases to 0.045, whereas with an increase in cycles (after 15,000 cycles), a gradual increase in friction coefficient (μ~0.15) was observed. In this case, an average property was measured over 50,000 cycles. It was also clear that with the addition of Au, the wear rate of Au or Au–MoS$_2$ coating was increased because of its ability to prevent MoS$_2$ to react with oxygen or moisture. As, MoS$_2$ reacts with oxygen and counterbody material that causes change in transfer of layer and mode of wear mechanism of deposited coatings and does not provides a lubrication effect, which results in loss of endurance of MoS$_2$. Thus, researchers have analysed that depositing a nanometre-thin Au layer over MoS$_2$ layer considerably avoids tribochemical or humidity impact on MoS$_2$ layer and enhances the wear life.

Huang and Xiong [49] studied the microsurface, phase, microhardness and tribological performance of MoS$_2$ coating with the addition of Al$_2$O$_3$ to develop Ni–MoS$_2$/Al$_2$O$_3$ coatings that have been deposited by pulse electrodeposition. The MoS$_2$ particles were coated with 5 wt. % of Al$_2$O$_3$ effectively by managing Al(NO$_3$)-3·9H$_2$O aqueous solution hydrolysis. Ni–MoS$_2$/Al$_2$O$_3$ composite electroplating was developed using the current method of pulse current electroplating. It is easy to incorporate the MoS$_2$/Al$_2$O$_3$ particles in Ni electroplating as compared to MoS$_2$ particles. Morphological studies show irregular Ni–MoS$_2$ coating deposition, since the nickel-coated MoS$_2$ seems as the protruding lines. Also, incorporation of particles of MoS$_2$ into a solution results in dendritic growth, while deposition of MoS$_2$/Al$_2$O$_3$ particles exhibits a uniform thickness and smooth surface. It was also observed that the Ni–MoS$_2$/Al$_2$O$_3$ composite shows better microhardness and resistance to wear than the Ni–MoS$_2$ composite. Zhang et al. [50] studied the effect of addition of nanographite on the tribological and anti-deliquescence performance of Ni/MoS$_2$ lubricating coating. It is a well-known fact that the lubricating performance of MoS$_2$ reduces in humid air, so to enhance its humid air, Ni/MoS$_2$–C coatings with 100 μm were deposited on GCr15 substrate by the nanocomposite electro-brush plating method. SEM, XPS and TEM were used to study the surface morphology, microstructure and elements of a deposited coating, and XRD was used to analyse the phase structure of a coating. The tribological performance of Ni/MoS$_2$–C and pure Ni/MoS$_2$ composite coatings was conducted after storage of samples for 12, 24 and 48 h in a

100% relative humidity atmosphere at room temperature, and chemical changes in an element were also analysed. Researchers observed that pure MoS_2 coatings that have been stored for 12 h in humid air have a small amount of MoS_2 due to the formation of MoO_3, which results in a decrease in tribological properties of pure MoS_2, whereas in similar conditions, Ni/MoS_2 coating with the addition of nanographite shows preferable tribological properties. It has also been observed that hardness of Ni/MoS_2 coating was enhanced with the addition of nanographite and combination plating displayed a well duration tribological performance as a result.

Cardinal et al. [51] studied the morphology, crystalline structure, composition, microhardness and frictional performance of $Ni-W-MoS_2$ nanostructured coatings. The pulse plating method was used for the development of $Ni-W-MoS_2$ coatings from a $Ni-W$ electrolyte that contains suspended particles of MoS_2. Results displayed that the addition of tungsten particles strongly influenced the properties of $Ni-W$ composite coating. It was observed that an increase in MoS_2 particles reduces the tungsten amount in the coating and results in an increase in average grain size. A high amount of MoS_2 in $Ni-W$ nanostructured coatings results in a porous sponge-like structure, high surface roughness and irregular frictional behaviour. However, with low amount of MoS_2 in $Ni-W$ coatings shows a 50% less friction coefficient than $Ni-W$ coatings.

Pimentel et al. [52] studied the mechanical, structural and tribological performance of solid lubricant $Mo-S-C$ solid coating. The DC magnetron sputtering technique was used for deposition of $Mo-S-C$ self-lubricating coatings. In order to get the composite coatings with carbon amount in the range of 0–55 at.%, the power ratio of targets was varied. It has been observed that with the addition of carbon content, the hardness of coating increases almost linearly. XRD suggests the presence of $Mo-C$ bonds, but it was predicted that molybdenum carbide grain size will be very small, since XRD does not show peaks that are related to $Mo-C$ phase. Tribological tests were conducted on pin-on-disc tribometer, and the result showed that in humid air, friction reduces with the increase in carbon content, whereas in nitrogen environment, all the coatings exhibit low friction of 0.02 except MoS_2 (0.04). $Mo-S-C$ coating shows better resistance to wear than MoS_2 coatings, but, on the other hand, exhibits low adhesion strength of the coatings with substrate. Such coatings were extremely susceptible to exposure to air and resulted in oxidation to the surface.

Tribological performances of $GNPs/MoS_2$ coating were studied by Meng et al. [53]. Due to good thermal conductivity and high stiffness and strength of graphene, it is considered to be a promising lubricating material. Therefore, it is feasible to add graphene nanoplatelets in MoS_2 to improve tribological performances of MoS_2 coating. Air spraying technique was used to deposit MoS_2 and $GNPs/MoS_2$ coatings on a GCr15 steel. The results show that tribological performance of MoS_2 coating increases with an increase in the addition of GNPs. Moreover, fluctuated wear track width, friction coefficient and wear rate were found in $GNPs/MoS_2$ coating with increasing applied load. However, the best anti-wear and friction-reduction performances were observed at an optimum applied normal load. More fine abrasive particles and furrows were also observed on the surface of $GNPs/MoS_2$ coating with increasing rotational speed. Therefore, such coating is more promising at low-rotational-speed working conditions.

7.2.3 HYBRIDIZED TRIBOLOGICAL COATINGS

Researchers were developing hybridized coatings to extend the tribological performance of coatings. The hybridized coatings incorporate the self-lubricating materials into either a coated or a composite structure to extend the tribological properties. Yongliang and Sunkyu [54] studied the effect of a combination of hard and solid lubricating coating on microstructural and tribological behaviour of $TiAlN/MoS_2$-Ti coatings. A coating system in which soft MoS_2–Ti layer was deposited on hard TiAlN layer exhibits an excellent combination of properties that improve the performance in high-wear, low-friction applications. The TiAN coating was deposited to provide high adhesion strength of coating with a substrate and high hardness, while MoS_2–Ti layer was to enhance the tribological properties without affecting the adhesion strength to the substrate. Researchers also observed that friction coefficient of TiAlN coating dropped by 48% after the MoS_2–Ti layer was deposited. Also, a remarkable decrease in the wear rate of TiAlN coating was observed due to the lubricating effect of MoS_2–Ti layer. Ren et al. [55] studied the tribological behaviour of Pb–Ti/MoS_2 for environmentally sensitive solid lubrication. MoS_2 coatings' lubricating properties decline in humid environments and limit its application as crystal MoS_2 is easily affected to form MoO_3 by water vapour, results in enhanced friction and wear rate. Hence, to enhance tribological performances of MoS_2 in high-humidity environment, Pb–Ti/MoS_2 coatings were deposited by unbalanced magnetron sputtering system. The doping element in MoS_2 coatings not only maintains the low-humidity sensitivity characteristic as Ti/MoS_2 coating, but also enhances the mechanical and tribological properties of coatings than that of single-doped coating. Moreover, in vacuum condition a very low coefficient of friction of 0.006 is measured or Pb–Ti/MoS_2 coating with 4.6 at.% Pb and main cause of ultra-low friction coefficient is due to dense coating structure. Interestingly, the wear rate of Pb–Ti/MoS_2 coatings is in accord with deviation in hardness from the elastic modulus ratio (H/E), as the coating having a high H/E value exhibits a low wear rate. The result shows that the lubricating performance of MoS_2 coatings can be enhanced in humid as well as vacuum conditions by doping of Pb and Ti, which is the main reason for MoS_2 coatings to be as environmentally adaptive lubricants. Shang et al. [56] also investigated MoS_2/Pb–Ti coating and multilayer MoS_2/Pb–Ti coating to enhance their resistance to corrosion in 3.5 wt. % NaCl solution and their tribological properties in a very high-humidity environment. The electrochemical impedance spectra and salt spray test revealed that MoS_2/Pb–Ti coating and multilayer coating can prevent penetration of oxygen (O_2) and other corrosion-causing elements and result in high resistance to corrosion. Moreover, tribological properties of MoS_2/Pb–Ti coating and multilayer MoS_2/Pb–Ti coating are also enhanced considerably, resulting in a compact structure and high mechanical properties than pure MoS_2 coatings. Furthermore, the heterogeneous interfaces in multilayer MoS_2/Pb–Ti coating are the main reason for the better resistance to corrosion and tribological properties of coatings. Overall, the dual doping and multilayer coating designs are favourable methods to develop environmentally adaptive lubricant MoS_2 coatings.

Aouadi et al. [57] investigated the tribological behaviour of Mo_2N/MoS_2/Ag nanocomposite coating deposited by unbalanced magnetron sputtering for aerospace applications. The coating composition was developed with 0–13 at.% S and 0–24 at.% Ag amount in order b-Mo_2N main phase in coatings with doping of Ag and MoS_2 phases. Tribological tests of this coating with Al_2O_3, Si_3N_4 and 440C steel as counterbodies were conducted at 25, 350 and 600°C. It was observed that at room temperature, high friction coefficients of 0.5–1.0 were obtained and the lowest friction coefficient was found against Al_2O_3 and the highest against Si_3N_4. Coatings with high sulphur content find a lower range of friction coefficient of 0.5 at room temperature; nevertheless, the total amount of amorphous MoS_2 was not sufficient in coating composition to provide hexagonal MoS_2 behaviour. Moreover, it was also observed that doping of both MoS_2 and Ag significantly reduces the friction coefficients due to the relocation of Ag to the surface at 350°C and results in lubricating silver molybdate phases at a temperature of 600°C over the wear track, which controlled the development of molybdenum oxides. The main cause of the lubricating behaviour and achieving the temperature-adaptive behaviour is the presence of sulphur in coatings. The lowest friction coefficient of 0.1 was measured for coating with S content of 5–14 at.% and Ag content of 16 at.%. Researchers concluded that the friction coefficients of nitride-based self-lubricating coatings stated in the literature were considerably higher. The $Mo_2N/MoS_2/Ag$ coatings show a two-order reduction in wear rate than single-phase Mo_2N coatings. The reduction in friction coefficients and wear rate of such self-lubricating adaptive composite coatings may be advantageous for the reduction of friction and wear in the case of hybrid bearings and other mechanical components that operate at high temperatures. Papp et al. [58] investigated the microstructure and tribological properties of multilayered nanocrystalline CrN/TiAlN/MoS_2 thin coating that has been developed from CrN, TiAlN and MoS_2 tri-layer systems on hardened high-speed steel and thin-oxide-covered Si (100) wafers using the reactive co-sputtering deposition method in Ar–N_2 atmosphere. The chemical composition analysis by Auger electron spectroscopy and the microstructure investigation were carried out by XTEM. The chemical composition analysis and microstructure investigation of the deposited coatings show that the CrN, TiAlN and MoS_2 phases form continuous layers with large gradient transition of composition. The developed CrN/(Al, Ti)N/MoS_2 coatings also show enhanced tribological properties in the dry sliding condition at room temperature. Xie et al. [59] studied the microstructure and performance of nitrogen ion implantation/AlN/CrAlN/MoS_2–phenolic resin duplex coatings deposited on magnesium alloys. The microstructural, electrochemical and mechanical properties of coating were analysed by X-ray photoelectron spectroscopy, XRD, SEM and electrochemical corrosion. Fourier transform infrared spectroscopy, wear tester and nanoindenter were used to understand the tribological properties of duplex coating. It has been analysed that MoS_2–phenolic resin coating possesses crystalline MoS_2 particles that have been embedded in an amorphous phenolic resin matrix with a two-phase microstructure. The presence of single-layer MoS_2–phenolic resin in magnesium alloys reduces the wear resistance and load-bearing property of a substrate, but at the same time increases the corrosion resistance. The load-bearing capacity of a substrate is significantly enhanced by doping of interlayer nitrogen ion implantation/AlN/CrAlN in MoS_2–phenolic resin/substrate system. In addition, it was also observed that interlayer nitrogen ion implantation/AlN significantly reduces the effect of galvanic corrosion due to its potential of being very similar to Mg alloys, whereas interlayer nitrogen ion implantation/AlN/CrAlN is ineffective in decreasing the galvanic corrosion because of the wide potential gap among CrN phase and a substrate. As a result, the nitrogen ion implantation/AlN/MoS$_2$–phenolic resin duplex coating displays excellent resistance to corrosion than nitrogen ion implantation/AlN/CrAlN/MoS$_2$–phenolic resin.

Sliney [60–65] developed a series of chromium carbide/Ag/BaF$_2$–CaF$_2$ coatings by plasma spraying, from PS100 to PS403. It has been observed that at high temperatures, among various coatings chromium carbide with metallic bond was used as wear resistance coating, while Ag and alkaline earth metal fluorides were used as low-friction coatings in different temperature ranges. Such coatings are effectively used at high temperatures, but at room temperature, they display a coefficient of friction above 0.2. Moreover, it was also observed that the plasma spray methods basically produce thick coating of about 100 µm and high roughness, thus restricting their application in precision components. Bidev et al. [66] studied adhesion strength and fatigue performance of Ti/TiB$_2$/MoS$_2$ coatings developed by closed-field unbalanced magnetron sputtering (CFUBMS). Ti/TiB$_2$/MoS$_2$ coatings' structural properties were analysed by EDS, SEM and XRD, and microhardness tester was utilized to measure the hardness of deposited coatings. Moreover, fatigue properties and the adhesion strength of deposited coatings with a substrate were investigated by scratch testing in two modes, involving sliding-fatigue multimode process in unidirectional sliding at varying fractions of critical load and progressive loading with a standard mode. The results show that Ti/TiB$_2$/MoS$_2$ coatings possess a thick and non-columnar structure. The XRD patterns of deposited coatings revealed the presence of MoS$_2$ (002) and TiB$_2$ (100). In addition, the thickness and hardness of coatings are predominantly affected by the target currents during the deposition procedure. The fatigue resistance and adhesion strength of the deposited coatings declined with S/Mo and Ti/Mo ratios, but improved with an increase in thickness and hardness. Baran et al. [67] also investigated the tribological performance of Ti/TiB$_2$/MoS$_2$ coatings developed by CFUBMS in air and vacuum environments. Various researchers have observed that in water vapour-free and vacuum environments, MoS$_2$ coatings are effectively used due to their high friction coefficient and low service life in humid conditions. In MoS$_2$ coatings, various alloy elements (e.g. Nb, Ti, Cr) and compounds (e.g. TiB$_2$, TiN) were added to improve the tribological properties. Characterization and tribological performance of Ti/TiB$_2$/MoS$_2$ coatings were studied in the same way as studied by the Bidev et al., and similar results were found. In addition, tribological properties of deposited coating are greatly affected by the thickness, hardness and stoichiometric ratio of elements in the structure under air and vacuum environments. The Ti/TiB$_2$/MoS$_2$ coatings exhibit a low friction coefficient and a wear rate under vacuum conditions as compared to air conditions.

Qin et al. [68] developed multilayered PEO/Ag/MoS$_2$ coatings on Ti6Al4V alloy by PEO treatment, an electroplating Ag coating and a burnished MoS$_2$ film and studied the adaptive-lubricating effect of the deposited coatings at elevated temperatures. Ball-on-disc tribometer was utilized to analyse the tribological properties over varying temperatures (RT–600°C). Results revealed that due to the lubricating effect of Ag and MoS$_2$, the PEO/Ag/MoS$_2$ multilayered coatings exhibit a low friction coefficient and better resistance to wear at room and medium temperatures (< 350°C), and they have high load-bearing ability due to the presence of hard PEO coating. At high temperatures (>350°C), it was analysed that oxidation of MoS_2 top layer occurred and due to the adhesion of softened Ag causes direct contact of Si_3N_4/exposed oxide under and results in a rapid increase in friction coefficient. However, COF decreased to 0.2 sharply as temperature increased to 600°C, which was due to the combined effect of MoOx and Ag_2MoO_4 high-temperature lubricants, the diffused Ag from discharged reservoirs, and development of lubricants/lubricating glaze layers contact. Yuan et al. [69] studied a simplistic and effective method for reducing the low dispersion of MoS_2 nanoflowers into a polyimide (PI) by cautiously embedding them on the surface of hollow carbon nanofibers (HCNF). The developed MoS_2@HCNF hybrid was used as a homogeneous filler to boost the tensile strength and lubricating properties of protective coating based on PI. The results display that tensile strength can be increased effectively by 46%, in addition to a slight decline (19%) in elongation with the addition of 2.0 wt. % MoS_2@HCNF. Moreover, MoS_2@HCNF/PI coatings also displayed better anti-wear and reduced friction behaviour under water lubricating conditions (0.5 wt. %, causing 72.5% lower wear rate) and liquid paraffin oil (1.5 wt. %, causing 56% lower wear rate), signifying that MoS_2@HCNF hybrid could jointly increase the wear created by friction shear force in the PI matrix. MoS_2@HCNF hybrid's improved tribological behaviours suggest their strong use as a novel filler in wear-resistant composite coatings.

Chen et al. [70] developed a hybrid of carbon nanotubes/graphene oxide/MoS_2 by the hydrothermal method and investigated its characterization and effect on epoxy tribological performances. Results displayed that EP-CNTs/GO/MoS_2 had a low friction coefficient of 0.042 and wear rate of 3.44×10^{-5} mm³/Nm, which is 90% and 95% less than that of EP and its other composite coatings embedded with single filler or binary hybrids. The reduction in friction coefficient and wear rate of EP-CNTs/GO/MoS_2 compared to pure EP is because of the uniform distribution of CNTs, MoS_2 and GO in CNTs/GO/MoS_2 hybrid, the self-lubricating effect of MoS_2, the load-carrying property of CNTs and GO and the development of transfer film onto counterbody surface.

Zhang et al. [71] investigated the microhardness and tribological performance of Cu–MoS_2–WC coating developed by cold spray and compared them with Cu–MoS_2 coating. Tribological tests showed that Cu–MoS_2–WC with alumina tribopair exhibits low and uniform friction all over the wear track in a dry nitrogen environment, and also low and homogeneous wear rates were measured. However, Cu–MoS_2 wear tracks displayed a large scale of detachment, while wear tracks in the case of Cu–MoS_2–WC coating displayed very small detachments. During testing, WC particles sit just beneath the wear track and form a situation of hard material with soft 'skin' and result in low friction. MoS_2 redistribution shows the same behaviour on a wear track for both the coatings; however, after 1,000 cycles more MoS_2 debris from Cu–MoS_2–WC gets adhered to counterbody and results in formation of tribofilm that avoids metal-to-metal contact and results in friction reduction. The subsurface wear track microstructure demonstrated dissimilarity among these two coatings and WC particle effect. First and foremost, microstructural modifications occurred only in nearby WC particles and slight modifications were analysed after large sliding distances, which recommends that WC particles have load-bearing capacity and help to develop a stable tribofilm, whereas in the case of Cu–MoS$_2$, the plastic deformation and microstructural refinement occurred more widely and led to formation of an unstable and thick tribofilm. This results in formation of more active material transfer between the metal-to-metal contacts and causes detachment, cracking and eventually high wear. Huang et al. [72] studied the tribological performance of ~23-μm-thick MoS$_2$–Ti(C, N)–TiO$_2$ coating deposited on Ti6Al4V by cathodic plasma electrolytic deposition (CPED) technique. SEM, EDS, optical microscopy and XRD were used to study morphologies, element composition, phase constituents and microstructure of the deposited coatings. Ball-on-disc tribometer was used to analyse the tribological performance of the deposited coating, and the hardness of the coating was also studied. The result shows that the hardness of the depositing coating is much higher (1024 ± 58 HV$_{0.025}$) than that of the coating free of MoS$_2$. It was also observed that both the coated samples show higher tribological performance. Incorporation of MoS$_2$ in coatings is the main cause of the lubricating effect and the reduction in wear rate. Additionally, the coated Ti6Al4V substrate shows an 11% and 25% reduction in wear rate and friction coefficient, respectively, compared to the uncoated Ti6Al4V substrate. Li et al. [73] studied the tribological performance and friction mechanism of YSZ/MoS$_2$ self-lubricating composite coatings deposited by both thermal spraying technology and hydrothermal reaction thermal on steel substrate. The results display that the developed MoS$_2$ powders consist of ultra-thin (7~8 nm) sheets of lamellar structure. The numbers of MoS$_2$ powders, after vacuum impregnation and hydrothermal reaction, look like flowers, formed within the YSZ coating sprayed by plasma. Furthermore, the intrinsic micropores of the YSZ coating are the growing point of the MoS$_2$ flower. Tribological tests show that composite coatings have an extremely long service life (>10^5 cycles) under high-vacuum conditions and have a low coefficient of friction (>0.1), which is 15% less than that of the YSZ coating. Moreover, the composite displays tremendous less wear rate of 2.30×10^{-7} mm^3/N/m and reduces wear damage of counterbody. The high wear-resistant and lubricating property is due to the development of transferred MoS$_2$ films and the ultra-smoothed worn surfaces of hybrid coatings.

7.3 CONCLUSION

Self-lubricating coatings or solid lubricant films have been utilized to reduce friction in sliding or rotating machine components, for example those working in ultra-high vacuum, in space or in applications where re-lubrication of tribopairs is not possible or any other conventional lubricant can't be used due to the lack of accessibility. Various surface coating methods, such as magnetron sputtering, plasma-assisted chemical vapour deposition, RF sputtering, molecular beam epitaxy, metal organic chemical vapour deposition, pulsed laser deposition, thermal spraying, ion implantation and fluidized bed, have been developed to enhance the surface properties. Among various coatings, MoS$_2$ coatings are most commonly used for various industrial applications. However, MoS$_2$ coatings show low lubricating properties in the humid air due to oxide formation, which in turn reduces coating lifetime and also increases the coefficient of friction. In order to reduce the deterioration of MoS$_2$ coatings in humid air, various metals, ceramics or mixed metals are added to the MoS$_2$

coatings. It is also clear from the above literature review that various tribological studies have been done to analyse the effect of addition of metals, ceramics or mixed metals such as Ti, Ni, Cr, Au, Pb, Nb, Ta, TiB_2 and TiN to MoS_2 coatings. However, the addition of metals or ceramics or mixed metals to MoS_2 enhances the coating hardness and corrosion resistance and reduces the sensitivity to water vapour, as compared to pure MoS_2 coating. Also, the hybridized tribological coatings combine the self-lubricating materials into either a composite or layered structure to extend the tribological properties. Moreover, it can be concluded that hybridized coatings show better mechanical, corrosion resistance and tribological properties as compared to the dichalcogenide lubricant coatings and adaptive tribological coatings.

REFERENCES

1. Bhushan B., 1999. *Principles and Applications of Tribology*, John Wiley & Sons, New York.
2. Handbook, ASM, 1992. *Friction, Lubrication, and Wear Technology*, Vol.18, ASM International, Materials Park, Ohio.
3. Fu, Y., Wei, J., & Batchelor, A.W., 2000. Some considerations on the mitigation of fretting damage by the application of surface-modification technologies. *Journal of Materials Processing Technology*, 99, 231–245.
4. Voevodin, A.A., O'neill, J. P., & Zabinski, J.S. (1999). Nanocomposite tribological coatings for aerospace applications. *Surface and Coatings Technology*, 116, 36–45.
5. Martin Jr, P. (1972). *A Survey of Solid Lubricant Technology (No. SWERR-TR-72–9)*. General Thomas J Rodman Lab, Army Weapons Command Rock Island, IL.
6. Grill, A. (1997). Tribology of diamond like carbon and related materials: An updated review. *Surface and Coatings Technology*, 94, 507–513.
7. Sherbiney, M.A., & Halling, J. (1977). Friction and wear of ion-plated soft metallic films. *Wear*, 45(2), 211–220.
8. Pauleau, Y., Juliet, P., & Gras, R. (1998). Tribological properties of calcium fluoride-based solid lubricant coatings at high temperatures. *Thin Solid Films*, 317(1–2), 481–485.
9. John, P.J., Prasad, S.V., Voevodin, A.A., & Zabinski, J.S. (1998). Calcium sulfate as a high temperature solid lubricant. *Wear*, 219(2), 155–161.
10. Kubart, T., Polcar, T., Kopecký, L., Novak, R., & Novakova, D. (2005). Temperature dependence of tribological properties of MoS_2 and $MoSe_2$ coatings. *Surface and Coatings Technology*, 193(1–3), 230–233.
11. Zhang, S.W. (1998). State-of-the-art of polymer tribology. *Tribology International*, 31(1–3), 49–60.
12. Gulbiński, W., & Suszko, T. (2006). Thin films of MoO_3–Ag_2O binary oxides–the high temperature lubricants. *Wear*, 261(7–8), 867–873.
13. Yang, J.-F., Parakash, B., Hardell, J., & Fang, Q.F. (2012). Tribological properties of transition metal di-chalcogenide based lubricant coatings. *Frontiers of Materials Science*, 6, 116–127.
14. Kaul, A.B. (2013). Graphene and two-dimensional layered materials for device applications. *Nanotechnology (IEEE-NANO), 2013 13th IEEE Conference on, IEEE*, 1–4.
15. Fleischauer, P.D., & Bauer, R. (1988, Jan 1) Chemical and structural effects on the lubrication properties of sputtered MoS_2 films. *Tribology Transactions,* 31(2), 239–50.
16. Wahl, K.J., & Singer, I.L. (1995, Jun 1). Quantification of a lubricant transfer process that enhances the sliding life of a MoS_2 coating. *Tribology Letters*, 1(1), 59–66.
17. Singer, I.L. (1996, Sep 18). Mechanics and chemistry of solids in sliding contact. *Langmuir*, 12(19), 4486–91.

18. Hilton, M.R., Bauer, R., & Fleischauer, P.D. (1990, Dec 15). Tribological performance and deformation of sputter-deposited MoS$_2$ solid lubricant films during sliding wear and indentation contact. *Aerospace Corp EL Segundo CA Lab Operations*, 188, 219–236.

19. Seitzman, L.E., Bolster, R.N., Singer, I.L., & Wegand, J.C. (1995, Jan 1). Relationship of endurance to microstructure of IBAD MoS$_2$ coatings. *Tribology transactions*, 38(2), 445–51.

20. Lauwerens, W., Wang, J., Navratil, J., Wieërs, E., D'haen, J., Stals, L.M., Celis, J.P., and Bruynseraede, Y. (2000, Sep 1). Humidity resistant MoSx films prepared by pulsed magnetron sputtering. *Surface and Coatings Technology*, 131(1–3), 216–21.

21. Zhang, X., Vitchev, R.G., Lauwerens, W., Stals, L., He, J., & Celis, J.P. (2001, Sep 21). Effect of crystallographic orientation on fretting wear behaviour of MoSx coatings in dry and humid air. *Thin Solid Films*, 396(1–2), 69–77.

22. Watanabe, S., Noshiro, J., & Miyake, S. (2004, May 24). Tribological characteristics of WS$_2$/MoS$_2$ solid lubricating multilayer films. *Surface and Coatings Technology*, 183 (2–3), 347–51.

23. Lansdown, A.R. (1999, May 28). *Molybdenum Disulphide Lubrication*. Tribology series, Vol. 35, Elsevier, Amsterdam.

24. Holmberg, K., & Matthews, A. (2009). *Coating Tribology*. Tribology and Interface Engineering series, Vol. 56, Elsevier, Amsterdam.

25. Roberts, E.W., & Price, W.B. (1988). In-vacuo, tribological properties of "high-rate" sputtered MoS2 applied to metal and ceramic substrates. *MRS Online Proceedings Library Archive,* 140.

26. Yukhno, T.P., Vvedensky, Y.V., & Sentyurikhina, L.N. (2001). Low temperature investigations on frictional behaviour and wear resistance of solid lubricant coatings. *Tribology International*, 34(4), 293–298.

27. Colbert, R.S., & Sawyer, W.G. (2010). Thermal dependence of the wear of molybdenum disulphide coatings. *Wear*, 269(11–12), 719–723.

28. Ostrovskaya, Y.L., Yukhno, T.P., Gamulya, G.D., Vvedenskij, Y.V., & Kuleba, V.I. (2001). Low temperature tribology at the B. Verkin Institute for low temperature physics & engineering (historical review). *Tribology International*, 34(4), 265–276.

29. Kohli, A.K., & Prakash, B. (2001). Contact pressure dependency in frictional behavior of burnished molybdenum disulphide coatings. *Tribology Transactions*, 44(1), 147–151.

30. Gamulya, G.D., Kopteva, T.A., Lebedeva, I.L., & Sentyurikhina, L.N. (1993). Effect of low temperatures on the wear mechanism of solid lubricant coatings in vacuum. *Wear*, 160(2), 351–359.

31. Brainard, W.A., & Buckley, D.H. (1969). The influence of ordering on the friction and wear of metals. *Vacuum* 4(2), 123–130.

32. Hu, T., Zhang, Y., & Hu, L. (2012). Tribological investigation of MoS$_2$ coatings deposited on the laser textured surface. *Wear*, 278, 77–82.

33. Donnet, C., Martin, J.M., Le Mogne, T., & Belin, M. (1996). Super-low friction of MoS$_2$ coatings in various environments. *Tribology International*, 29(2), 123–128.

34. Vierneusel, B., Schneider, T., Tremmel, S., Wartzack, S., & Gradt, T. (2013). Humidity resistant MoS$_2$ coatings deposited by unbalanced magnetron sputtering. *Surface and Coatings Technology*, 235, 97–107.

35. Zabinski, J.S., Donley, M.S., Dyhouse, V.J., & McDevitt, N.T. (1992). Chemical and tribological characterization of PbO/MoS$_2$ films grown by pulsed laser deposition. *Thin Solid Films*, 214(2), 156–163.

36. Renevier, N.M., Fox, V.C., Teer, D.G., Hampshire, J. (2000). Performance of low friction MoS$_2$/ titanium composite coatings used in forming applications, *Materials and Design*, 21, 337–343.

37. Martins, R.C., Moura, P.S., & Seabra, J.O. (2006). MoS_2/Ti low-friction coating for gears. *Tribology International*, 39(12), 1686–1697.

38. Stoyanov, P., Strauss, H.W., & Chromik, R.R. (2012). Scaling effects between micro- and macro-tribology for a Ti–MoS_2 coating. *Wear*, 274, 149–161.

39. Banday, S., & Wani, M.F. (2019). Nanoscratch resistance and nanotribological performance of Ti/MoS_2 coating on Al-Si alloy deposited by pulse laser deposition technique. *Journal of Tribology*, 141(2).

40. Banday, S., & Wani, M.F. (2019). Nanomechanical and nanotribological properties of self-lubricating Ti/MoS_2 nanocoating at nanoscale level. *Journal of Surface Science and Engineering*, 14, 89–104.

41. Jing, Y., Luo, J., & Pang, S. (2004). Effect of Ti or TiN codeposition on the performance of MoS_2-based composite coatings. *Thin Solid Films*, 461(2), 288–293.

42. Ding, X.Z., Zeng, X.T., He, X.Y., & Chen, Z. (2010). Tribological properties of Cr-and Ti-doped MoS_2 composite coatings under different humidity atmosphere. *Surface and Coatings Technology*, 205(1), 224–231.

43. Kao, W.H. (2005). Tribological properties and high speed drilling application of MoS_2–Cr coatings. *Wear*, 258(5–6), 812–825.

44. Baker, C.C., Hu, J.J., & Voevodin, A.A. (2006). Preparation of Al_2O_3/DLC/Au/MoS_2 chameleon coatings for space and ambient environments. *Surface and Coatings Technology*, 201(7), 4224–4229.

45. Lince, J.R., Kim, H.I., Adams, P.M., Dickrell, D.J., & Dugger, M.T. (2009). Nanostructural, electrical, and tribological properties of composite Au–MoS_2 coatings. *Thin Solid Films*, 517(18), 5516–5522.

46. Stoyanov, P., Chromik, R.R., Gupta, S., & Lince, J.R. (2010). Micro-scale sliding contacts on Au and Au-MoS_2 coatings. *Surface and Coatings Technology*, 205(5), 1449–1454.

47. Stoyanov, P., Gupta, S., Chromik, R.R., & Lince, J.R. (2012). Microtribological performance of Au–MoS_2 nanocomposite and Au/MoS_2 bilayer coatings. *Tribology International*, 52, 144–152.

48. Chien, H.H., Ma, K.J., Vattikuti, S.P., Kuo, C.H., Huo, C.B., & Chao, C.L. (2010). Tribological behaviour of MoS_2/Au coatings. *Thin Solid Films*, 518(24), 7532–7534.

49. Huang, Z.J., & Xiong, D.S. (2008). MoS_2 coated with Al_2O_3 for Ni–MoS_2/Al_2O_3 composite coatings by pulse electrodeposition. *Surface and Coatings Technology*, 202(14), 3208–3214.

50. Zhang, S., Li, G.L., Wang, H.D., Xu, B.S., & Ma, G.Z. (2013). Impact of nanometer graphite addition on the anti-deliquescence and tribological properties of Ni/MoS_2 lubricating coating. *Physics Procedia*, 50, 199–205.

51. Cardinal, M.F., Castro, P.A., Baxi, J., Liang, H., & Williams, F.J. (2009). Characterization and frictional behavior of nanostructured Ni–W–MoS_2 composite coatings. *Surface and Coatings Technology*, 204(1–2), 85–90.

52. Pimentel, J.V., Polcar, T., & Cavaleiro, A. (2011). Structural, mechanical and tribological properties of Mo–S–C solid lubricant coating. *Surface and Coatings Technology*, 205(10), 3274–3279.

53. Meng, F., Han, H., Gao, X., Yang, C., & Zheng, Z. (2018). Experiment study on tribological performances of GNPs/MoS_2 coating. *Tribology International*, 118, 400–407.

54. Yongliang, L.I., & Sunkyu, K.I.M. (2006). Microstructural and tribological behavior of TiAlN/MoS_2-Ti coatings. *Rare Metals*, 25(4), 326–330.

55. Ren, S., Li, H., Cui, M., Wang, L., & Pu, J. (2017). Functional regulation of Pb-Ti/MoS_2 composite coatings for environmentally adaptive solid lubrication. *Applied Surface Science*, 401, 362–372.

56. Shang, K., Zheng, S., Ren, S., Pu, J., He, D., & Liu, S. (2018). Improving the tribological and corrosive properties of MoS_2-based coatings by dual-doping and multilayer construction. *Applied Surface Science*, 437, 233–244.

57. Aouadi, S.M., Paudel, Y., Luster, B., Stadler, S., Kohli, P., Muratore, C., & Voevodin, A.A. (2008). Adaptive Mo$_2$N/MoS$_2$/Ag tribological nanocomposite coatings for aerospace applications. *Tribology Letters*, 29(2), 95–103.

58. Papp, S., Kelemen, A., Jakab-Farkas, L., Vida-Simiti, I., & Biró, D. (2013). Multilayered nanocrystalline CrN/TiAlN/MoS$_2$ tribological thin film coatings: Preparation and characterization. In *IOP Conference Series: Materials Science and Engineering* (Vol. 47, No. 1, p. 012016). IOP Publishing.

59. Xie, Z., Chen, Q., Chen, T., Gao, X., Yu, X., Song, H., & Feng, Y. (2015). Microstructure and properties of nitrogen ion implantation/AlN/CrAlN/MoS$_2$-phenolic resin duplex coatings on magnesium alloys. *Materials Chemistry and Physics*, 160, 212–220.

60. Sliney, H. E. (1960). *Lubricating Properties of Some Bonded Fluoride and Oxide Coatings for Temperature to 1500 F* (Vol. 478). NASA-TN-D-478 (12 Pages).

61. Sliney, H.E. (1974). High temperature solid lubricants, Part 1: Layer lattice compounds and graphite. *ASME Journal of Mechanical Engineering*, 96(2), 18–22.

62. Sliney, H.E. (1979). Wide temperature spectrum self-lubricating coatings prepared by plasma spraying. *Thin Solid Films*, 64(1–2), 211–217.

63. Amato, I., & Martinengo, P.C. (1973). Some improvements in solid lubricants coatings for high temperature operations. *ASLE Transactions*, 16(1), 42–49.

64. Sliney, H.E. (1986). The use of silver in self-lubricating coatings for extreme temperatures. *ASLE Transactions*, 29(3), 370–376.

65. Sliney, H.E. (1987). Coatings for friction and wear control at high temperatures. *Surface and Coatings Technology*, 33, 243–244.

66. Bidev, F., Baran, Ö., Arslan, E., Totik, Y., & Efeoğlu, İ. (2013). Adhesion and fatigue properties of Ti/TiB$_2$/MoS$_2$ graded-composite coatings deposited by closed-field unbalanced magnetron sputtering. *Surface and Coatings Technology*, 215, 266–271.

67. Baran, Ö., Bidev, F., Çiçek, H., Kara, L., Efeoğlu, İ., & Küçükömeroğlu, T. (2014). Investigation of the friction and wear properties of Ti/TiB$_2$/MoS$_2$ graded-composite coatings deposited by CFUBMS under air and vacuum conditions. *Surface and Coatings Technology*, 260, 310–315.

68. Qin, Y., Xiong, D., Li, J., Jin, Q., He, Y., Zhang, R., & Zou, Y. (2016). Adaptive-lubricating PEO/Ag/MoS$_2$ multilayered coatings for Ti6Al4V alloy at elevated temperature. *Materials & Design*, 107, 311–321.

69. Yuan, H., Yang, S., Liu, X., Wang, Z., Ma, L., Hou, K., & Wang, J. (2017). Polyimide-based lubricating coatings synergistically enhanced by MoS$_2$@ HCNF hybrid. *Composites Part A: Applied Science and Manufacturing*, 102, 9–17.

70. Chen, B., Li, X., Jia, Y., Xu, L., Liang, H., Li, X., & Yan, F. (2018). Fabrication of ternary hybrid of carbon nanotubes/graphene oxide/MoS$_2$ and its enhancement on the tribological properties of epoxy composite coatings. *Composites Part A: Applied Science and Manufacturing*, 115, 157–165.

71. Zhang, Y., Epshteyn, Y., & Chromik, R.R. (2018). Dry sliding wear behaviour of cold-sprayed Cu-MoS$_2$ and Cu-MoS$_2$-WC composite coatings: The influence of WC. *Tribology International*, 123, 296–306.

72. Huang, J., Zhu, J., Fan, X., Xiong, D., & Li, J. (2018). Preparation of MoS$_2$-Ti (C, N)-TiO$_2$ coating by cathodic plasma electrolytic deposition and its tribological properties. *Surface and Coatings Technology*, 347, 76–83.

73. Li, S., Zhao, X., An, Y., Liu, D., Zhou, H., & Chen, J. (2018). YSZ/MoS$_2$ self-lubricating coating fabricated by thermal spraying and hydrothermal reaction. *Ceramics International*, 44(15), 17864–17872.

8 Frictional Behaviour of Gelatin Based Soft Lubricants

Shraddha Gondane and Arun K. Singh
Visvesvaraya National Institute of Technology

Nitish Sinha
G H Raisoni Institute of Business Management

CONTENTS

8.1 INTRODUCTION

Lubricants are the substances that basically reduce friction and wear of sliding surfaces (Mang et al., 2011; Bart et al., 2013). These are the important class of materials for reducing energy consumption in a variety of mechanical, physical and biological systems as well as other moving systems (Mang et al., 2011; Bart et al., 2013). A large number of lubricants have been developed for a variety of applications (Mang et al., 2011; Bart et al., 2013). In recent times, lubricants made of soft materials such as gelatin have also gained importance for practical applications (Bart et al., 2013). Synovial fluid is an example of a natural soft lubricant which is present at the joints of human body (Lee et al., 2013). Soft lubricants are biocompatible and non-degradable over a long period of time (Lee et al., 2013). Further, when two contacting surfaces slid past each other, generally two types of motion, namely stick-slip and steady sliding, occur (Gong, 2006; Persson, 2000; Baumberger et al., 2003). Stick-slip is a frictional instability that encounters in the form of jerky motion of the sliding surfaces up to a threshold shear velocity, and this is known as the critical velocity of the sliding interface (Persson, 2000; Baumberger et al., 2003; Thakre & Singh, 2018; Berman et al., 1996). For stick-slip motion, a necessary condition is that friction must reduce with slip velocity, and this is characterized as velocity-weakening (VW) effect (Persson, 2000; Thakre & Singh, 2018; Berman et al., 1996). However, if friction increases with sliding velocity, this is defined as velocity-strengthening (VS) effect. Steady sliding generally occurs in VS regime of friction (Persson, 2000; Berman et al., 1996; Gupta & Singh, 2016). More interestingly, when motion of a steadily sliding block is brought to zero, frictional stress from steady sliding may or may not reduce to a zero value during the relaxation time (Gupta & Singh, 2016). This relaxation time is also known as hold time, waiting time, time of stationary contact or ageing time, etc. (Berman et al., 1996; Gupta & Singh, 2016). Although steady dynamic stress decreases during relaxation time, static stress of the sliding interface increases simultaneously (Baumberger et al., 2003; Gupta & Singh, 2016). In other words, friction and adhesion are just the opposite and their effectiveness depends on shear velocity. For instance, in the case of static friction, the rate of rupture dominates over the rate of adhesion, while during steady motion, both the rates become equal (Berman et al., 1996; Gupta & Singh, 2016). At residual stress, the rate of adhesion becomes important, but the rate of rupture results in almost zero (Gupta & Singh, 2016). In stick-slip regime too, when the rate of adhesion exceeds the rate of rupture, this results in unstable static and dynamic frictions. It has been established through the experiments that it is the shear velocity that determines whether a sliding mass undergoes unsteady (stick-slip) or steady sliding (Thakre & Singh, 2018). Friction models for steady stress as well as stress relaxation of soft solids have been proposed, and these models are also validated with the friction experiments (Gupta & Singh, 2016, 2018; Singh & Juvekar, 2011).

Viscosity is also important for the frictional properties of a lubricant (Persson, 2000; Gong, 2006; Mang et al., 2007; Bart et al., 2013; Lee et al., 2013). There is extensive study on the frictional behaviour of different lubricants under the condition of boundary layer having a very small thickness (Drummond & Israelachvili, 2001; Drummond et al., 2002). A theory based on shear stress-induced solid–liquid transition has been proposed to elucidate the frictional properties of the thin layer of lubricants (Drummond & Israelachvili, 2001; Drummond et al., 2002). Noting that earlier friction experiments were mainly carried out on gelatin hydrogels in solid form (Baumberger et al., 2003; Thakre & Singh, 2018; Gupta & Singh, 2016, 2018, 2019). Present study, in contrast, is focused on investigating the frictional properties of soft gelatin hydrogels in liquid form. Formation and rupture of molecular chains submerged in viscous liquid result in adhesion and friction of such a lubricant (Persson, 2000; Baumberger et al., 2003). These results are, ultimately, analysed in terms of static, dynamic and residual stresses in stable sliding regime, that is in VS regime.

Generally, two types of friction tests, namely slide-hold-slide (SHS) and slide-free-slide (SFS), are carried out to characterize the frictional properties of sliding solids (Gupta & Singh, 2019). In SHS test, residual stress appears because of strengthening of the sliding interface during the relaxation or hold period (Gupta & Singh, 2018). However, there is no residual stress in SFS test (Gupta & Singh, 2019). This is because of the external force on the sliding block; if that is removed, residual stress also becomes zero (Gupta & Singh, 2019). In other words, friction is a force that opposes the relative motion of the contacting surfaces upon application of external force (Persson, 2000). Since there is no external force during the free period of SFS test, friction force is also zero. Nevertheless, SHS test is more useful than SFS test (Gupta & Singh, 2019). In the present study, all experiments are carried out in stable sliding regime.

Coulomb's law of friction is used along the soft lubricant. According to this law, frictional stress τ is related to cohesion c_0 and $\mu_0\sigma_0$ as $\tau = c_0 + \mu_0\sigma_0$, where σ_0 is the effective normal stress and μ_0 is the coefficient of friction (Persson, 2000; Jaeger et al., 2009). Generally, μ_0 is expressed as $\mu_0 = \tan\phi_0$, where ϕ_0 is the internal friction angle (Jaeger et al., 2009). A gel lubricant/glass sliding system consists of different materials, and their interaction should be characterized by the adhesion process (Jaeger et al., 2009). Consequently, in the case of gel/glass system, c_0 is replaced with a_0, that is adhesive stress. Accordingly, Coulomb's law of friction takes the form as $\tau = a_0 + \mu_0\sigma_0$. Further, the physical significance of a_0 is that still remains present at sliding interface at $\sigma_0 = 0$ (Jaeger et al., 2009). Adhesive friction develops because of bond formation and rupture of molecular chains with the substrate (Drummond et al., 2002). While the other term, $\mu_0\sigma_0$ is related to normal stress-controlled friction (Drummond & Israelachvili, 2001). Baumberger et al. (2003) have reported in their experiments on soft gelatin hydrogels that there is a negative normal stress at which shear stress becomes zero. They have also studied dynamic stress τ_d of gelatin hydrogel as $\tau_d = \mu_0\sigma_0$ to show that μ_d depends on shear velocity. In recent times, the scaling laws have been proposed for adhesive and viscous components of friction of the hydrogels (Shoaib & Espinosa-Marzal, 2018; Urueña et al., 2018; Leroy et al., 2009). For instance, the scaling law for Gemini hydrogels shows negative dependence of coefficient of friction on normal force (Urueña et al., 2018). Motivated by these results, it becomes interesting to extend that study further to understand the effect of gelatin concentrations on friction properties of the gel lubricants.

8.2   EXPERIMENTAL METHODS

The present method of sample preparation and experiments are the same as reported in previous studies (Gupta & Singh, 2016, 2018, 2019). In order to maintain the thickness of the gel lubricant, a fixed amount of gel lubricant was spread along the interface of two smooth glass surfaces. The area of the slider was $6 \times 6 \text{cm}^2$. The thickness of the lubricant was very thin, so there is hardly any leakage of the lubricant from the sides of the sliding interface. The concentration of gelatin in the lubricant was also varied to study its effect on the frictional properties. In SHS test, hold time was fixed at 60 s, while external shear velocity V_0 was varied from 0.1 to 1.0 mms^{-1}. The normal load was also varied in the experiment from 1 to 6 N. A load cell (5 N) connected with data acquisition system (DAS) is used for measuring the friction force at the sampling rate of 10. A constant temperature of $20 \pm 1°C$ as well as constant humidity (60%) was maintained in the test chamber. The aim of the present study is to understand the effect of gelatin concentration, normal stress and sliding velocity on frictional properties of the gelatin based liquid lubricants.

8.3   RESULTS AND DISCUSSION

Figure 8.1a presents a frictional stress τ vs. sliding time t obtained from a typical slide-hold-slide (SHS) test. It is seen from the plot that frictional stress initially

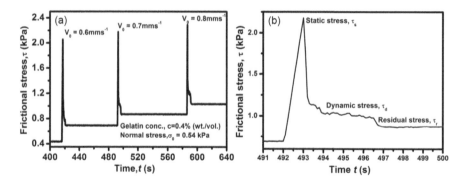

FIGURE 8.1 (a) A typical slide-hold-slide test between frictional stress τ vs. time t at a fixed normal stress $\sigma_0 = 0.54$ kPa and a fixed hold time of 60 s. (b) A portion of Figure 8.1a, which shows static, dynamic and residual stresses at fixed shear velocity $V_0 = 0.7$ mms^{-1} for $c = 0.4$ % (wt./vol.).

increases with time and reaches a threshold value. This is generally characterized with static stress τ_s of the sliding interface, as in Figure 8.1b. Static stress depends on hold time as well as shear rate of sliding and varies as logarithm of hold time as well as logarithm of shear rate (Gong, 2006; Gupta & Singh, 2019). After the threshold in Figure 8.1a, frictional stress decreases in a regime in which frictional stress practically remains constant with time, and this is generally characterized as dynamic stress τ_d in Figure 8.1b. Although steady stress depends on shear rate, that does not depend on history of sliding (Gupta & Singh, 2019). Further, the steadily sliding block is brought to standstill; that is, $V_0 = 0$. It is seen in Figure 8.1b that frictional stress decreases with time before that becomes constant, and that is known as residual stress τ_r. Figure 8.1a also presents the evidence that residual stress increases with shear velocity. The presence of residual stress is attributed to the competition between bond strengthening at the sliding interface and stiffness of the overall sliding system (Gupta & Singh, 2018). It is also concluded from Figure 8.1a and b that frictional properties of the soft gelatin lubricant are qualitatively similar to the soft gelatin hydrogel, yet their magnitudes are smaller in comparison with the soft solid friction (Gupta & Singh, 2016, 2018).

Figure 8.2a shows the effect of sliding velocity V_0 on static stress τ_s for different gelatin concentrations $c\%$ (wt./vol.) in the lubricant. It is concluded from the plot that τ_s increases with V_0. It is also observed that τ_s increases with $c\%$ in the soft lubricant. These results are qualitatively similar to soft gelatin hydrogels (Gupta & Singh, 2016). Figure 8.2b, on the other hand, indicates that τ_d increases linearly with V_0. It is important to note that static friction is a history as well as rate dependent property (Persson, 2000; Gupta & Singh, 2019). Figure 8.2c illustrates the shear velocity dependence of residual stress τ_r, and that increases with V_0. This observation is, in contrast, to the earlier study in which τ_r remains independent of shear velocity (Gupta & Singh, 2019). A possible reason for this contradictory observation may be attributed to high stiffness of the sliding interface, and this does not allow the relaxation process to the same level of residual stress.

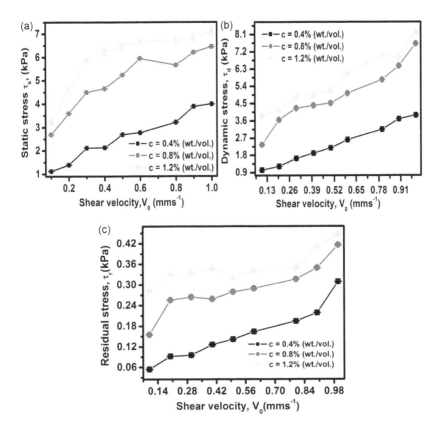

FIGURE 8.2 (a) Static stress vs. shear velocity; (b) dynamic stress vs. shear velocity; and (c) residual stress vs. shear velocity, for different gelatin concentrations $c = 0.4\%$, 0.8% and 1.2% (wt./vol.).

Figure 8.3a presents τ_s vs. σ_0 for different c% of the gelatin lubricant. This is owing to the fact that increase in normal stress on the gel lubricant results in increased area of contact at the sliding interface (Persson, 2000). Figure 8.3b shows τ_d vs. σ_0, and in this case too, τ_d increases with an increase in σ_0 due to enhanced adhesive interaction between the lubrication and glass interfaces. Baumberger et al. (2003) have also reported a similar observation in the case soft gelatin hydrogels. Figure 8.3c presents τ_r vs. σ_0. The reason for increasing τ_r is the same as the effect of σ_0 on τ_s and τ_d.

8.4 DEVELOPMENT OF SCALING LAWS FOR ADHESIVE STRESS AND COEFFICIENT OF FRICTION

Aiming to develop scaling laws, we have plotted all three frictional stresses τ_s, τ_d and τ_r vs. σ_0 for different c%. Figure 8.3a–c presents the evidence that all three frictional stresses vary almost linearly with σ_0. Further, while slope of the straight lines results in coefficient of friction μ_0, their intercept measures adhesive stress a_0 of friction. Further analysis of the results in Figure 8.4a shows the scaling law between static

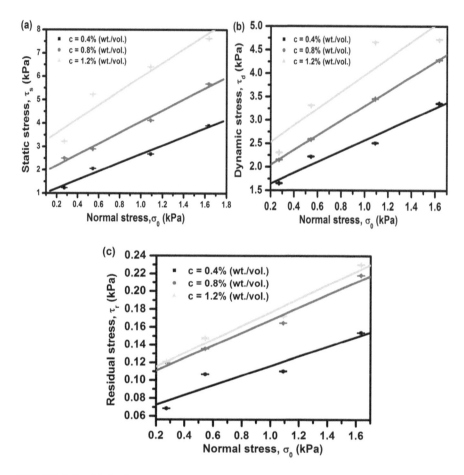

FIGURE 8.3   (a) Static stress τ_s vs. normal stress σ_0; (b) dynamic stress τ_d vs. σ_0; and (c) residual stress τ_r vs. σ_0 at fixed gelatin concentrations $c = 0.4\%$, 0.8% and 1.2% (wt./vol.).

adhesive stress a_s and gelatin concentration $c\%$ for τ_s as $a_s \sim c^{1.14}$. Figure 8.4b, on the other hand, establishes the scaling relation between coefficient of friction μ_s and $c\%$ as $\mu_s \sim c^{0.43}$. As a result, exponent of a_s is larger in magnitude than corresponding μ_s. This is due to the fact that gelatin is a soft solid that possesses low elasticity but high surface energy; thus, the gelatin lubricant has more tendency to adhere with the glass surface than due to normal stress.

Figure 8.5a demonstrates the scaling law between dynamic adhesive stress a_d and gelatin concentration $c\%$ for dynamic stress τ_d as $a_d \sim c^{0.38}$. Figure 8.5b shows the scaling law between coefficient of friction μ_d and gelatin concentration $c\%$ for τ_d as $\mu_d \sim c^{0.40}$. But in this case, the exponent of μ_d is slightly larger than the corresponding a_d. This contradictory observation could be attributed to the dominance of normal stress-controlled friction over the molecular controlled adhesive friction.

Figure 8.6a presents the plot between residual adhesive stress a_r and $c\%$. The scaling relation is found to be $a_r \sim c^{0.26}$. In the same way, Figure 8.6b shows the scaling

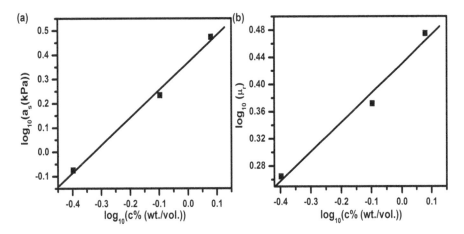

FIGURE 8.4 (a) The scaling law between static adhesive stress a_s and gelatin concentration $c\%$ as $a_s \sim c^{1.14}$. (b) The scaling law between static coefficient of friction μ_s and $c\%$ as $\mu_s \sim c^{0.43}$ for static stress.

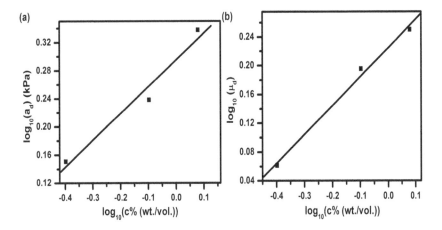

FIGURE 8.5 (a) The scaling law between adhesive dynamic stress a_d and gelatin concentration $c\%$ for dynamic stress τ_d as $a_d \sim c^{0.38}$. (b) The scaling law between dynamic coefficient of friction μ_d and $c\%$ as $\mu_d \sim c^{0.40}$.

law between residual coefficient of friction μ_r and $c\%$ as $\mu_r \sim c^{0.31}$, and noting that exponent of μ_r is larger than a_r. Thus, it is concluded that in this case too, the normal stress-controlled friction dominates over the molecular-controlled adhesive friction.

Based on the above scaling laws between components of friction and concentration of the gelatin lubricants, their numeric values are summarized in Table 8.1.

A comparison of power laws from Table 8.1 for τ_s, τ_d and τ_r shows that the exponent of both a_0 and μ_0 follows the trend $\tau_s > \tau_d > \tau_r$. At the same time, it is also obvious that exponent of a_0 is larger in magnitude than μ_0 in the case of static friction τ_s. However, the same is opposite in the case of τ_d and τ_r. The reason for these

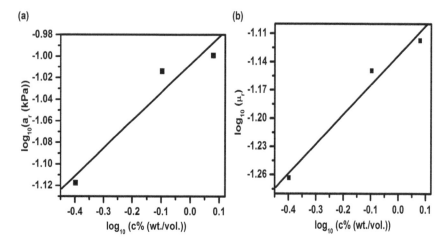

FIGURE 8.6 (a) The scaling law between adhesive stress a_r and gelatin concentration $c\%$ for residual stress as $a_r \sim c^{0.26}$. (b) The scaling law between residual coefficient of friction μ_r and $c\%$ as $\mu_r \sim c^{0.31}$.

TABLE 8.1

Summary of the Scaling Laws for Static, Dynamic and Residual Stresses in Terms of Adhesive Stress, Coefficient of Friction and Gelatin Concentration $c\%$

Frictional Stress/Components of Friction	Static Stress, τ_s	Dynamic Stress, τ_d	Residual Stress, τ_r
Adhesive stress, a_0 (a_s, a_d, & a_r)	1.14	0.38	0.26
Coeff. of friction, μ_0 (μ_s, μ_d, & μ_r)	0.43	0.40	0.31

---

observations is attributed to the competition between normal stress and molecular chain-controlled mechanisms of friction (Drummond & Israelachvili, 2001; Berman et al., 1998). For instance, in the case of τ_s, it is believed that molecular chains rupture from the glass surfaces more strongly than the contacts between glass–glass surfaces during the static friction process. However, during dynamic sliding and relaxation processes, it is the normal stress-controlled friction that dominates over the molecular chain-controlled friction. Nevertheless, further study is needed on this issue for a better understanding of the whole process. A future study concerning the shear velocity dependence of scaling laws will be interesting.

8.5 CONCLUSIONS

Present experimental study shows that frictional behaviour of a soft gelatin lubricant on a glass substrate is quite similar to the solid gelatin. It is observed that all three static, dynamic and residual stresses increase with gelatin concentrations in the lubricant. Magnitude of adhesive stress and coefficient of friction shows an increase ng trend with shear velocity. The scaling law study reveals that exponent of adhesive stress is larger in magnitude than corresponding coefficient of friction for static friction. However, the trend is opposite in the case of dynamic and residual stresses. In addition, both adhesive stress and coefficient of friction show the following trend of friction as static>dynamic>residual stress. These observations are explained in the light of normal stress and molecular controlled friction at the sliding interface.

ACKNOWLEDGEMENTS

We are thankful to MHRD-GOI for providing fund through TEQIP-II to develop the experimental set-up at VNIT Nagpur for the present study.

REFERENCES

Berman, A. D., Ducker, W. A., & Israelachvili, J. N. (1996). Origin and characterization of different stick– slip friction mechanisms. *Langmuir*, *12*(19), 4559–4563.

Berman, A., Drummond, C., & Israelachvili, J. (1998). Amontons' law at the molecular level. *Tribology Letters*, *4*(2), 95.

Baumberger, T., Caroli, C., & Ronsin, O. (2003). Self-healing slip pulses and the friction of gelatin gels. *The European Physical Journal E*, *11*(1), 85–93.

Bart, J. C., Gucciardi, E., & Cavallaro, S. (2013). *Biolubricants: Science and Technology*. Woodhead Publishing Ltd., Cambridge, UK.

Drummond, C., Elezgaray, J., & Richetti, P. (2002). Behavior of adhesive boundary lubricated surfaces under shear: A new dynamic transition. *EPL (Europhysics Letters)*, *58*(4), 503.

Drummond, C., & Israelachvili, J. (2001). Dynamic phase transitions in confined lubricant fluids under shear. *Physical Review E*, *63*(4), 041506.

Gong, J. P. (2006). Friction and lubrication of hydrogels—its richness and complexity. *Soft Matter*, *2*(7), 544–552.

Gupta, V., & Singh, A. K. (2016). Scaling laws of gelatin hydrogels for steady dynamic friction. *International Journal of Modern Physics B*, *30*(26), 1650198.

Gupta, V., & Singh, A. K. (2018). Stress relaxation at a gelatin hydrogel–glass interface in direct shear sliding. *Modern Physics Letters B*, *32*(02), 1750345.

Gupta, V., & Singh, A. K. (2019). Effect of residual strength on frictional properties of a soft and hard solid interface. *Materials Research Express*, *6*(8), 085317.

Gondane, S., Singh, A. K., Sinha, N., & Vijayakumar, R. P. (2019). Experimental study on steady dynamic friction of MWCNTs mixed lubricants. *Surface Review and Letters*, *27*, 1950172.

Jaeger, J.C. Cook, N.G. & Zimmerman, R. (2009). *Fundamentals of Rock Mechanics*. John Wiley & Sons, New Delhi.

Leroy, S., Steinberger, A., Cottin-Bizonne, C., Trunfio-Sfarghiu, A. M., & Charlaix, E. (2009). Probing biolubrication with a nanoscale flow. *Soft Matter*, *5*(24), 4997–5002.

Lee, D. W., Banquy, X., & Israelachvili, J. N. (2013). Stick-slip friction and wear of articular joints. *Proceedings of the National Academy of Sciences*, *110*(7), E567–E574.

Mang, T., Bobzin, K., & Bartels, T. (2011). *Industrial Tribology: Tribosystems, Friction, Wear and Surface Engineering, Lubrication*. John Wiley & Sons.

Persson, B. N. (2000). Historical Note. In *Sliding Friction* (pp. 9–16). Springer, Berlin, Heidelberg.

Singh, A. K., & Juvekar, V. A. (2011). Steady dynamic friction at elastomer–hard solid interface: a model based on population balance of bonds. *Soft Matter*, *7*(22), 10601–10611.

Shoaib, T., & Espinosa-Marzal, R. M. (2018). Insight into the viscous and adhesive contributions to hydrogel friction. *Tribology Letters*, *66*(3), 96.

Thakre, A. A., & Singh, A. K. (2018). An experimental study on stick-slip of gelatin hydrogels using fracture mechanics. *Materials Research Express*, *5*(8), 085301.

Urueña, J. M., McGhee, E. O., Angelini, T. E., Dowson, D., Sawyer, W. G., & Pitenis, A. A. (2018). Normal load scaling of friction in gemini hydrogels. *Biotribology*, *13*, 30–35.

Section II

Sustainable Lubrication

9 Recent Progress in Vegetable Oil-Based Lubricants for Tribological Applications

Sooraj Singh Rawat and A. P. Harsha
Indian Institute of Technology (Banaras Hindu University)

CONTENTS

9.1 INTRODUCTION

The lubricants are natural lubricious substances. They protect the tribosurfaces against wear and minimize the risk of frequent breakdown. They also facilitate for smooth slide or roll of machine parts over each other and reduce the noise developed during the operation. The lubricant decreases the friction and direct interaction of asperities by establishing a tribofilm amid the interacting surfaces. The lubricants have many different functions, which are given in the later part (Section 9.1.1). Lubricant is an amalgamation of base stock and a package of additives. The percentage of additives in the lubricant is varied according to the severity of applications.

Typically, a liquid lubricant is constituted of base oil and additives in the range 1–25 wt. % (Rizvi and Syed, 2009).

The mineral oil is a refined, purified and processed product of crude oil via the fractional distillation process. In conventional lubricants, mineral oils are preferred as base stock due to their low cost, readily available in abundance, excellent oxidation stability and availability in a variety of viscosities. Besides, crude oil has a finite source in the world, and it can be utilized until available. The dramatic growth in the industrial sector in the 21st century caused the increase in the depletion rate of crude oil, which results in an increase in the price. Furthermore, mineral oils are toxic, and their biodegradability is very poor (Panchal et al., 2017). Therefore, the disposal of mineral oil is hazardous to both terrestrial and aquatic ecosystems. The environmental threat in consequence of the disposal of mineral oil has arisen the interest in developing a suitable substitute for conventional lubricants and eco-friendly too. In view of the above, the lubricants are formulated with vegetable oil, which exhibit high viscosity index, low evaporation losses, high flash point, renewability, high biodegradability and the excellent lubrication performance in comparison with mineral oils (Fox and Stachowiak, 2007; Jayadas et al., 2007; Panchal et al., 2017; Shashidhara and Jayaram, 2010; Zulkifli et al., 2013a). The long fatty acid chains of vegetable oil and the presence of polar groups furnish to establish an excellent tribofilm at the interacting surfaces under boundary and hydrodynamic conditions (Sahoo and Biswas, 2009; Zachariah et al., 2020).

9.1.1 Function of Lubricants

The function of lubricants is varied according to the tribological situations. The prime purpose of the lubricants is to protect the interacting surfaces and control friction and wear. Nevertheless, an extensive variety of lubricants encompasses the following functions.

- Cleaning and sweep away the debris and contaminations
- Protect against corrosion
- Reduce friction and wear
- Reduce vibrations and noise
- Seals for gases
- Separate interacting surfaces
- To dissipate heat
- Transfer of power

9.1.2 Classification of Lubricants

The lubricants are chiefly categorized in conformity with their (a) physical state, (b) origin of base oil and (c) applications (Mobarak et al., 2014). Figure 9.1 shows the classification of lubricants. By their physical state, the lubricants are further categorized as solid, semi-solid and liquid. In some exceptional cases, gases are also available as lubricants. As per the derived origin of base oil, the lubricants are assorted as mineral oil (oil derived from crude oil), vegetable oil (oil derived from plant seeds or

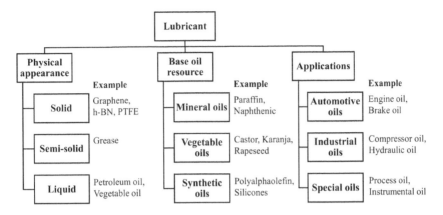

FIGURE 9.1 Classification of lubricants.

TABLE 9.1

Category of Base Oil as per API Guidelines (Rizvi and Syed, 2009; Torbacke et al., 2014)

Category	Saturates (%)	Sulphur (%)	Viscosity Index
Group I	< 90	> 0.03	80–120
Group II	≥ 90	≤ 0.03	80–120
Group III	≥ 90	≤ 0.03	≥ 120
Group IV	PAOs		–
Group V	All others not included in Group I, Group II, Group III and Group IV		–

kernels) and synthetic oil (oils are tailored as per end requirement). Apart from that, lubricants are classified based on their applications as industrial oils, automotive oils and special-purpose oils.

Further, the American Petroleum Institute (API) has segmented the base oils into five categories according to their physicochemical properties, which are presented in Table 9.1.

9.1.3 CURRENT STATUS AND LUBRICANT MARKET

Approximately 1,700 small and large lubricants' manufacturers exist worldwide. Around 60% volume of the global demand for the lubricants is produced by less than 2% of lubricant manufacturers (Nagendramma and Kaul, 2012). BP, Chevron, Conoco Phillips, Exxon Mobil, Fuchs, Lukoil, Nippon oil, Petro China, Shell and Valvoline are the major lubricant manufacturers in the world. In India, Hindustan Petroleum Corporation Ltd. (HPCL), Bharat Petroleum Corporation Ltd. (BPCL) and Indian Oil Corporation Ltd. (IOCL) are the leading lubricant manufactures. Indian

TABLE 9.2
Forecast Medium-Term Oil Demand for Period
2018–2024 (OPEC-Report, 2019)

Year	World (mb/d)	India (mb/d)
2018	98.7	4.7
2019	99.9	4.9
2020	101.0	5.1
2021	102.0	5.3
2022	103.0	5.5
2023	103.9	5.7
2024	104.8	6.0
Growth 2018–2024	**6.1**	**1.2**

Institute of Petroleum, Dehradun, Indian Oil Corporation, Faridabad, and Indian Institute of Chemical Technology, Hyderabad, are premier research and development laboratories in India.

In the last 20 years, the global market of lubricants has experienced significant demand due to the globalization of industries. In 1991, the worldwide demand for lubricants was estimated at around 35 million tonnes/year. In 2004, the global consumption of lubricants remained around 37.4 million tonnes/year (Nagendramma and Kaul, 2012). In 2005, approximately 37.9 million tonnes/year of lubricants were consumed worldwide (Mobarak et al., 2014). According to the Organization of the Petroleum Exporting Countries (OPEC) report 2014, world and India's medium-term oil demand is increased from 90 to 96 mb/d (million barrels per day) and 3.7–4.6 mb/d, respectively. However, in 2018, the demand for lubricants is reached up to 4.7 in India and 98.7 mb/d worldwide. At the end of the year 2024, the growth in demand for lubricants is expected to be 1.2 mb/d in India and 6.1 mb/d worldwide, and the long-term global oil demand is expected to increase by about 12 mb/d, rising from 98.7 mb/d in 2018 to 110.6 mb/d in 2040 (OPEC-Report, 2019). The forecast growth of medium-term oil demand for the period 2018–2024 is summarized in Table 9.2.

Environmental Lubricants Manufacturing, Inc. (ELM) in Iowa, United States, Lubrizol Corporation of Ohio, US research centre and National Ag-Based Lubricants (NABL) centre have introduced various lubricants such as hydraulic fluids, metalworking fluids and greases formulated with vegetable oil. IOCL, BPCL, Balmer Lawrie & Corporation Ltd., and Standard Greases and Specialties are major grease manufactures. The total production of greases in the world and in India is around 1174 TMT and 82 TMT in the year 2018, respectively (Singh, 2020). Table 9.3 exhibits the percentage excerpt of different kinds of base oils used for the production of lubricating greases.

9.2 BASIC CHEMISTRY OF HYDROCARBON

Carbon and hydrogen are the primary critical elements of a hydrocarbon. The hydrogen element represents with a symbol 'H', consisting of a proton in its nucleus and a single

TABLE 9.3

The Percentage Excerpt of Different Kinds of Base Oils Used for the Production of Lubricating Greases (Singh, 2020)

Type of Base Oil	World (%)	India (%)
Mineral	88.97	97.53
Synthetic	5.77	1.79
Semi-synthetic	4.64	0.44
Bio-based	0.63	0.23

electron in its orbit. Hydrogen is having atomic number 1; therefore, it is the lightest element and abundantly available in the universe. Carbon is a Latin invented word 'carbo', which means 'coal'. It also abundantly exists in the earth's crust. It is represented with a symbol 'C' and atomic number 6. It consists of six protons, six neutrons (sometimes seven or eight) in its nucleus and six electrons in its orbit. The atomic mass number of an atom is assigned based on the atomic mass of the nucleus, so the carbon atom has an atomic mass of 12, 13 and 14 amu. The atoms having the same atomic number and having different atomic mass are referred to as 'isotopes'. Therefore, carbon 12 (C^{12}), carbon 13 (C^{13}) and carbon 14 (C^{14}) are the isotopes of the carbon. C^{14} is radiocarbon, and it is used in estimating the age of organic materials based on radiocarbon dating. C^{14} is more concentrated on bio-based oils than on crude products.

The C atom has four valencies that prefer to bond with the H atom through the sharing of the electron(s). A pair of electrons form a single bond. In a pair, one electron is being shared by each atom, which is bonded together. Hydrocarbons comprised a chain length of C and H atoms. Crude oil is a complex blend of hydrocarbons molecules having a large assortment of molecular mass. The structure of few hydrocarbon chain lengths is visualized in Figure 9.2. Further, it can be subdivided into the following:

- **Alkanes** – the hydrocarbon chain length forms a single bond between C and C atoms. Single-bonded hydrocarbons are also referred to as saturated. The empirical relationship for such hydrocarbon is C_nH_{2n+2}, where n represents the number of C atoms in a chain length. It is also known as *paraffin*, comprised of linear (n-paraffin) or branched (iso-paraffin) chain hydrocarbons.
- **Alkenes** – this hydrocarbon chain length contains one or more double bonds between C and C atoms. The single bond is more chemically stable than a double bond. Therefore, it is referred to as unsaturated. The empirical relationship for alkenes is C_nH_{2n}. It is also known as *olefins*.
- **Alkynes** – if hydrocarbon chain length is bonded with at least a triple bond, it is referred to as alkynes. The empirical relationship for alkynes is C_nH_{2n-2}. The triple bond is more unstable than double and single bonds.
- **Alicyclics** – it is also known as *naphthenes*. In alicyclics, the six or five C atoms are arranged in a saturated cyclic ring. Therefore, it is referred to as cycloalkenes. The chain length contains more than one or more saturated cyclic ring.

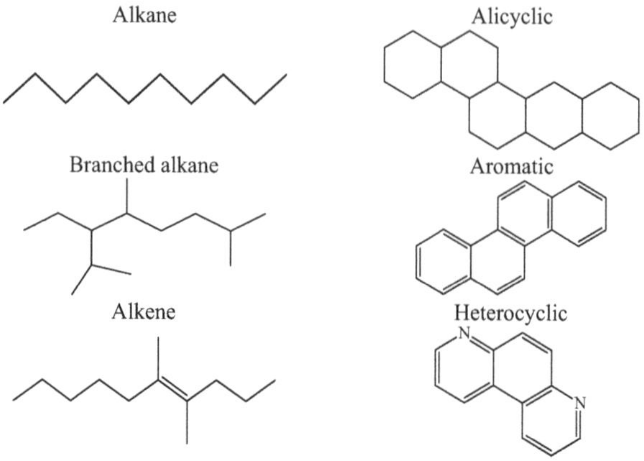

FIGURE 9.2   Few hydrocarbon structures of straight and branched paraffin, olefin, alicyclic, aromatic and heterocyclic.

TABLE 9.4

Comparison of Basic Properties of the Paraffins, Naphthenic and Aromatic Hydrocarbons (Torbacke et al., 2014; Totten, 1992)

Property	Paraffins	Naphthenic	Aromatic
Density	Low	Low	High
Elastomer compatibility	Shrink	Swell	Swell
Flash point	High	Low	Low/medium
Oxidation stability	High	High	Low
Pour point	High	Low	Low
Thermal stability	Low	Low/medium	High
Toxicity	Low	Low	Medium
Viscosity index	High	Low	Low
Volatility	Low	Medium	High

- **Aromatics** – it also has a six-membered cyclic structure with alternate double and single bonds like benzene ring.
- **Non-hydrocarbons** – some compounds that are present in crude oil have elements other than C and H, in either cyclic or non-cyclic chain structures. Generally, sulphur, oxygen, nitrogen and other elements are available in cyclic structure in the crude; it is referred to as *heterocyclic*.

The chain length of hydrocarbon has a significant effect on its physical properties. The longer chain length of hydrocarbons has demonstrated higher viscosity and boiling or melting point. For example, methane (CH_4) boils at $-164°C$, and octane (C_8H_{18}) boils at 125°C. The basic properties of hydrocarbons are based on their chemical structure and molecular weight. A comparative study of fundamental properties of the paraffins, naphthenes and aromatics is summarized in Table 9.4.

If hydrocarbon molecules are chemically bonded with oxygen molecules, then parent hydrocarbon is transformed into an alcohol (—OH) group. The alcohols are more chemically reactive than hydrocarbons. When alcohol is oxidized in the presence of oxygen, it formed a functional (—COOH) group. For example, vinegar (acetic acid) is the oxidized product of ethanol and oxygen. The presence of (—COOH) group is referred to as organic acid (fatty acid) and responsible for acidity. The plant seeds oil is the primary source of fatty acid (FA). The FA chain length varies from four C atoms to 36 C atoms (Syahir et al., 2017). Further, organic acid reacts with organic alcohol, then esters are formed, and water is the by-product. The esters have natural, pleasant fragrances. For example, wintergreen is an ester, which is a reaction product of salicylic acid and menthol.

9.3 BASICS OF VEGETABLE OIL

The seeds or kernels of the plants are crushed between the rollers to extract the vegetable oil. These oils are mainly of two types: non-edible and edible oils. The vegetable oils are comprised of triglycerides (also referred to as 'natural oil'), primarily a glycerol molecule which consists of hydroxyl groups attached with three long FA chains through ester linkages, as shown in Figure 9.3. If hydroxyl group of primary glycerol molecules is replaced with a FA, then it is referred to as 'monoglycerides'. Similarly, if any two hydroxyl groups are replaced from glycerol molecules, then it is called 'diglycerides'. The triglycerides present as a major component (98%) and some minor components such as free fatty acids (0.1%), tocopherols (0.1%), sterols (0.3%) and diglycerol (0.5%) are also available in vegetable oil (Zainal et al., 2018).

In a FA chain, if each C atom is bonded with a single bond to an adjacent C atom, then it referred to as 'saturated FA'. In other words, the saturated FA chain is ruled out for the presence of double- or triple-bonded C–C atoms. In a saturated FA chain, each C atom can bond with two H atoms. Suppose H atom(s) is missing from adjacent C atom, at that point, C atom shares a double bond or triple bond in place of a single bond. The presence of a double bond or triple bond in a FA chain is referred to as 'unsaturated FA'. If the count of the double bond in a FA chain is one and more than one, then it is termed 'monounsaturated FA' and 'polyunsaturated FA', respectively.

Vegetable oils are a mixture of saturated, monounsaturated and polyunsaturated FAs. The FA compositions of some vegetable oils are abridged in Table 9.5. The FA

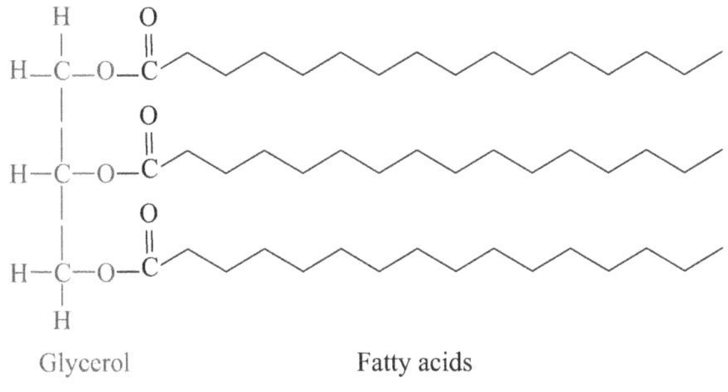

FIGURE 9.3 Schematic representation of triglyceride structure of vegetable oil.

TABLE 9.5

FA Compositions of Some Vegetable Oils

Vegetable Oils	Myristic Acid (C14:0)	Palmitic Acid (C16:0)	Palmitoleic Acid (C16:1)	Stearic Acid (C18:0)	Oleic Acid (C18:1)	Linoleic Acid (C18:2)	Linolenic Acid (C18:3)	Arachidic Acid (C20:0)	Eicosenoic Acid (C20:1)	Behenic Acid (C22:0)	Erucic Acid (C22:1)	Reference
Cottonseed oil	–	11.67	–	0.89	13.27	57.51	–	–	–	–	–	Sahoo et al. (2007)
Hydnocarpus wightianus	0.56	22.91	0.44	2.71	25.77	6.39	–	–	–	–	–	Salaji and Jayadas (2020)
Jatropha oil	–	16.0	–	6.5	43.5	34.4	0.80	–	–	–	–	Sahoo et al. (2007)
Karanja oil	–	11.65	–	7.50	51.59	16.46	2.65	–	–	–	–	Sahoo et al. (2007)
Mahua oil	–	19.2	–	23.4	58.6	12.4	–	2.1	–	–	–	Kumar et al. (2020a)
Rapeseed oil	–	4.0	0.3	1.6	18.4	14.7	10.5	0.6	7.9	0.5	41.2	Zareh-desari and Davoodi (2016)
Soya bean oil	–	10.4	0.3	2.3	22.5	56.8	7.1	0.4	–	–	–	Zareh-desari and Davoodi (2016)
Sunflower oil	–	1.6	6.1	4.2	25.6	60.8	0.8	–	–	0.9	–	Siniawski et al. (2007)

Note: (C A:B) is a notation where A signifies the number of C atoms in a given FA chain and B describes the number of double bonds in a given FA chain.

composition of each vegetable is unique, and their physicochemical properties are different. The oxidation stability of unsaturated FAs is poor in contrast to saturated FAs. Nevertheless, the saturated FAs have lower liquidity (higher melting point) than unsaturated FAs. For example, the pour point of coconut oil is ~21°C because it contains more than 90% saturated fats (Jayadas and Nair, 2006), while castor oil demonstrates the pour point as −10°C because of the presence of unsaturation (Singh, 2011). C–C single bonds are more chemically stable than C–C double or triple bonds. As the degree of unsaturation increases, it leads to a decrease in pour point (higher liquidity) with an increase in chemical reactivity. Therefore, unsaturated FAs are more chemically reactive and, in consequence, exhibited poor oxidation stability.

9.3.1 BENEFITS AND DRAWBACKS OF VEGETABLE OILS OVER MINERAL OILS

Vegetable oils have several advantages and few disadvantages over mineral oils. A comparative study on the physicochemical characteristics of vegetable oils and mineral oils is given in Table 9.6. Mineral oils are non-renewable and finite in resource, while vegetable oils are renewable and infinite in resource. Additionally, the vegetable oils are comprised of FAs, which are non-toxic, and their biodegradability is much superior to mineral oils. Therefore, the disposal is much more effortless and safer for the environment and aquatic life. Vegetable oils have better fire resistance property than mineral oil due to their high fire point and flash point. The vegetable oil-based lubricants effectively reduce the emission of smoke (i.e. toxic hydrocarbons and carbon monoxide) (Singh, 2011). Further, the vegetable oils are good solvents and have excellent lubricity due to their polar nature, which provides them with a high

---

TABLE 9.6

Comparison of Physicochemical Properties of Vegetable Oil and Mineral Oil (Rudnick, 2006)

Properties	Vegetable Oil	Mineral Oil
Biodegradability (%)	80–100	10–30
Cold flow behaviour	Poor	Good
Density (@ 20°C, kg/m³)	890–970	840–920
Evaporation loss	Lower	Higher
Flash point, °C	Higher	Lower
Fire point, °C	Higher	Lower
Hydrolytic stability	Poor	Good
Load-bearing capacity	High	Low
Oxidation stability	Fair	Good
Pour point, °C	−22 to +12	−15
Shear stability	Minimum	More than vegetable oil
Toxicity	Non-toxic	Toxic
Viscosity index	100–200	100
Volatility	Low	High

affinity with metal surfaces (Mannekote and Kailas, 2009). These properties suggest that lubricants formulated with vegetable oils are alternative to petroleum oil-derived lubricants. Apart from that, the vegetable oils in their natural form have few limitations (i.e. oxidation stability and fluidity at low temperature), which confines the use as a lubricant. These limitations can be rectified by various chemical treatments (Syahir et al., 2017).

9.4 PROCESSING OF VEGETABLE OILS

The geographical location and climatic condition have a substantial effect on the FA composition of vegetable oils. The geographical location comprehends the altitude, latitude and geographical area. On the other hand, the mean climatic conditions are exposed to sunlight and its intensity and weather. The Zarazi olive oil is obtained from the four different geographical locations in the south of Tunisia, and its static analyses exposed that the quality and sterolic and acidic composition of olive oil were notably altered according to the geographical locations (El-gharbi et al., 2018). The following steps are involved in extracting vegetable oil from oilseed.

a. Planting and harvesting
b. Preparation of beans
c. Cleaning
d. Extraction of oil (expeller crushing)
e. Reclamation

Refining is a vital process for any crude oil. The primary goal of the refining process is to separate major and minor impurities in crude oil and produce a high-quality base stock that shows satisfactory results during engineering applications. It is a treatment for purification, which is used to remove unsaponifiable materials, insoluble matter, free FAs, gums and phosphatides. The oil that passes through the purifying treatment process is referred to as refined oil. Refined soya bean oil is commonly used as cooking oil.

Degumming is a process that involves the removal of gums. The gums are composed of non-triglycerides and phospholipids, which are insoluble, and soluble in vegetable oils. The content of gums in vegetable oils is measured about parts per million (ppm) of phosphorus. The presence of phosphorus content in vegetable oil has a pronounced effect on colour, oxidation and hydrolytic stability but also reduces refining yield. Insoluble gums are settled down at the bottom of the storage drums; it is referred to as natural degumming. Soluble gums are removed from the oils through a chemical treatment. Lecithin, a by-product of degumming, is preferred as an emulsifying agent (Honary and Richter, 2011).

Bleaching is another crucial step used to absorb soaps, metal traces, oxidation products, pigments and decomposition products by an absorbent or bleaching agent. Activated natural clay or Fuller's earth, activated carbon, and silica are used as bleaching agents. The process of bleaching is very simple in which neutralized oil is added with stoichiometric quality of clay and the solution is heated at bleaching temperature and followed by filtration. The effective adsorption of impurities from crude oil depends on some factors such as bleaching temperature, degree of intimate contact amid oil and bleaching agent.

Deodorization process is used to remove the available volatile substances and odoriferous compounds from the vegetable oil. In this process, the steps involved are heating of vegetable oil followed by steam stripping and finally cooling. The purpose of the steam stripping step is to increase the rate of the process. This process also reduces the content of free FAs to <0.05%.

9.5   OXIDATION STABILITY

When vegetable oils are exposed to oxygen, some degree of chemical reaction occurs, which results in oxidation. The degree of unsaturation of a FA chain increases the chemical reactivity. Therefore, unsaturated FAs are more prone to get oxidized. Oxidation stability specifies the proficiency of the oil to restrict oxidation. The oxidation stability of unsaturated FAs is deficient as compared to saturated FAs. Several factors, such as pressure, temperature, water content and contaminations, govern the oxidation rate of oil. Also, the oxidation rate increases with rise in temperature. As a general thumb rule, the rate of oxidation doubles with every 10°C escalation in operating temperature (Honary and Richter, 2011). The low oxidation stability of a lubricant is undesirable. Therefore, the lubricants should have high oxidation stability, which is a crucial criterion for them. The free radicals available in oils react with oxygen and form peroxides, further decomposing into new radicals, aldehydes and ketones. The aldehydes have a strong odour and rancidity taste. When the oil undergoes oxidation degradation, consequently, it polymerizes and thickens. This polymerization process increases the acidity of oil and viscosity and initiates the formation of an insoluble deposit (Nosonovsky and Bhushan, 2012).

There are several methods to amend the oxidation stability of bio-based oil. These methods are frequently used in industrial lubricants, which include:

a. Addition of antioxidants additive
b. Preventing direct exposure of oil to the air by nitrogen dosing
c. Chemical treatment to remove unsaturation

A wide variety of antioxidants additives have been developed over the years for engine oils, transmission fluids, turbine oils, compressor oils, hydraulic fluids, metalworking fluids, gear oils and greases. Some common examples of antioxidants are methyl cyclopentadiene dimer, zinc dialkyldithiophosphate (ZDDP), triaryl phosphites, t-butylhydroxytoluene, diisodecyl pentaerythritol diphosphite, tertiary butyl hydroquinone and propyl gallate (Rudnick, 2017). Nitrogen is heavier than air, so the containers containing vegetable oil-based product are topped up with nitrogen to avoid oxidation. This process is referred to as *nitrogen dosing*. The nitrogen dosing is an effective barrier to prevent oxidation. The presence of C=C bonds in the FA chain length is the leading cause of unsaturation in the triglyceride structure of vegetable oil. The C=C bonds are the potential sites that can be chemically modified and transform the unsaturated FA chain into a saturated FA chain. Some chemical modification treatments can ameliorate the oxidation stability of the vegetable oils and refurbish them into a suitable lubricant application. These chemical treatments include:

a. Epoxidation
b. Hydrogenation
c. Transesterification

Epoxidation is a chemical treatment process in which epoxidized vegetable oils are produced by joining an oxygen atom at olefinically unsaturated C=C bonds to form a three-membered cyclic ring, which is also referred to as *oxirane ring*. The final product of the epoxidation process is called *epoxides* or *oxirane compounds*. In this method, the vegetable oil is reacted with peroxy acid (i.e. concentrated hydrogen peroxide) in the presence of acetic acid or formic acid, and a small quantity of concentrated sulphuric acid is added as a catalyst (Kashyap and Harsha, 2016). They have found better oxidation stability and better friction-reducing characteristics as compared to unmodified rapeseed oil.

Hydrogenation is a process of joining hydrogen atom to the C=C bonds in the structure of triglycerides of a vegetable oil molecule. The final product of this process is referred to as *hydrogenated oil*. In this process, the vegetable oils in the presence of supported or Raney Ni catalyst are hydrogenated under pressure between 69 and 413 kPa and temperature ranges from 150°C to 225°C (Zainal et al., 2018). If the hydrogenation process transforms polyunsaturated FAs into monounsaturated FAs without affecting the saturated component of the FAs, then it is referred to as *partial hydrogenation*. This process does not deteriorate the pour point of the vegetable oil. The partial hydrogenation of palm oil is studied in the presence of $Pd/\gamma-Al_2O_3$ catalyst (Shomchoam and Yoosuk, 2014). The results showed that the oxidation stability of palm oil is augmented from 13.8 to 22.8 h by transforming the polyunsaturated FAs into monounsaturated FAs.

Transesterification is another multistep chemical treatment process in which triglyceride molecule reacts with short-chain alcohol in the presence of an acidic or basic catalyst to produce FA alkyl esters. Further, the resulting FA alkyl esters are reacted with various types of alcohol with acid or base catalyst to yield triesters. The transesterified FA alkyl esters lead to an improvement in oxidation stability and increased low-temperature fluidity while preserving the beneficial viscosity and tribological performance of vegetable oils (McNutt and He, 2016). The palm oil polyol esters are synthesized by transesterification under the temperature range between 12°C and 150°C using sodium methoxide as a catalyst to yield trimethylolpropane (TMP) triesters (Yunus et al., 2005). They have found the low pour point (−37°C) of the obtained product.

9.6 ROLE OF NANOADDITIVES

Chemicals, solid particles and/or fillers are deliberately added to a base stock to enhance its physicochemical and lubrication properties, which are referred to as additives. Commercial lubricants are comprised of a package of additives such as extreme pressure (EP) agent, viscosity modifiers (VM), oxidation inhibitors, friction modifiers (FM), antiwear (AW) agents, corrosion inhibitors, pour point depressants and foam inhibitors. Conventional organic additives such as ZDDP and trixylyl phosphate are the most promising AW/EP characteristics. They comprised active elements such as chlorine (Cl), sulphur (S) or phosphorus (P), and the presence of the polar group facilitates the strong adsorption on friction surfaces. The environmental threat arises from the use of P and S elements in additives (Chen et al., 2015).

In recent years, nanomaterials have shown huge interest as an alternative to conventional additives. Enormous materials have been explored as nanoadditives in lubricants and greases (Gulzar et al., 2016; Rawat and Harsha, 2019). The lubricant dispersed with additives of nanometer size is referred to as *nanolubricants*. The high surface energy of nanomaterials tends to agglomerate, which results in poor dispersion stability. Therefore, achieving the long period dispersion of nanoadditives in the lubricant is a great challenge. In this context, the surface modification of nanomaterials using surface-active molecules is a wonderful approach (Chouhan et al., 2018; Kumari et al., 2017). Nanocomposites comprise more than one nanomaterial, and their synergistic effect shows superior tribological performance than the singular effect (Sun and Du, 2019; Verma et al., 2019). Therefore, in recent times, nanocomposites are gaining immense attraction as an additive in lubricants. The engineered nanomaterials are broadly classified into seven categories according to their characteristic chemical elements: metal oxides, sulphides, metals, carbon and its derivatives, nanocomposites, rare earth compounds and others (Dai et al., 2016).

Nanomaterials have several advantages of using as an additive in lubricants. The solubility of nanomaterials in non-polar base oils is very poor, and its reactivity with other additives present in the lubricants is low. The nanodimension of materials eases to enter tribointerfaces and increases the probability of film formation. Further, the real area of contact decreases, owing to the entrapped nanomaterials at the interacting surfaces (Rawat et al., 2020). The non-volatility of inorganic nanomaterials is better than organic nanomaterials that make them favourable to endure for high-temperature applications. Nanomaterials exhibit versatile potential that solo nanomaterials served multiuse as FM, AW and EP additive (Gupta and Harsha, 2017; Kumar and Harsha, 2020).

The effectiveness of nanomaterials in augmenting the tribological performance of lubricants is dependent on various attributes such as their size, crystal structure, concentration, morphology and compatibility with base stock (Rawat et al., 2019). It is found that the optimum concentration and size of nanomaterials show excellent tribological performance and beyond which tribological performance starts to deteriorate. Apart from that, the friction and wear preventive characteristics of a nanolubricant depend on ambient and working conditions such as temperature, applied load, velocity and surface roughness of interacting surfaces. The lubrication mechanism of lubricants doped with nanoadditives is a crucial parameter to understand the active mechanism involved in the improvement of tribological performance. Four lubrication mechanisms have been proposed by the researchers to understand the role of nanolubrication and to explain the tribological performance of nanolubricants: (1) mending effect, (2) rolling mechanism, (3) protective film and (4) polishing mechanism. Different lubrication mechanisms of nanolubricants are schematically illustrated in Figure 9.4.

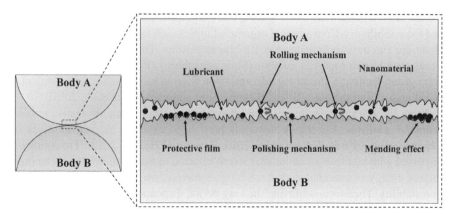

FIGURE 9.4 Schematic representation of lubrication mechanism of nanolubricants.

Rolling Mechanism, in this mechanism, the quasi-spherical and spherical nano-materials would roll over the asperities and impersonate as nanobearing. It is believed that these nanobearings transform the sliding friction into rolling friction or partially rolling friction, which leads to minimize friction and wear. This phenomenon is also called the *ball-bearing effect*. The type of loading and hardness of tribopair are the other factors for favouring the rolling mechanism. At stable low-load conditions, the shape and rigidity of nanomaterials are maintained between the shearing surfaces, resulting in rolling friction. This rolling mechanism has been reported by several researchers (Asrul et al., 2013; Kashyap and Harsha, 2016; Rawat et al., 2019a).

Mending effect is a surface repairing technique by the deposition of nanomaterial in the grooves and scars of the tribosurfaces. The deposition of nanomaterials reduces the abrasion phenomenon, which results in compensating the material loss at the interacting surfaces. The nanodimension of materials furnishes the ability to patch the scars and grooves developed on worn surfaces. Therefore, it is also referred to as a *self-repairing effect*. A majority of investigations have affirmed the deposition of nanomaterials on the worn scars through energy-dispersive X-ray spectroscopy (EDX) (Gupta and Harsha, 2017; Talib et al., 2017).

Polishing mechanism is also another surface improvement technique. In this phenomenon, the nanomaterials behave as three-body abrasive, resulting in reducing surface roughness. The reduction in surface roughness is ascribed to the polishing of surfaces, and it is also termed a *smoothing effect* or *artificial smoothing*. This smoothing effect improves the tribological performance (Sajeeb and Rajendrakumar, 2020). The surface topography characterization technique, i.e. atomic force micros-copy (AFM), is beneficial to observe the effect of nanomaterials on the roughness of tribopairs (Rawat et al., 2019b; Thottackkad et al., 2012).

Protective film of nanomaterials is formed on the friction surfaces by the phy-sisorption and chemisorption interactions. This protective film is also termed *tri-bofilm*. The hardness and chemical reactivity of nanomaterials play a vital role in the formation of tribofilm (Shafi and Charoo, 2020). The tribofilm minimizes the direct interactions of asperities, so it protects interacting surfaces against friction and wear. The X-ray photoelectron spectroscopy (XPS) and Raman spectroscopy are very informative analytical tools to verify the formation of tribofilm (Chen et al., 2015; Gulzar et al., 2015).

9.7 APPLICATION OF VEGETABLE OIL-BASED LUBRICANTS

The application of lubricants is mainly categorized into closed systems and open systems (Syahir et al., 2017). In a closed lubrication system, the lubricant does not cross the system boundary and keeps within the system, such as compressors and engines. After the accomplishment of the usefulness period of the lubricant, it is replaced. In an open lubrication system, typically, spray system, atomization systems, drip feed or gravity feed systems are used to lubricate the tribopairs such as metalworking processes, metal machining and chainsaw oiling. During the operation, a quantitatively large amount of lubricant is required, and it is a total loss of the lubricant. The used lubricants are drained into the environment over time. Therefore, the demand for environmentally friendly lubricants is gaining immense attention. Enormous investigations are carried out with vegetable oil-based as lubricants/greases, and their tribological performances showed promising results (Bhaumik et al., 2020; Delgado et al., 2020; Jayadas and Nair, 2006; Panchal et al., 2017).

Nowadays, numerous bio-based lubricants are developed and used in various industrial applications. The United Nations have permitted (Honary and Richter, 2011) commercialized soya bean oil-based greases to lubricate rail curve and flanges of locomotive wheels. Further, bio-based greases have applications in the greasing of hitch point (also called the fifth wheel) of the tractor. The bio-lubricants are widely used in engine oils, hydraulic fluids, compressors oils, metalworking fluids, gear oils, chainsaw lubricants, greases, asphalt release agent, concrete mould release oil, food industry, corrosion preventive oils, chain and conveyor lubricants, and insulating fluids. The specific application of distinctive vegetable oils is listed in Table 9.7.

9.8 EVALUATION OF LUBRICANTS

The American Society of Testing Materials (ASTM) has developed various technical standards to evaluate the performance of the lubricants. The performance evaluation of lubricants is broadly classified into two categories:

a. Physicochemical properties
b. Tribological properties

9.8.1 EVALUATION OF PHYSICOCHEMICAL PROPERTIES

The performance of a lubricant is based on its physicochemical properties. The physicochemical properties of some non-edible and edible vegetable oils are listed in Table 9.8. The effect of degree of unsaturation, chain length and chain branching on the physicochemical properties of vegetable oils is exhibited in Table 9.9. Therefore,

TABLE 9.7

Various Applications of Different Vegetable Oils

Bio-Based oil	Applications
Canola oil	Metalworking fluids, chain oil bar lubes, tractor transmission fluids, hydraulic oils, food-grade lubes, penetrating oils
Castor oil	Gear lubricants, greases
Coconut oil	Gas engine oils
Crambe oil	Surfactants, intermediate chemicals, grease
Cuphea oil	Motor oil, cosmetics
Jojoba oil	Lubricant, cosmetic, grease
Linseed oil	Varnishes, stains, lacquers, paints, coating
Olive oil	Automotive lubricants
Palm oil	Grease, rolling lubricant
Rapeseed oil	Greases, air compressor-farm equipment, chain saw bar lubricants
Safflower oil	Diesel fuel, light-coloured paints, enamels, resins
Soya bean oil	Hydraulic oil, plasticizers, disinfectants, pesticides, detergents, shampoos, soaps, coating, paints, printing inks, metal casting/metalworking, biodiesel fuel, lubricants
Tallow oil	Soaps, steam cylinder oils, cosmetics, plastics, lubricants

Adapted from Panchal et al. (2017) with permission.

it is imperative to determine the limiting points of vegetable oil that can be modified and formulated as a lubricant suitable for its earmarked use. The list of ASTM standards is required to evaluate the physicochemical properties of vegetable oils as summarized in Table 9.10.

Once the vegetable oil-based lubricants are formulated, further, it is evaluated for their impact on terrestrial and aquatic ecosystems. In this context, the final product is characterized by one or all of the environmental performance tests listed in Table 9.11.

9.8.2 EVALUATION OF TRIBOLOGICAL PERFORMANCE

Each tribosystem confronts with wear and friction, so the prime objective of the lubricant is to minimize friction and enhance wear prevention characteristics by furnishing a tribofilm in between tribopairs. This tribofilm enables metal-to-metal separation. Tribological tests are executed to determine the friction coefficient, wear prevention characteristics and EP behaviour of the lubricant. Typically, the machines have flat on flat, round on round or round on flat tribocontacts. In this context, several performance test rigs such as pin-on-disc/ball-on-disc, SRV test machine, four-ball tester and FZG gear are designed to examine the tribological performance of the lubricants. The tribological behaviour of the lubricants has been assessed in a specific tribometer under the prescribed test conditions furnished by ASTM standard, which is listed in Table 9.12.

TABLE 9.8
The Physicochemical Properties of Some Non-Edible and Edible Vegetable Oils

Vegetable Oil	Density (g/cm³)	Flash Point (°C)	Pour Point (°C)	Viscosity (mm²/s) 40°C	Viscosity (mm²/s) 100°C	Viscosity Index	Reference
Canola oil methyl ester (COME)	0.881	136	-12	4.95	1.8	236	Nagabhooshanam et al. (2020)
Castor oil	–	>300	-10	252	19	90	Singh (2011)
Chemically modified palm oil (CMPO)	0.901	–	–	40.03	9.15	221	Gulzar et al. (2015)
Coconut oil	–	–	–	27.82	7.07	224	Sajeeb and Rajendrakumar (2020)
Epoxidized mahua oil	0.871	153.4	2.3	123.22	11.7	187	Kumar et al. (2020a)
Hydnocarpus wightianus oil	–	244	11	50.5	9.2	166	Salaji and Jayadas (2020)
Jatropha oil	–	240	0	35.8	8.04	208	Joseph et al. (2007)
Mahua oil	0.925	47.5	-3.1	39.2	8.2	109	Kumar et al. (2020a)
Neem oil	0.920	100	10	35.83	–	–	Wakil et al. (2015)
Palm oil	0.899	254.5	12	41.93	8.49	185	Wakil et al. (2015)
Rapeseed oil	0.913	193	-12	45.26	–	–	Kashyap and Harsha (2016)
Rapeseed oil	0.913	314	-11	38.6	13.7	–	Zareh-desari and Davoodi (2016)
Sesame oil	0.906	280	-4	34.08	7.63	202.9	Wakil et al. (2015)
Soya bean oil	–	325	-9	28.86	7.55	246	Siniawski et al. (2007)
Soya bean oil	0.926	324	-12	31.4	8.2	–	Zareh-desari and Davoodi (2016)
Sunflower oil	–	252	-12	40.05	8.65	206	Siniawski et al. (2007)

TABLE 9.9
The Effect of Degree of Unsaturation, Chain Branching and Chain Length on Physicochemical Properties of Vegetable Oils

Physicochemical Properties	Low Temperature Properties	Oxidation Stability	Kinematic Viscosity	Viscosity Index
Higher degree of unsaturation	Positive	Negative	Negative	Positive
Increase of chain length	–	–	Positive	Positive
Increase of chain branching	Positive	Positive	Negative	Negative

TABLE 9.10
List of Physicochemical Characterizations of the Lubricant

ASTM Standard	Purpose
D 92-05	This standard method is used to evaluate the fire and flash points of the lubricants by cleave land open cup tester
D 93-02a	This standard method is used to evaluate flash points of the lubricants by Pensky–Martens closed cup tester
D 94-07	The purpose of this standard is to quantify the saponification number of the lubricants
D 97-05	This standard test practice is used to evaluate the low-temperature fluidity (pour point) of the lubricants
D 130-12	The purpose of this standard is to detect the corrosiveness of copper strip by lubricating oil whose vapour pressure is not greater than 124 kPa at 37.8°C
D 445-06	This standard test method is used to determine the kinematic viscosity of opaque and transparent lubricants
D 611-12	This standard test practice is used to assess the aniline point of hydrocarbon solvents and petroleum products.
D 664-04	The purpose of this standard is to quantify the acid number of the lubricants by potentiometric titration
D 892-03	This standard test practice is used to assess the foam characteristics of lubricants at 24°C and 93.5°C
D 943-19	This standard test method is used to determine the oxidation stability of inhibited lubricants under the environment of oxygen at an elevated temperature.
D 972-16	This standard practice is used to measure the evaporation loss of lubricants and greases at any temperature in the range between 100°C and 150°C
D 974-14	The purpose of this standard is to quantify the acid and base number of the lubricant by colour-indicator titration
D 1147-09	This standard test method is used to determine the refractive indexes of light-coloured and transparent viscous hydrocarbon liquids and melted solid that have refractive indexes in the range of 1.33 and 1.60 and at a temperature from 80°C to 100°C.
D 1401-19	This standard test practice is used to assess the water separation ability of petroleum oils and synthetic fluids. It indicates the demulsibility of the oil.
D 1748-10	This standard practice is used to determine the rust protection by metal preservatives under the conditions of high humidity
D 1838-16	The purpose of this standard is to detect the corrosiveness to copper by liquefied petroleum gases whose vapour pressure is greater than 124 kPa at 37.8°C

(Continued)

TABLE 9.10 (*Continued*)
List of Physicochemical Characterizations of the Lubricant

ASTM Standard	Purpose
D 2070-16	This standard test method is used to evaluate the thermal stability of hydrocarbon-based hydraulic oils
D 2140-08	The purpose of this test method is to characterize the carbon-type composition of an oil
D 2270-93	This standard practice is used to calculate the viscosity index from kinematic viscosity at 40°C and 100°C
D 2272-14a	This standard test practice is used to assess the oxidation stability of lubricants at 150°C by rotating oxygen-pressured vessel.
D 2500-17a	This standard test practice is used to measure the cloud point of petroleum products biodiesel fuels
D 2501-14	This standard test practice is used to calculate the viscosity gravity constant (VGC) of petroleum oils having viscosities in excess of 5.5 mm²/s at 40°C and excess of 0.8 mm²/s at 100°C. VGC indicates the solubility behaviour of the fluid.
D 2619-09	This standard test practice is used to assess the hydrolytic stability of hydraulic fluids.
D 3339-12	This standard test method is used to determine the acid number of lubricants by semi-micro colour indicator titration
D 4052-96	This standard practice is used to determine the density or relative density of viscous oils and petroleum distillates in the temperature range between 15°C and 35°C and vapour pressures below 80 kPa
D 4289-19	This standard test practice is used to assess the elastomer compatibility of lubricating grease and fluids
D 5355-95	This test method is used to determine the specific gravity of oils and lubricants at 25°C
D 5554-15	This test practice is used to evaluate the iodine value of fats and oils. It is a parameter that indicates the degree of unsaturation in the oils and fats and is expressed with reference to the number of centigrams of iodine absorbed per gram of sample.
D 5555-95	This test practice is used to quantify the free FA contents of fats and oils.
D 5558-95	This standard test practice is used to assess the saponification value of fats and oils.
D 5800-19a	This test procedure is used to determine the evaporation loss of lubricants by the Noack method
D 6082-12	This test practice is used to assess the foaming characteristic of lubricants at 150°C.
D 6304-16	This test practice is used to quantify the water content in petroleum products, lubricants and additives by coulometric Karl Fischer titration.
D 6375-09	This standard test practice is used to assess the Noack evaporation loss of lubricants using a thermogravimetric analysis (TGA) test
D 6749-02	This test practice is used to evaluate the pour point of lubricants by an automatic air pressure method

9.8.3 TRIBOLOGICAL PERFORMANCE OF VEGETABLE OIL-BASED LUBRICANTS

The tribological performance of the bio-lubricants is susceptible to the composition of vegetable oils. These oils present a great diversity in their composition due to variation in percentage share of saturated, monounsaturated and polyunsaturated FAs. The degree of unsaturation, chain branching and chain length also governs the lubrication performance of FAs. The tribological test on eight natural oils

TABLE 9.11

List of Environmental Performance Characterizations of the Lubricant

ASTM Standard	Purpose
D 5864-18	This test standard is used to evaluate the aerobic aquatic biodegradation of lubricants or their components within the controlled laboratory conditions.
D 6081-98	This standard practice is used in the formulation of lubricants or their components for acute or chronic aquatic toxicity tests.
D 6866-20	The purpose of this method is to quantify the bio-based content of gaseous, liquid and solid samples using radiocarbon analysis

TABLE 9.12

List of Tribological Characterizations of the Lubricant

ASTM Standard	Purpose
D 2509-14	This test standard is used to measure the load-carrying capacity of lubricants by the Timken EP tester
D 2783-03	This standard test practice is used to assess the EP properties of lubricants by the four-ball method
D 3233-19	This standard test practice is used to assess the EP properties of lubricants by Falex pin and vee block methods
D 4172-94	This standard test practice is used to assess the wear preventive characteristics of lubricants by the four-ball method
D 5182-19	This test practice is used to determine the scuffing load-carrying capacity of oils by Forschungsstelle für Zähnräder und Getriebebau (FZG) visual method
D 6425-11	This standard test practice is used to assess the wear and friction properties of EP lubricants using high-frequency, linear oscillation (SRV) test machine.
D 7043-17	This standard practice is used to evaluate the wear characteristics of petroleum and non-petroleum hydraulic fluids operating in a constant volume vane pump.
D 7421-19	This standard test practice is used to determine the load-carrying capacity of lubricant via SRV test machine.
G 99-05	This test method is used to determine friction and wear characteristics of lubricants via pin-on-disc apparatus.

(i.e. soya bean, sesame, safflower, peanut, olive, corn, canola and avocado oils) has been conducted (Reeves et al., 2015) to develop a clear understanding of the effect of FA composition. These oils have a large variety in their compositions. Figure 9.5a shows the friction profiles of various vegetable oils concerning sliding distance under ambient conditions. The steady friction profiles are maintained after sliding approximately 2,000 m, and it is a similar trend for all oil samples. Figure 9.5b depicts the average coefficient of friction (COF) of each natural oil lubricant. The experimental results demonstrate that avocado, olive, canola and peanut oils maintain the average COF in the range of between 0.02 and 0.10. However, sesame, soya bean, corn and

(a)

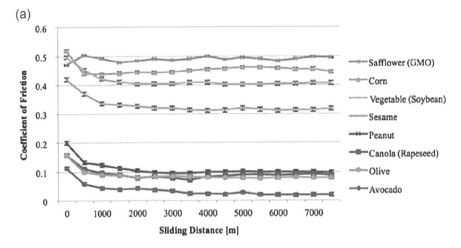

(b)

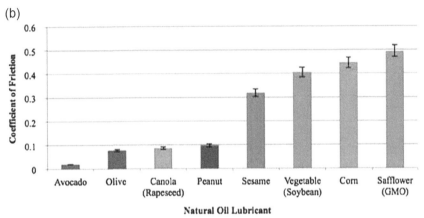

FIGURE 9.5 Variation of COF during ambient conditions for various vegetable oils with (a) sliding distance and (b) average COF (Adapted from Reeves et al. (2015) with permission.)

safflower oils uphold the average COF in the range of 0.30–0.50. Figure 9.6a exhibits the variation in wear volume (WV) of the pin of various vegetable oils concerning sliding distance. There is a rapid increase in WV of the pin within the first 50–100 m, and this increment persists until approximately 1,000 m. Subsequently, the variation in WV decreases concerning sliding distance. Figure 9.6a depicts the average WV of each natural oil samples. As shown in the figure, soya bean and corn oils have the highest WV, and avocado is achieved with the lowest WV.

Fox and Stachowiak (2007) reported that the organic polar molecules are bonded to the triglyceride structure of the vegetable oil, which imparts an excellent adherence to the metallic surfaces. A molecule of FA contains a carboxyl group ($-COOH$) at one end and an alkyl group ($-CH_3$) at another end. The $-COOH$ group has a strong affinity with metallic surfaces. However, $-CH_3$ group behaves as a repellent to

(a)

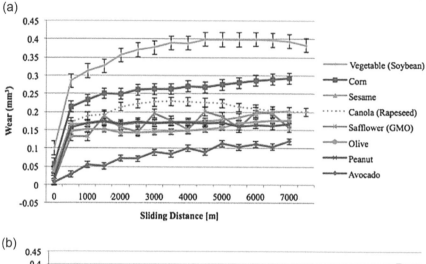

(b)

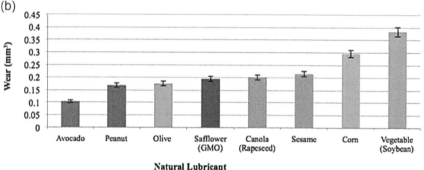

FIGURE 9.6 Variation of WV during ambient conditions for various vegetable oils with (a) sliding distance and (b) average WV (Adapted from Reeves et al. (2015) with permission.)

another material. A physisorption monolayer of organic groups is formed on metallic surfaces. This monolayer corroborates to form tribofilm results in reducing the COF (Reeves et al., 2015; Zainal et al., 2018).

Zulkifli et al. (2016) synthesized the pentaerythritol ester (PE) and TMP ester by the transesterification of palm oil methyl ester (POME) and compared their load-carrying capacity and lubrication behaviour with fully formulated lubricant (FFL) and paraffin. The results showed that the PE and TMP ester lubricants have comparable tribological characteristics and load-carrying capacity as commercial FFL. Jayadas et al. (2007) evaluated the vegetable oil for automotive application. The plain coconut oil shows lower COF and larger wear scar diameter (WSD) in comparison with 2T oil. Further, coconut oil is doped with AW/EP additive (2 wt. %), and the tribological performance is substantially improved. These findings suggest that coconut oil in its unmodified form cannot be recommended as 2T oil for automotive applications.

In the metal forming process, vegetable oil has shown potential candidate as bio-based lubricants. Zareh-desari and Davoodi (2016) formulated the metal forming oil with rapeseed and soya bean oils blended with SiO_2 and CuO in variable concentration and compared their tribological performance with two conventional metal forming lubricants (CM202A press drawing oil and solid lubricant of zinc phosphate plus sodium soap). The experimental results exhibited that SiO_2 and CuO nanoadditives have enhanced the friction reduction ability of bio-lubricants (21%–31%) similar to conventional metal forming lubricants. Similarly, bio-based nanofluids have shown their potential as cutting fluids in machining operations. Padmini et al. (2016) in their minimum quantity lubrication (MQL) investigated canola oil (CAN), coconut oil (CC) and sesame oil (SS) dispersed with nano molybdenum disulphide ($nMoS_2$) used as cutting fluid in turning of AISI 1040 steel. The test results showed that the machining parameters such as cutting forces, tool wear, surface roughness and temperature were significantly reduced by 44%, 37%, 39% and 21%, respectively, using 0.5 wt. % of $nMoS_2$ in CC as compared to dry machining.

Enormous nanosized materials of various sizes and morphologies have been explored as nanoadditives to improve the antiwear and antifriction characteristics of bio-lubricants. Several research articles show the positive effect of nanoadditives on the lubrication behaviour of bio-lubricants. The effect of some nanoadditives on the tribological performance of bio-lubricants is summarized in Table 9.13.

Many researchers have also used vegetable oil (edible or non-edible) as a base oil in the formulation of bio-greases. The palm grease formulated with lithium and calcium soap demonstrated the superior AW performance than mineral grease, and it is attributed to the presence of active functional polar groups (Sukirno et al., 2010, 2009). Panchal et al. (2015) formulated lithium grease using transesterified Karanja oil as base oil. The tribological results of Karanja oil esters grease exhibited competitive performance as compared to SN-500 mineral grease. Similarly, Jatropha oil thickened with lithium soap and blended with a multifunctional additive exhibited superior tribological performance as compared to commercial grease (Nagendramma and Kumar, 2015).

9.9   SUMMARY

This chapter has introduced the lubricants, their function, classification and current status in the market. In brief, the basic chemistry of hydrocarbons and vegetable oils is discussed. Besides, the advantages of bio-lubricants over conventional lubricants are also covered. Further, this chapter develops a clear understanding of the processing of vegetable oils and the remedies for their limiting characteristics. Moreover, a small section is dedicated to comprehending the lubrication mechanism of nanolubricants. Finally, the chapter unwraps the physicochemical and tribological properties evaluation techniques and tribological investigation of vegetable oil-based lubricants. This overview suggests that bio-lubricants have a substantial superior capability to protect the interacting surfaces as compared to conventional lubricants. The uses of bio-lubricants are increasing day by day for sustainable development. In the coming decades, it is believed that bio-lubricants would replace conventional lubricants.

TABLE 9.13

Tribological Performance of Bio-Lubricants Enriched with Different Nanoadditives

Base Oil	Nanoparticles	Optimum Concentration of Nanoparticles	COF (μ)	WSD/SWR/WV	EP	Reference
Coconut oil	CuO (d = 20–150 nm)	0.34 wt. %	μ↓	WSD↓	–	Thottackkad et al. (2012)
Palm oil-based TMP ester	TiO$_2$	1.0 wt. %	μ↓15%	WSD↓11%	–	Zulkifli et al. (2013b)
Rapeseed oil	CuO (d = 20 nm)	0.5 wt. %	μ↓40%	WSD↑	–	Kashyap and Harsha (2016)
	CeO$_2$ (d = 9 nm)	0.1 wt. %	μ↓54%	WSD↑		
Avocado	hBN (70 nm, 0.5 μm, 1.5 μm and 5.0 μm)	5.0 wt. % of 70 nm	μ↓64%	WV↓72%	–	Reeves and Menezes (2017)
Castor oil	ZnO	0.1 wt. %	μ↓	WSD↓	–	Bhaumik et al. (2018)
Castor oil	CeO$_2$ (d = 90 nm)	0.25 w/v%	μ↓43.3%	WSD↓ 37.4%	EP↑	Gupta and Harsha (2018a)
	PTFE (d = 150 nm)	0.5 w/w%	μ↓64.0%	WSD↓ 35.3%	–	
Castor oil	CuO (80 nm)	0.1 wt. %	μ↓34.6%	WSD↓28.3%	EP↑	Gupta and Harsha (2018b)
Blend of coconut oil and mustard oil	CuO (30–50 nm)	0.2 wt. %	μ↓	WSD↓	–	Sajeeb and Rajendrakumar (2019)
Castor oil	MoS$_2$ (d = 50 nm)	0.1 wt. %	μ↓	WSD↓	–	Yu et al. (2019)
Canola	CuO	0.1 wt. %	μ↓61.15%	SWR↓ 64.7%	–	Kumar et al. (2020b)
Castor oil + 40% cashew nut shell liquid	rGO	0.5 wt. %	μ↓	WSD↓	–	Bhaumik et al. (2020)
Neem	SiO$_2$	0.3 wt. %	–	WSD↓	–	Mahara and Singh (2020)
Epoxidized mahua oil	CuO	0.4 wt. %	μ↓	WSD↓	–	Kumar et al. (2020a)
Coconut oil	CeO$_2$/CuO 50:50	0.25 wt. %	μ↓15.7%	WSD↓23.4%	–	Sajeeb and Rajendrakumar (2020)

COF, coefficient of friction; WSD, wear scar diameter; SWR, specific wear rate; WV, wear volume; EP, extreme pressure; TMP, trimethylolpropane; ↓, decrease; ↑, increase; d, diameter.

REFERENCES

Asrul, M., Zulkifli, N.W.M., Masjuki, H.H., Kalam, M.A., 2013. Tribological properties and lubricant mechanism of nanoparticle in engine oil. *Procedia Eng.* 68, 320–325. Doi: 10.1016/j.proeng.2013.12.186.

Bhaumik, S., Kamaraj, M., Paleu, V., 2020. Tribological analyses of a new optimized gearbox biodegradable lubricant blended with reduced graphene oxide nanoparticles. *Proc. Inst. Mech. Eng. Part J J. Eng. Tribol.* Doi: 10.1177/1350650120925590.

Bhaumik, S., Maggirwar, R., Datta, S., Pathak, S.D., 2018. Analyses of anti-wear and extreme pressure properties of castor oil with zinc oxide nano friction modifiers. *Appl. Surf. Sci.* 449, 277–286. Doi: 10.1016/j.apsusc.2017.12.131.

Chen, B., Gu, K., Fang, J., Jiang, W., Jiu, W., Nan, Z., 2015. Tribological characteristics of monodispersed cerium borate nanospheres in biodegradable rapeseed oil lubricant. *Appl. Surf. Sci.* 353, 326–332. Doi: 10.1016/j.apsusc.2015.06.107.

Chouhan, A., Mungse, H.P., Sharma, O.P., Singh, R.K., Khatri, O.P., 2018. Chemically functionalized graphene for lubricant applications: Microscopic and spectroscopic studies of contact interfaces to probe the role of graphene for enhanced tribo-performance. *J. Colloid Interface Sci.* 513, 666–676. Doi: 10.1016/j.jcis.2017.11.072.

Dai, W., Kheireddin, B., Gao, H., Liang, H., 2016. Roles of nanoparticles in oil lubrication. *Tribol. Int.* 102, 88–98. Doi: 10.1016/j.triboint.2016.05.020.

Delgado, M.A., Cortes-Trivino, E., Valencia, C., Franco, J.M., 2020. Tribological study of epoxide-functionalized alkali lignin-based gel-like biogreases. *Tribol. Int.* 146, 106231. Doi: 10.1016/j.triboint.2020.106231.

El-gharbi, S., Tekaya, M., Bendini, A., Valli, E., Palagano, R., Hammami, M., Toschi, T.G., Mechri, B., 2018. Effects of geographical location on chemical properties of Zarazi virgin olive oil produced in the South of Tunisia. *Am. J. Food Sci. Technol.* 6, 228–236. Doi: 10.12691/ajfst-6-6-1.

Fox, N.J., Stachowiak, G.W., 2007. Vegetable oil-based lubricants-A review of oxidation. *Tribol. Int.* 40, 1035–1046. Doi: 10.1016/j.triboint.2006.10.001.

Gulzar, M., Masjuki, H.H., Kalam, M.A., Varman, M., Zulkifli, N.W.M., Mufti, R.A., Zahid, R., 2016. Tribological performance of nanoparticles as lubricating oil additives. *J. Nanoparticle Res.* 18, 223. Doi: 10.1007/s11051-016-3537-4.

Gulzar, M., Masjuki, H., Varman, M., Kalam, M., Mufti, R.A., Zulkifli, N., Yunus, R., Zahid, R., 2015. Improving the AW/EP ability of chemically modified palm oil by adding CuO and MoS_2 nanoparticles. *Tribol. Int.* 88, 271–279. Doi: 10.1016/j.triboint.2015.03.035.

Gupta, R.N., Harsha, A.P., 2017. Synthesis, characterization, and tribological studies of calcium-copper-titanate nanoparticles as a biolubricant additive. J. Tribol. 139, 021801–021811. Doi: 10.1115/1.4033714.

Gupta, R.N., Harsha, A.P., 2018a. Antiwear and extreme pressure performance of castor oil with nano-additives. *Proc. Inst. Mech. Eng. Part J J. Eng. Tribol.* 232, 1055–1067. Doi: 10.1177/1350650117739159.

Gupta, R.N., Harsha, A.P., 2018b. Tribological study of castor oil with surface-modi fi ed CuO nanoparticles in boundary lubrication. *Ind. Lubr. Tribol.* 4, 700–710. Doi: 10.1108/ILT-02-2017-0030.

Honary, L.A.T., Richter, E., 2011. *Biobased Lubricants and Greases.* John Wiley & Sons, Ltd. Doi: 10.1002/9780470971956.ch7.

Jayadas, N.H., Nair, K.P., 2006. Coconut oil as base oil for industrial lubricants — evaluation and modification of thermal, oxidative and low temperature properties. *Tribiology Int.* 39, 873–878. Doi: 10.1016/j.triboint.2005.06.006.

Jayadas, N.H., Prabhakaran Nair, K., G, A., 2007. Tribological evaluation of coconut oil as an environment-friendly lubricant. *Tribol. Int.* 40, 350–354. Doi: 10.1016/j.triboint.2005.09.021.

Joseph, P. V., Saxena, D., Sharma, D.K., 2007. Study of some non-edible vegetable oils of Indian origin for lubricant application. *J. Synth. Lubr.* 24, 181–197. Doi: 10.1002/jsl.39.

Kashyap, A., Harsha, A.P., 2016. Tribological studies on chemically modified rapeseed oil with CuO and CeO$_2$ nanoparticles. *Proc. Inst. Mech. Eng. Part J J. Eng. Tribol.* 230, 1562–1571. Doi: 10.1177/1350650116641328.

Kumar, H., Harsha, A.P., 2020. Investigation on friction, anti-wear and extreme pressure properties of different grades of polyalphaolefins with functionalized multi-walled carbon nanotubes as an additive. *J. Tribol.* 142. Doi: 10.1115/1.4046571.

Kumar, S., Anuj, C., Sehgal, K., Kumar, N., 2020a. Improved lubrication mechanism of chemically modified mahua (Madhuca indica) oil with addition of copper oxide nanoparticles. *J. Bio- Tribo-Corrosion* 6, 1–10. Doi: 10.1007/s40735-020-00387-2.

Kumar, V., Dhanola, A., Garg, H.C., Kumar, G., 2020b. Improving the tribological performance of canola oil by adding CuO nanoadditives for steel / steel contact. *Mater. Today Proc.* Doi: 10.1016/j.matpr.2020.04.807.

Kumari, S., Mungse, H.P., Gusain, R., Kumar, N., Sugimura, H., Khatri, O.P., 2017. Octadecanethiol-grafted molybdenum disulfide nanosheets as oil-dispersible additive for reduction of friction and wear. *Flat. Chem* 3, 16–25. Doi: 10.1016/j.flatc.2017.06.004.

Mahara, M., Singh, Y., 2020. Tribological analysis of the neem oil during the addition of SiO$_2$ nanoparticles at different loads. *Mater. Today Proc.* Doi: 10.1016/j.matpr.2020.04.813.

Mannekote, J.K., Kailas, S.V., 2009. Studies on boundary lubrication properties of oxidised coconut and soy bean oils. *Lubr. Sci.* 21, 355–365. Doi: 10.1002/ls.101.

McNutt, J., He, Q.S., 2016. Development of biolubricants from vegetable oils via chemical modification. *J. Ind. Eng. Chem.* 36, 1–12. Doi: 10.1016/j.jiec.2016.02.008.

Mobarak, H.M., Mohamad, E.N., Masjuki, H.H., Kalam, M.A., Mahmud, K.A., Habibullah, M., Ashraful, A.M., 2014. The prospects of biolubricants as alternatives in automotive applications. *Renew. Sustain. Energy Rev.* 33, 34–43. Doi: 10.1016/j.rser.2014.01.062.

Nagabhooshanam, N., Baskar, S., Prabhu, T.R., Arumugam, S., 2020. Evaluation of tribological characteristics of nano zirconia dispersed biodegradable canola oil methyl ester metalworking fluid. *Tribol. Int.* 106510. Doi: 10.1016/j.triboint.2020.106510.

Nagendramma, P., Kaul, S., 2012. Development of ecofriendly/biodegradable lubricants: An overview. *Renew. Sustain. Energy* Rev. 16, 764–774. Doi: 10.1016/j.rser.2011.09.002.

Nagendramma, P., Kumar, P., 2015. Eco-friendly multipurpose lubricating greases from vegetable residual oils. Lubricants 3, 628–636. Doi: 10.3390/lubricants3040628.

Nosonovsky, M., Bhushan, B. (Eds.), 2012. *Green Tribology Biomimetics, Energy Conservation and Sustainability.* Springer. Doi: 10.1007/978-3-642-01150-4.

OPEC-Report, 2019. Organization of the petroleum exporting countries world oil outlook. ISBN 978-3-9503936-9-9

Padmini, R., Krishna, P.V., Mohana, G.K., 2016. Effectiveness of vegetable oil based nano fl uids as potential cutting fl uids in turning AISI 1040 steel. *Tribol. Int.* 94, 490–501. Doi: 10.1016/j.triboint.2015.10.006.

Panchal, T., Chauhan, D., Thomas, M., Patel, J., 2015. Bio based grease a value added product from renewable resources. *Ind. Crops Prod.* 63, 48–52. Doi: 10.1016/j.indcrop.2014.09.030.

Panchal, T.M., Patel, A., Chauhan, D.D., Thomas, M., Patel, J. V., 2017. A methodological review on bio-lubricants from vegetable oil based resources. *Renew. Sustain. Energy Rev.* 70, 65–70. Doi: 10.1016/j.rser.2016.11.105.

Rawat, S.S., Harsha, A.P., 2019. Current and future trends in grease lubrication, in: Katiyar, J.K., Bhattacharya, S., Patel, V.K., Kumar, V. (Eds.), *Automotive Tribology.* Springer Nature, Singapore, pp. 147–182.

Rawat, S.S., Harsha, A.P., Agarwal, D.P., Kumari, S., Khatri, O.P., 2019. Pristine and alkylated MoS$_2$ nanosheets for enhancement of tribological performance of paraffin grease under boundary lubrication regime. *J. Tribol.* 141, 072102–072112. Doi: 10.1115/1.4043606.

Rawat, S.S., Harsha, A.P., Chouhan, A., Khatri, O.P., 2020. Effect of graphene-based nanoad-ditives on the tribological and rheological perfeormance of paraffin grease. *J. Mater. Eng. Perform.* 29, 2235–2257.

Rawat, S.S., Harsha, A.P., Das, S., Deepak, A.P., 2019a. Effect of CuO and ZnO nano-additive on the tribological performance of paraffin oil-based lithium grease. *Tribol. Trans.* 63, 90–100. Doi: 10.1080/10402004.2019.1664684.

Rawat, S.S., Harsha, A.P., Deepak, A.P., 2019b. Tribological performance of paraf-fin grease with silica nanoparticles as an additive. *Appl. Nanosci.* 9, 305–315. Doi: 10.1007/s13204-018-0911-9.

Reeves, C.J., Menezes, P.L., 2017. Evaluation of boron nitride particles on the tribological performance of avocado and canola oil for energy conservation and sustainability. *Int. J. Adv. Manuf. Technol.* 89, 3475–3486. Doi: 10.1007/s00170-016-9354-1.

Reeves, C.J., Menezes, P.L., Jen, T., Lovell, M.R., 2015. The influence of fatty acids on tri-bological and thermal properties of natural oils as sustainable biolubricants. *Tribiology Int.* 90, 123–134. Doi: 10.1016/j.triboint.2015.04.021.

Rizvi, A., Syed, Q.A., 2009. *A Comprehensive Review of Lubricant Chemistry, Technology, Selection and Design.* ASTM International. West Conshohocken, PA.

Rudnick, L.R., 2006. *Synthetics, Mineral Oils, and Bio-Based Lubricants Chemistry and Technology.* CRC Press, Taylor & Francis Group, LLC. Boca Raton.

Rudnick, L.R. (Ed.), 2017. *Lubricants Additives Chemistry and Applications*, Third ed. CRC Press. Boca Raton.

Sahoo, R.R., Biswas, S.K., 2009. Frictional response of fatty acids on steel. *J. Colloid Interface Sci.* 333, 707–718. Doi: 10.1016/j.jcis.2009.01.046.

Sahoo, P.K., Das, L.M., Babu, M.K.G., Naik, S.N., 2007. Biodiesel development from high acid value polanga seed oil and performance evaluation in a CI engine. *Fuel* 86, 448–454. Doi: 10.1016/j.fuel.2006.07.025.

Sajeeb, A., Rajendrakumar, P.K., 2019. Experimental studies on viscosity and tribological characteristics of blends of vegetable oils with CuO nanoparticles as additive. *Micro Nano Lett.* 14, 1121–1125. Doi: 10.1049/mnl.2018.5595.

Sajeeb, A., Rajendrakumar, P.K., 2020. Tribological assessment of vegetable oil based CeO2/CuO hybrid nano-lubricant. *Proc. Inst. Mech. Eng. Part J J. Eng. Tribol.* Doi: 10.1177/1350650119899208.

Salaji, S., Jayadas, N.H., 2020. Evaluation of physicochemical and tribological properties of chaulmoogra (Hydnocarpus wightianus) oil as green lubricant base stock. *Proc. Inst. Mech. Eng. Part J J. Eng. Tribol.* Doi: 10.1177/1350650119899529.

Shafi, W.K., Charoo, M.S., 2020. An overall review on the tribological, thermal and rheo-logical properties of nanolubricants. *Tribol. - Mater. Surfaces Interfaces*, 1–35. Doi: 10.1080/17515831.2020.1785233

Shashidhara, Y.M., Jayaram, S.R., 2010. Vegetable oils as a potential cutting fluid — An evo-lution. *Tribiology Int.* 43, 1073–1081. Doi: 10.1016/j.triboint.2009.12.065.

Shomchoam, B., Yoosuk, B., 2014. Eco-friendly lubricant by partial hydrogenation of palm oil over Pd/γ -Al$_2$O$_3$ catalyst. *Ind. Crop. Prod.* 62, 395–399. Doi: 10.1016/j.indcrop.2014.09.022.

Singh, A.K., 2011. Castor oil-based lubricant reduces smoke emission in two-stroke engines. *Ind. Crop. Prod.* 33, 287–295. Doi: 10.1016/j.indcrop.2010.12.014.

Singh, T., 2020. Grease production survey report, in: *22nd Lubricating Grease Conference, NLGI India-Chapter.* NLGI, Missouri.

Siniawski, M.T., Saniei, N., Adhikari, B., Doezema, L.A., 2007. Influence of fatty acid com-position on the tribological performance of two vegetable-based lubricants. *J. Synth. Lubr.* 24, 101–110. Doi: 10.1002/jsl.32.

Sukirno, R. F., Bismo, S., Nasikin, M., 2009. Biogrease based on palm oil and lithium soap thickener: Evaluation of antiwear. *World Appl. Sci. J.* 6, 401–407.

Sukirno, L., Fajar, R., Bismo, N., 2010. Anti-wear properties of bio-grease from modified palm oil and calcium soap thickener. *Agric. Eng. Int. CIGR J.* 12, 64–69.

Sun, J., Du, S., 2019. Application of graphene derivatives and their nanocomposites in tribology and lubrication. *RSC Adv.* 9, 40642–40661. Doi: 10.1039/c9ra05679c.

Syahir, A.Z., Zulkifli, N.W.M., Masjuki, H.H., Kalam, M.A., Alabdulkarem, A., Gulzar, M., Khuong, L.S., Harith, M.H., 2017. A review on bio-based lubricants and their applications. *J. Clean. Prod.* 168, 997–1016. Doi: 10.1016/j.jclepro.2017.09.106.

Talib, N., Nasir, R.M., Rahim, E.A., 2017. Tribological behaviour of modified jatropha oil by mixing hexagonal boron nitride nanoparticles as a bio-based lubricant for machining processes. *J. Clean. Prod.* 147, 360–378. Doi: 10.1016/j.jclepro.2017.01.086

Thottackkad, M.V., Perikinalil, R.K., Kumarapillai, P.N., 2012. Experimental evaluation on the tribological properties of coconut oil by the addition of CuO nanoparticles. *Int. J. Precis. Eng. Manuf.* 13, 111–116. Doi: 10.1007/s12541-012-0015-5.

Torbacke, M., Rudolphi, Å.K., Kassfeldt, E., 2014. *Lubricants: Introduction to Properties and Performance*, 1st ed. John Wiley & Sons, Ltd. West Sussex, PO19 8SQ, United Kingdom.

Totten, G.E., 1992. *ASM Handbook, Volume 18: Friction, Lubrication, and Wear Technology.* ASM International. Materials Park, Ohio.

Verma, D.K., Kumar, B., Kavita, Rastogi, R.B., 2019. Zinc oxide- and magnesium-doped zinc oxide-decorated nanocomposites of reduced graphene oxide as friction and wear modifiers. *ACS Appl. Mater. Interf.* 11, 2418–2430. Doi: 10.1021/acsami.8b20103.

Wakil, M.A., Kalam, M.A., Masjuki, H.H., Atabani, A.E., Fattah, I.M.R., 2015. Influence of biodiesel blending on physicochemical properties and importance of mathematical model for predicting the properties of biodiesel blend. *Energy Convers. Manag.* 94, 51–67. Doi: 10.1016/j.enconman.2015.01.043

Yu, R., Liu, J., Zhou, Y., 2019. Experimental study on tribological property of MoS_2 nanoparticle in castor oil. *J. Tribol.* 141, 102001–102005. Doi: 10.1115/1.4044294.

Yunus, R., Fakhru, A., Ooi, T.L., Omar, R., Idris, A., 2005. Synthesis of palm oil based trimethylolpropane esters with improved pour points. *Ind. Eng. Chem. Res.* 44, 8178–8183.

Zachariah, Z., Nalam, P.C., Ravindra, A., Raju, A., Mohanlal, A., Wang, K., 2020. Correlation between the adsorption and the nanotribological performance of fatty acid - based organic friction modifiers on stainless steel. *Tribol. Lett.* 68, 1–16. Doi: 10.1007/s11249-019-1250-z.

Zainal, N A, Zulki, N.W.M., Gulzar, M., Masjuki, H.H., 2018a. A review on the chemistry, production, and technological potential of bio- based lubricants. *Renew. Sustain. Energy Rev.* 82, 80–102. Doi: 10.1016/j.rser.2017.09.004.

Zareh-desari, B., Davoodi, B., 2016. Assessing the lubrication performance of vegetable oil-based nano-lubricants for environmentally conscious metal forming processes. *J. Clean. Prod.* 135, 1198–1209. Doi: 10.1016/j.jclepro.2016.07.040.

Zulkifli, N.W.M., Azman, S.S.N., Kalam, M.A., Masjuki, H.H., Yunus, R., Gulzar, M., 2016. Lubricity of bio-based lubricant derived from different chemically modified fatty acid methyl ester. *Tribol. Int.* 93, 555–562. Doi: 10.1016/j.triboint.2015.03.024.

Zulkifli, N.W.M., Kalam, M.A., Masjuki, H.H., Shahabuddin, M., Yunus, R., 2013a. Wear prevention characteristics of a palm oil-based TMP (trimethylolpropane) ester as an engine lubricant. *Energy* 54, 167–173. Doi: 10.1016/j.energy.2013.01.038.

Zulkifli, N.W.M., Kalam, M.A., Masjuki, H.H., Yunus, R., 2013b. Experimental analysis of tribological properties of biolubricant with nanoparticle additive. *Procedia Eng.* 68, 152–157. Doi: 10.1016/j.proeng.2013.12.161.

10 Biolubricants

T. Mohanraj and N. Radhika
Amrita Vishwa Vidyapeetham

CONTENTS

10.1 PRINCIPLES OF LUBRICATION

Tribology is an extremely interdisciplinary research area. Tribological variables consist of different types including motion load, sliding velocity, disc rotational speed, temperature, mode of lubrication and nature of lubricant (Suganeswaran et al., 2020). The main intention is to reduce wear, friction and heat between the adjoining surfaces in a relative movement. At the same time, as wear and high temperature could not be entirely reduced, together consequences of wear and heat can be condensed to satisfactory or bring down to slight levels. This can be decreased by putting in a material (lubricant) with less coefficient of friction (COF) between the contacting materials. Lubricants are employed for controlling the friction and wear to prevent direct contact between the asperities of the surface and decrease the heat (Kozma, 1997).

Tribosystem necessitates satisfactory material, protection against wear through unique coatings, lubricants and surface modifications. Lubrication plays a considerable role to prevent wear in mechanical systems, and they have to effectively operate for an extended time. Lubrication may be employed for further intentions (Bart et al., 2013), like

- decrease oxidation
- avoid rust

- decrease energy utilization
- act as an insulator
- power transformation in hydraulic systems
- Closure against dirt and dust.

With technology boon, increased number of industries, machinery and automobiles, the demand for lubricant oils is approximately 40 million tons per annum. Even though petroleum-based lubricants occupy 90% of applications, the basic stock is non-renewable. The concerns of their inherent toxicity and non-biodegradable nature lead to the fetch of alternate resources. The biolubricants also known as vegetable oils are found to be the potential replacement. From literature and market surveys, it was found that the biolubricants' market growth rate is approximately 15% (Panchal et al., 2017).

The biolubricants started to replace mineral lubricants as they are renewable raw materials, biodegradable and non-toxic. Also, they exhibit better antiwear property, friction modifying property because of their strong interactions with the surface during the contact with the surfaces, in particular metallic surfaces (Shankar et al., 2017, Mohanraj et al., 2019, Shankar et al., 2019a, 2019b). The performance of biolubricants can be enhanced by adding appropriate additives or viscosity modifiers (Li et al., 2019).

10.2   PREPARATION OF BIOLUBRICANTS

Biolubricants have been prepared from esterification of vegetable oil that comprises oleic/other long-chain fatty acids (FAs) with long-chain alcohol with an acid reagent. The essential property of a lubricant is biodegradability. A lubricant with the nature of high biodegradablity is essential for the ecosystem. The biodegradability of petroleum-based lubricants (20%–30%) is poor than the lubricants produced from plant-based oils (90%–100%) and they will not cause any harmful consequences to the environment (Rani, 2017). Thus, the harmful consequences of lubricants on the ecosystem are decreased to a better scope, which is a need of an hour. The advantages of biolubricants over petroleum-based lubricants are listed below.

1. Biodegradation
2. Negligible environmental pollution
3. Non-toxic
4. Related characteristics like lubricity and viscosity.

Table 10.1 presents the evaluation of commercial lubricants with biolubricants. It is vibrant from Table 10.1 that pollutants emitted from commercial lubricants are the major source to pollute the environment, whereas the biolubricants do not emit any harmful pollutants and decrease the environmental pollution. Alternative sources like non-edible/edible oils could be used for producing the biolubricants domestically, whereas commercial lubricants might not be produced domestically, and on the edge of exhaustion subsequently, the price of petroleum products also rising. Commercial lubricants are hazardous to handle and store in the event of spillage, outflow and owing to their poor biodegradation and high harmfulness; however, these issues are not caused in biolubricants (Jain and Suhane, 2013).

TABLE 10.1
Evaluation of Commercial Lubricants with Biolubricants

Feature	Commercial Lubricants	Biolubricants
Biodegradation	No	Yes
Poisonous	Yes	No
Source of global warming	Yes	No
Dangerous emission of by-products	Yes	No
Environmental pollution	Yes	No
Economic improvement	No	Yes
Hazardous to use and storage	Yes	No
Commercial manufacturing	No	Yes
Another source	No	Yes

10.3 PREPARATION OF BIOLUBRICANTS WITH CONVENTIONAL CATALYST

Biolubricants can be developed from vegetable oils via chemical modifications like transesterification, epoxidation and the esterification process (Sharma et al., 2015). The esterification response among FA and polyol is a technologically significant method to obtain a preferred polyhydric ester. Usually, homogeneous acids and alkali are employed as a catalyst for the esterification process, though alkali catalysts possibly will originate saponification and inorganic acids are accountable for products obtained through this process. Essentially, they are exhaustible and might end up with a challenging separation process, which reasons ecological pollution. Heterogeneous catalysts work as an alternate option to prepare the biolubricants. Though it is easy to recycle heterogeneous catalysts, they are suffered from the deactivation from coking and expensive, and few catalysts still have the affinity to initiate the saponification process (Bondioli, 2004).

A specified batch container (~1.5 L) fabricated from stainless steel was used for performing the esterification reaction in the prescribed reaction surroundings. The container was furnished with a motorized agitator that is operated through the electric motor with a specified speed, to continuously monitor and maintain the fluid at the specified temperature, a thermocouple that is employed with the accuracy of ±0.5°C, an inlet channel to pour the sample and an outlet channel with a valve. Figure 10.1 shows the schematic representation of the esterification reaction (Chowdhury et al., 2013).

The esterification process was performed in the designed container by adding 0.5 L of FA produced from different hydrolysis processes. The molar ratio of FA to octanol was changed from 4:1 to 1:1. Adding the catalyst initiated the process. The quantity of catalyst was also altered from 0.2 to 3 g as per the requirement and the temperature was maintained from 60°C to 150°C. The influence of stirrer speed was analysed from 450 to 650 rpm. Desiccants such as magnesium sulphate, sodium sulphate and silica gel powder were further mixed with the reactor at 50 w/w % of free fatty acid (FFA) to examine the influence on reaction rate. Proceeding to desiccant usage, magnesium sulphate and sodium sulphate were dehydrated at 110°C consecutively, and then, silica gel was dehydrated at 180°C (Chowdhury et al., 2013).

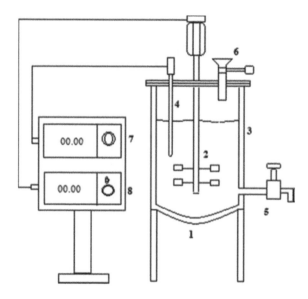

FIGURE 10.1 Graphical illustration of esterification process: 1. container, 2. agitator, 3. container jacket, 4. temperature sensor, 5. outlet channel, 6. inlet channel, 7. temperature display unit and 8. stirring speed display unit

10.4 PREPARATION OF BIOLUBRICANTS WITH BIOCATALYSTS

Biocatalysts are eco-friendly and attain more concentration towards the preparation of biolubricants. Recent enzymology has been attained enhancements in the progress and the usage of biocatalysts. The employment of biocatalyst in industrial application is the same as conventional catalysts used for heterogeneous reactions owing to the progress and development of novel control technologies. For instance, esters arranged from the response of FAs with alcohols have an essential application (Dossat et al., 2002, Zaidi et al., 2002).

Evaluating traditional chemical production from alcohols and FAs by employing mineral acid as a catalyst, employment of enzymes like lipases for preparing FA esters without solvents may be the substantial merits (Yadav and Lathi, 2003). These contain the usage of any hydrophobic substance, high selectivity, slight operating conditions, easiness to isolate the product and recycling of protein. Conversely, it requires extensive time for activity, inactivation from polyols and quantity of lipases equated with early catalysts restricted the wide application of proteins (Kiriliauskaitė et al., 2011, Linko et al., 1997).

10.5 PREPARATION OF BIOLUBRICANTS THROUGH CHEMICAL MODIFICATION

The viscosity index of vegetable oils is enormously low and restricts the usage of vegetable oils at low temperatures, especially in automobiles and industries. Vegetable oils have an affinity to shape macrocrystalline anatomical structure at lower operating temperature via uniform piling of 'bend' in the triacylglycerol spine. Such macrocrystals confine the flow owing to reduce the energy in the course of self-stacking. Numerous diester compounds were produced from existing oleic acid (OA) and FAs (Salimon and Salih, 2009b). Chemical modifications, namely epoxidation, the formation of estolides and transesterification, were revealed to enhance the oxidization stability of biolubricants and to succeed optimum performance for industrial applications (Sharma et al., 2006).

10.5.1 EPOXIDATION

The significant stages in the three-step production of oleochemical diesters comprise epoxidation and ring-opening reaction of epoxidized OA with diverse FAs using p-toluenesulphonic acid (PTSA) as a catalyst for harvesting the mono-ester. The reaction of mono-ester compounds with aromatics was performed in the manifestation of $10\,mol\%$ H_2SO_4, making the compounds of anticipated diester (Salimon and Salih, 2009a). Figure 10.2 shows the epoxidation of OA, followed by ring-opening acylation of epoxidized OA employing PTSA as a promoter.

As expected, the dimension of midchain rise, a subsequent enhancement in the performance was witnessed at less temperature. This sensation is owing to the augmented capability of long-chain esters to interrupt the macrocrystalline foundation at small temperatures. The additional comment is the constructive influence of splitting at the end of the chain on the low-temperature behaviour of subsequent products, which leads to the establishment of the microcrystalline anatomical structure to a certain extent than macrocrystalline anatomical structure.

10.5.2 ESTOLIDES FORMATION

Estolides are a type of esters derived from vegetable oils and then prepared through the establishment of carbocation at the place of unsaturation. Carbocation may experience nucleophilic outbreak by other FAs, with/without carbocation immigration beside the length of chain to develop an ester. Estolides were established to overwhelm a few drawbacks associated with vegetable oils, like poor thermal oxidization steadiness (Cermak and Isbell, 2003) and limited temperature characteristics (Asadauskas and Erhan, 1999). Few drawbacks might be enhanced by the usage of additives with the compromise of biodegradability, harmfulness and price.

Cermak and Isbell produced the saturated mono-estolide esters and enhanced saturated mono-estolide 2-ethylhexyl esters from OA as well as saturated FAs by employing three dissimilar methods (Cermak and Isbell, 2009). Figure 10.3 illustrates the graphics of the synthesis of oleic estolide esters. The estolide numbers (ENs), a mean number of FA units added to the base acid, are varied and are reliant on operating surroundings. The physical characteristics specified that equally chain length and EN influence the properties of less temperature.

10.5.3 TRANSESTERIFICATION/ESTERIFICATION

Biodegradable organic polyesters obtained from transesterification/esterification of vegetable oils and divided into neopolyols like trimethylolpropane (TMP) and

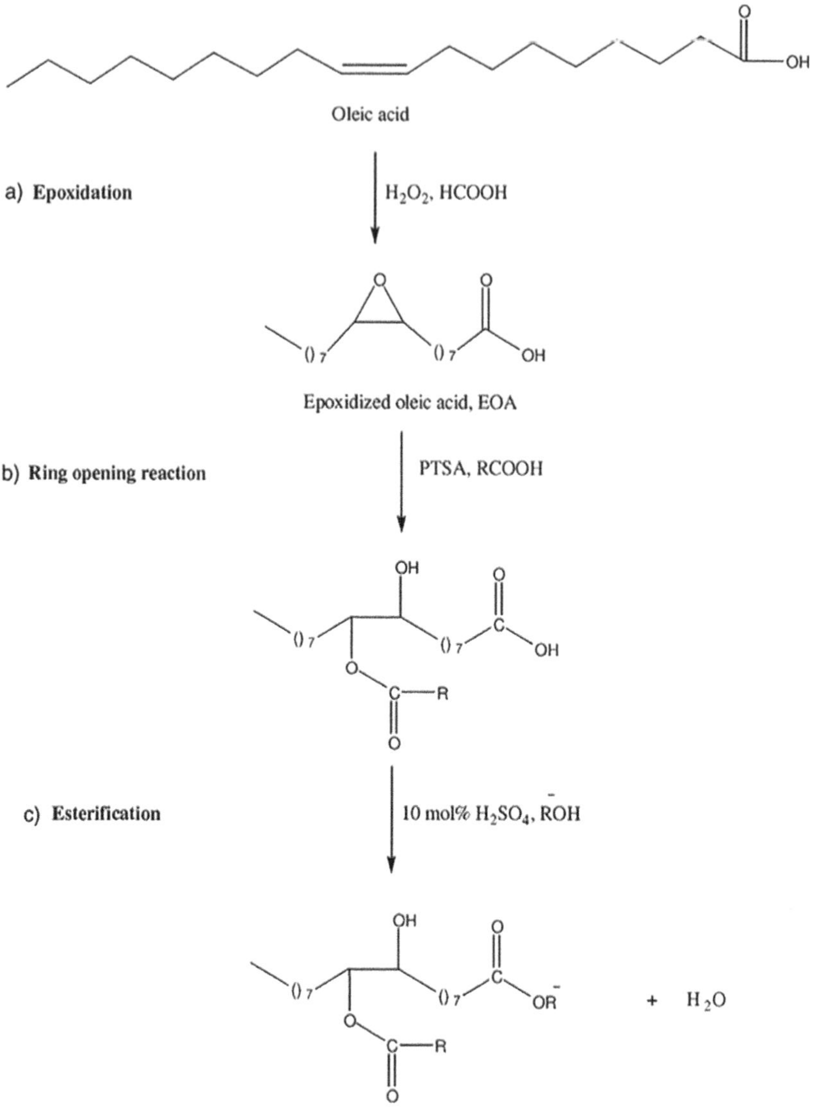

a) Epoxidation   H$_2$O$_2$, HCOOH

Oleic acid

Epoxidized oleic acid, EOA

b) Ring opening reaction   PTSA, RCOOH

c) Esterification   10 mol% H$_2$SO$_4$, ROH

+   H$_2$O

FIGURE 10.2   (a) Epoxidation of OA, (b) Ring-opening reaction, and (c) Esterification

PE were produced for numerous applications. Figure 10.4 shows the transesteri-fication process of TMP. The preparation of biodegradable TMP of rapeseed oil (RSO) FAs utilizing enzymatic and chemical routes were described. 0.5% w/w of sodium methylate was used as a promoter, and the response blend was flowing in low pressure. Around 99% transformation was attained at 110°C–120°C in 10 h. With 40% w/w Candida rugosa lipase, 64% of TMP was transformed to tries-ters in 24 h at 5.3 kPa. By restrained Rhizomucor miehei (50% w/w), maximum

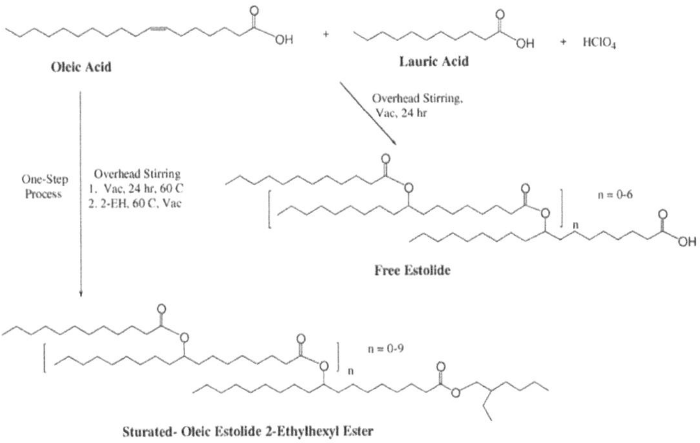

FIGURE 10.3 Graphics of synthesis of oleic estolide esters

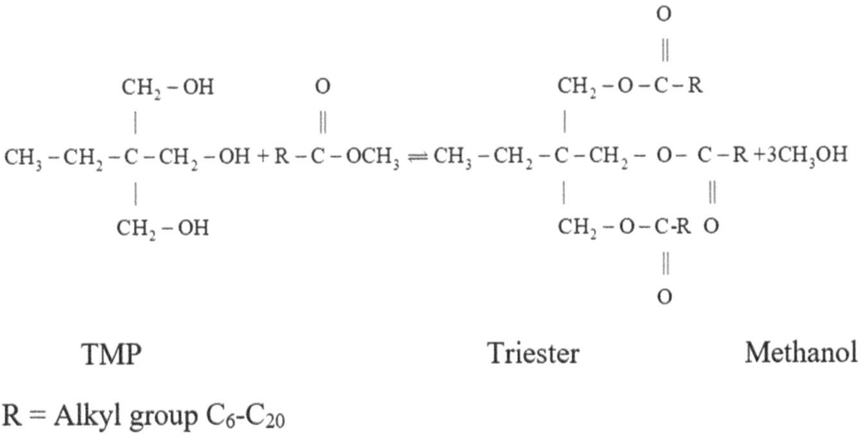

TMP   Triester   Methanol

R = Alkyl group C$_6$-C$_{20}$

FIGURE 10.4 TMP transesterification

transformation to TMP triester was around 90% and attained in 66 h (Uosukainen et al., 1998).

10.6 CHARACTERIZATION OF BIOLUBRICANTS

According to tribological tests, biolubricants are characterized as lubricants having less flashpoint, less pour point, larger viscosity index and lesser acid number. Hence, it is recommended that the biolubricants are employed at a high-speed and low-load tribosystem. In industry, they can be employed to cool the lubricant compounds for the material removal process, bearings and also in the specific process where the loss of lubricant might be happening, e.g. chain lubrication.

10.7   TRIBOLOGICAL PERFORMANCE OF BIOLUBRICANTS

The tribological response in terms of wear rate and COF was discussed. Adhvaryu and Erhan (2002) investigated the characteristics of epoxidized soybean oil (ESBO) over soybean oil (SBO) and genetically modified high-oleic soybean oil (HOSBO) for extreme temperature lubrication. The addition of 0.6 M alkylated phenolic compound in SBO decreases the COF up to 66%. It was identified that the removal of several unsaturations in the FA chain reduced the forming propensity of ESBO. The establishment of the steady polymeric film on the metal surface through boundary lubrication offers a significant reduction in COF.

COF and wear performance of SBO and sunflower oil (SO) were compared with the performance of base mineral oil (MO). They found that wear and COF were least for SBO followed by SO. SBO decreased COF by 68.53% as compared to MO. They found that reason for the least wear and friction in SBO was the chemical structure and presence of the least amount of linoleic acid and OA (Siniawski et al., 2007).

Engine oil was developed via palm oil (PO) as feedstock and tested with four-stroke engines. The biodegradable lubricant was formulated by blending PO and MO. The blended mixture consists of 50.6% (wt.) PO, 41.6% MO and 7.8% additives. The selected mixture decreases the COF by 43.33% as compared to MO. Better characteristics were observed in the formulated blend than conventional oil in terms of thermal and tribological properties (Cheenkachorn and Fungtammasan, 2010).

The tribological characteristics of raw SBO and SO with MO at a different level of humidity were compared. They found that at higher humidity levels, SBO and SO retain their friction and wear reduction ability than MOs due to the inherent ability to act in response to surfaces and develop tribofilms. At maximum humidity, SBO decreases the COF by 81.25% as compared to MO. Further, they found that SBO had elevated wear resistance than SO due to its least viscosity and its chemical structure (Siniawski et al., 2011).

A tribological test on AISI 1040 steel under raw, altered forms of pongam and jatropha (JO) and MO was executed. At 70 N load and 7,500 m sliding distance, pongam raw oil decreased the COF by 27.27% and epoxidized JO raw oil decreased the wear by 91.66% than MO. The authors found a remarkable reduction in COF and wear under pongam and JO raw oil as compared to MO (Shashidhara and Jayaram, 2012). The development of biolubricants from vegetable oil via various chemical modification routes was reviewed. They suggested as biolubricants will be a better replacement for MO-based lubricants (McNutt, 2016).

The following vegetable oils are identified as lubricant oil (Panchal et al., 2017). Canola oil, castor oil, coconut oil, olive oil, PO, RSO, safflower oil, linseed oil, SBO, jojoba oil, crambe oil, SO, cuphea oil, tallow oil, JO, PO methyl ester, RSO methyl ester, karanja oil, methyl ester, stearic acid, OA and linoleic acid. The tribological performance of PO for hip implant material was investigated. They found that the lubricant with higher viscosity has the least friction and can be used to enhance the lifespan of implants. They suggested that the PO can be employed as biolubricants in the upcoming days (Sapawe et al., 2017).

10.8   TRIBOLOGICAL PERFORMANCE OF BIOLUBRICANTS WITH ADDITIVES

The tribological characteristics of vegetable-based oils without any additives were reviewed. They found a great variation in the properties of vegetable oils than MOs. They found that sesame oil and wheat germ oil had possible properties for industrial lubricant. Under abrasive wear conditions, the castor, sesame and soybean oils exhibited the best results (Gerbig et al., 2004). The relationship between end-to-end carbon chain distance in FA and its efficiency at higher temperature was evaluated. The authors added the additives like stearic, arachidic and behenic FA with SO. They found the least wear rate and COF for all the FAs and the additives had a significant effect up to 100°C (Baumgart et al., 2010).

The lubrication, wear and thermal conductivity of pure SBO, soybean methyl ester (SBME), SBME + 1.5% of Cu nanopowder and MO (4T) were presented. They found that wear scar diameter was decreased by 12% for SBME+1.5% Cu equated to SBME and 9.5% as compared to base MO (Kanagasabapathi et al., 2012). The tribological performance of chemically modified RSO (CMRSO) with and without nano- and microparticles of TiO_2 was compared. They found that RSO with TiO_2 nanoparticles (NPs) provides excellent friction and wear properties than RSO with micro-TiO_2 particles and RSO without additive. They concluded that TiO_2 NPs decreased the wear scar by 15% as compared to RSO with TiO_2 microparticles (Arumugam and Sriram, 2013).

The vegetable oil-based lubricants from the epoxidation of SO and SBO were developed. They investigated the tribological performance of biolubricants with the adding up of ZnO and CuO NPs as additives. They found that adding up of NPs in vegetable oils considerably enhanced the tribological properties. The SO and SBO decreased the COF by 50.9% and 49.03%, respectively, than MO. The friction and wear were decreased considerably owing to the development of smooth and compact tribofilm on the surfaces (Alves et al., 2013).

The effect of ethylene-vinyl acetate (EVA) copolymer and ethyl cellulose (EC) as an additive with high-oleic sunflower oil (HOSO), SBO and castor oil (CO) blended with 4% (wt.) of EVA and 1% (wt.) of EC was reported. They found that CO had better lubricant properties like good film-forming properties, enhanced friction and wear properties than HOSO and SBO. Compared to SBO, HOSO and CO decreased the wear scar by 14.81% and 40.74%, respectively. At higher temperatures, CO decreased the COF by 50% as compared to other lubricants. They observed that EVA exercised the least influence on the development of tribofilm and helps to reduce wear and friction. On the contrary, EC was found as effective for enhancing the mixed and boundary lubrication with CO (Quinchia et al., 2014). Biolubricant (SBO and SO) with ZnO and CuO NPs was prepared to investigate the tribological properties. They analysed the tribological performance of biolubricants with a high-frequency reciprocating rig (HFRR). The addition of ZnO and CuO NPs in SBO increases the COF by 16% and 20%, respectively (Trajano et al., 2014).

The tribological characteristics of coconut oil with unmodified and surfactant-modified MoS_2 NPs were examined and compared the performance with mineral-base oil. The optimal concentration of 0.53 and 0.58 wt. % NP was obtained for coconut- and mineral-base oil. The coconut oil with surfactant-modified MoS_2 NPs decrease the COF by 47.95% and 38.39% was decreased for base oil. The experimental results revealed that the performance of coconut oil with NPs was better than base oil in terms of COF and wear properties and coconut oil was found as an environmentally friendly substitute for MO (Koshy et al., 2015).

The tribological performance of rice bran oil (RBO), SO and coconut oil using a four-ball tester (FBT) was assessed. They found that the physicochemical and thermal properties of RBO were found as superior compared to other vegetable oils and SAE20W40. The COF of RBO, sunflower and coconut oil was decreased by 37.60%, 48.71% and 13.67%, respectively, and wear scar of them was increased by 6.55%, 12.20% and 9.47%, respectively. They concluded that RBO had more wear scar than SAE20W40 and it can be enhanced by adding appropriate antiwear additives with RBO (Rani et al., 2015).

The tribological characteristics of CMRSO with CuO, WS_2 and TiO_2 NPs as additives were described. They used the FBT to perform the tribological tests. The change in viscosity of the nanolubricants with different additives was measured according to ASTM D 445. They compared the performance of nanolubricants with the performance of SAE20W40 lubricant. They found that CMRSO with CuO NPs had enhanced tribological performance, smooth wear scar and high viscosity than SAE20W40 and other nanobased biolubricants (Baskar et al., 2015).

The experiments with biodegradable nanolubricant to study their tribological properties were conducted. The base oil used was a mixture of palm olein and SBO mixed with hybrid additives. The hybrid additives used zinc dithiophosphate (ZDDP) and CuO NPs. The formulated palm olein lubricant with ZDDP and CuO NPs at 0.75% concentration reduced 54.6% wear, 29.2% COF reduction and smooth surface roughness compared with base palm olein and SBO (Darminesh et al., 2017). The lubricant properties of RBO were compared with SAE20W40 oil. They found that RBO was highly biodegradable, less corrosive and non-toxic than MO and had lower COF than SAE20W40 oil (Rani, 2017).

The tribological performance of SBO with nanographite additive (0.25, 0.50 and 1 wt. %) was presented. They performed the tribological test using FBT and found that COF was increased for lubricant with NP additives. The addition of 0.25 wt. % of graphite in SBO decreases the COF by 66.67%. The wear scar diameter for NP-added lubricant was increased than lubricant without NP additive. The concentration of NP additive had an insignificant effect on the tribological properties of SBO (Cristea et al., 2017).

The effect of the addition of black carbon nanopowder (mean size of 13 nm) with different concentrations (0.25, 0.50, and 1 wt. %) in SBO was analysed and tribological performance was presented. They used the FBT and performed the experiments with different loads and speeds. They found that the performance of additive added SBO was substantial than base SBO. They suggested that the additives should be physically or chemically bonded with the oil for better performance (Cristea et al., 2018a). The nature and concentration of additives (nanoamorphous carbon, nanographite and nanographene) in SBO were focused on and examined its rheological and tribological performance (Cristea et al., 2018b).

The addition of NPs with vegetable oils for minimum quantity lubrication (MQL) in the end milling operation was focused. When the results compared for nanofluid of Cu, SiC and diamond particles with dry lubrication, about 15% reduction on cutting fore and friction was observed under canola oil-based diamond nanofluid MQL. They found that NPs with greater size could reduce the milling force but increase the surface roughness (Yuan et al., 2018). The friction and wear performance of vegetable oil with NPs for sustainable lubrication were characterized. They suggested that vegetable oils have the remarkable potential to employ as lubricating oils (Shafi et al., 2018).

The above reviews revealed that the addition of NPs in biolubricants enhances tribological performance in terms of wear and friction characteristics. The NPs like graphite, MoS_2, SiO_2, TiO_2, Al_2O_3 and multiwalled carbon nanotube (MWCNT) can be supplemented to increase the performance. Table 10.2 presents the tribological characteristics of various biolubricants with different additives in different operating conditions.

10.9   CONCLUSIONS

A remarkable requirement for biolubricants is anticipated over the upcoming days since vegetable oils are normal, inexhaustible, non-noxious, non-contaminating and cost-effective than petroleum-based lubricants. Biolubricants become essential base stocks for lubricant development owing to their desirable characteristics. Owing to the development of environmental concerns, biolubricants are identifying their way to lubricants for engineering applications. Biolubricants, in comparison with synthetic oils, have diverse characteristics owing to their exclusive chemical structures. The distinct features of biolubricants are owing to the privileged attraction of vegetable oils to metallic surfaces. Besides, higher flashpoints of vegetable oils become non-flammable liquids. Further, the tribological performance can be improved by adding additives. Biolubricants can be employed as a lubricant in almost all automobile and engineering applications.

TABLE 10.2

Summary of Tribological Performance of Various Biolubricants

S. No.	Reference	Base Lubricant	Tribosystem	Wear Scare	Wear Rate	COF	Viscosity
	Shahabuddin et al. (2013b)	SAE 40 + 10% - JO	Pin-on-disc (POD): aluminium and cast iron, FBT	0.35	–	0.045	–
	Reeves et al. (2013)	Canola oil + boron nitride NPs	POD tribometer	–	0.19	0.06	–
	Shahabuddin et al. (2013a)	SAE 40+ 10% - JO	POD tribometer, 30 N load, 200 rpm speed	–	0.02	0.22	125 cSt
	Baskar et al. (2015)	CMRSO with CuO NPs	HFRR	0.3546 mm	–	0.09	15.6 cSt
	Reeves et al. (2015)	SBO	POD: Pin (Cu 6.35 mm φ, 50 mm length, hemispherical tip); disc (Al 2024 alloy, 70 mm φ, 6.35 mm thickness)	0.384 mm	–	0.406	25.9 cP
		Corn oil		0.297 mm	–	0.446	26.8 cP
		Sesame oil		0.215 mm	–	0318	28.6 cP
		Canola oil		0.201 mm	–	0.087	29.9 cP
		Peanut oil		0.168 mm	–	0.098	31.9 cP
		Genetically modified safflower oil		0.195 mm	–	0.494	31.9 cP
		Avocado oil		0.104 mm	–	0.020	31.9 cP
		Olive oil		0.175 mm	–	0.078	32.7 cP
	Rani et al. (2015)	Coconut oil	FBT: ball (AISI E-52100 Cr alloy steel, 12.7 mm φ, extra polish EP grade 25, 64–66 Hardness Rockwell C (HRC))	0.601 mm	–	0.101	24.8 cSt
		SO		0.616 mm	–	0.060	27.8 cSt
		RBO		0.585 mm	–	0.073	40.6 cSt
		MO		0.549 mm	–	0.117	105 cSt

(Continued)

TABLE 10.2 (Continued)
Summary of Tribological Performance of Various Biolubricants

S. No.	Reference	Base Lubricant	Tribosystem	Wear Scare	Wear Rate	COF	Viscosity
	Quinchia et al. (2014)	HOSO	HFRR, Mini Traction Machine technique	0.22 mm	–	0.06	38.50 cSt
		HOSO+EVA 4%		0.19 mm	–	0.05	168.14 cSt
		HOSO+EC 1%		0.20 mm	–	0.05-1	50.60 cSt
		CO		0.32 mm	–	0.02	242.50 cSt
		CO+EVA 4%		0.16 mm	–	0.021	427.10 cSt
		CO+EC 1%		0.17 mm	–	0.02	546.70 cSt
		Soy Bean Oil (SYO)		0.24 mm	–	0.07	33.60 cSt
		SYO+EVA 4%		–	–		174.00 cSt
	Guezmil et al. (2016)	Sesame oil	POD: Pin (SS); disc Ultra-high-molecular-weight polyethylene (UHMWP)	0.024 mm	–	0.028	58.0 cP
		Nigella sativa oil		0.028 mm	–	0.032	64.5 cP
		Dry		0.051 mm	–	0.290	–
	Kalam et al. (2017)	Olive oil	HFRR: ball (chrome alloy steel, 62 HRC)	0.621 mm	–	0.052	37.042 cSt
	Ruggiero et al. (2017)	JO	HFRR: ball (chrome alloy steel, 62 HRC)	<0.001 mm	–	0.126	36.61 cSt
		Hydrotreated RSO		<0.001 mm	–	0.164	4.68 cSt
	Silva et al. (2015)	Moringa oil	HFRR: ball (AISI 52-100 steel, 570–750 HV, 6 mm φ); disc (AISI 52100 steel, 190–210 HV, 10 mm φ)	0.15 mm	–	0.084	44.9 cSt
		Passion fruit oil		0.076 mm	–	0.060	31.8 cSt
		Epoxidized moringa oil		0.146 mm	–	0.080	80.4 cSt
		Epoxidized passion fruit oil		0.131 mm	–	0.068	185.7 cSt
	Zulkifli et al. (2013)	Lunaria annua oil-based TMP ester	HFRR	0.269 mm	–	0.079	29.97 cSt
		PO-based TMP ester		0.8053 mm	–	0.081	40.03 cSt

(Continued)

TABLE 10.2 (Continued)
Summary of Tribological Performance of Various Biolubricants

S. No.	Reference	Base Lubricant	Tribosystem	Wear Scare	Wear Rate	COF	Viscosity
	Zulkifli et al. (2016)	PO-based TMP ester	FBT: ball (AISI 52-100 steel, 12.7 mm φ, 64-66 HRC)	2.47mm	–	0.25	40.03 cSt
		PO-based PE ester		2.54mm	–	0.06	68.40 cSt
		PO-based NPG ester	FBT: ball (AISI 52–100 steel, 12.7 mm φ, 64-66 HRC)	0.659mm	–	0.092	27.92 cSt
		PO-based PE ester		0.669mm	–	0.094	72.78 cSt
		PO-based PE ester with antiwear and anticorrosion additives		0.377mm	–	0.063	75.37 cSt
	Suresha et al. (2020)	Pure neem oil (NO)	FBT: ball (Cr alloy steel, 12.7 mm φ, 0.1 μm conjugated linoleic acid (CLA), 62–66 HRC)	1.486μm	–	0.115	53.3±0.35 cSt
		NO+0.25 wt. % GNPs		1.424μm	–	0.11	55.5±0.61 cSt
		NO+0.50 wt. % GNPs		1.302μm	–	0.085	58.7±0.19 cSt
		NO+0.75 wt. % GNPs		1.256μm	–	0.080	62.3±0.53 cSt
		NO+1.0 wt. % GNPs		1.144μm	–	0.067	67.193±0.4 cSt
	Suresh et al. (2020)	SAE20W40 lubricant+5% JO	POD (1-2000N load, 0.49–9.5 m/s sliding speed, 200–200 rpm speed)	–	11.5μm	0.026	–
		SAE20W40 lubricant+10% JO		–	12.5μm	0.031	–
		SAE20W40 lubricant+15% JO		–	17μm	0.040	–
		SAE20W40 lubricant+5% Cardanol oil		–	5μm	0.175	–
		SAE20W40 lubricant+10% Cardanol oil		–	25μm	0.06	–
		SAE20W40 lubricant+15% Cardanol oil		–	26μm	0.63	–

(Continued)

TABLE 10.2 (*Continued*)

Summary of Tribological Performance of Various Biolubricants

S. No.	Reference	Base Lubricant	Tribosystem	Wear Scare	Wear Rate	COF	Viscosity
	Habibullah et al. (2014)	SAE40	FBT: ball EN31 steel, 12.7 mm φ, surface finish 0.1 μm CLA	0.49 mm	–	–	158 cSt
		SAE40+1% JO		0.51 mm	–	–	118 cSt
		SAE40+2 % JO		0.48 mm	–	–	100 cSt
		SAE40+ 3% JO		0.5 mm	–	–	102 cSt
		SAE40+ 4% JO		0.47 mm	–	–	101 cSt
		SAE40+ 5% JO		0.46 mm	–	–	130 cSt
	Sneha et al. (2019)	RBO	FBT: ball (Cr alloy steel, 12.7 mm φ, 0.1 μm CLA, 62–66 HRC)	0.547 mm	–	0.0898	35.26 cP
		JO		0.584 mm	–	0.0624	30.141 cP
		Karanja oil		0.571 mm	–	0.0673	38.995 cP
	Tobón and Chaparro (2019)	Sesame oil	FBT: ball (Cr alloy steel, 12.7 mm φ, 0.1 μm CLA, 62–66 HRC)	2.390 mm	–	0.05311	–
		Sesame oil + 1 wt. % polyethylene wax		2.0186 mm	–	0.051507	–
		Sesame oil+ 1 wt. % paraffin		2.2196 mm	–	0.051661	–
	Sharma and Sachan (2019)	Karanja oil	FBT: ball (Cr alloy steel, 12.7 mm φ, 0.1 μm CLA, 62–66 HRC)	0.44 mm	–	0.042	–

REFERENCES

Adhvaryu, A. & Erhan, S. 2002. Epoxidized soybean oil as a potential source of high-temperature lubricants. *Industrial Crops and Products*, 15, 247–254.

Alves, S., Barros, B., Trajano, M., Ribeiro, K. & Moura, E. 2013. Tribological behavior of vegetable oil-based lubricants with nanoparticles of oxides in boundary lubrication conditions. *Tribology International*, 65, 28–36.

Arumugam, S. & Sriram, G. 2013. Preliminary study of nano-and microscale TiO_2 additives on tribological behavior of chemically modified rapeseed oil. *Tribology Transactions*, 56, 797–805.

Asadauskas, S. & Erhan, S. Z. 1999. Depression of pour points of vegetable oils by blending with diluents used for biodegradable lubricants. *Journal of the American Oil Chemists' Society*, 76, 313–316.

Bart, J. C. J., Gucciardi, E. & Cavallaro, S. 2013. 2- Principles of lubrication. In: Bart, J. C. J., Gucciardi, E. & Cavallaro, S. (eds.) *Biolubricants*. Woodhead Publishing Ltd. Cambridge, UK.

Baskar, S., Sriram, G. & Arumugam, S. 2015. Experimental analysis on tribological behavior of nano based bio-lubricants using four ball tribometer. *Tribology in Industry*, 37, 449–454.

Baumgart, P., Canzi, G., Hanashiro, T., Doezema, L. A. & Siniawski, M. T. 2010. Influence of fatty acid additives on the tribological performance of sunflower oil. *Lubrication Science*, 22, 393–403.

Bondioli, P. 2004. The preparation of fatty acid esters by means of catalytic reactions. *Topics in Catalysis*, 27, 77–82.

Cermak, S. C. & Isbell, T. A. 2003. Improved oxidative stability of estolide esters. *Industrial Crops and Products*, 18, 223–230.

Cermak, S. C. & Isbell, T. A. 2009. Synthesis and physical properties of mono-estolides with varying chain lengths. *Industrial Crops and Products*, 29, 205–213.

Cheenkachorn, K. & Fungtammasan, B. 2010. Development of engine oil using palm oil as a base stock for four-stroke engines. *Energy*, 35, 2552–2556.

Chowdhury, A., Mitra, D. & Biswas, D. 2013. Biolubricant synthesis from waste cooking oil via enzymatic hydrolysis followed by chemical esterification. *Journal of Chemical Technology & Biotechnology*, 88, 139–144.

Cristea, G. C., Dima, C., Dima, D., Georgescu, C. & Deleanu, L. 2017. Nano graphite as additive in soybean oil. *MATEC Web of Conferences*, EDP Sciences, 04023.

Cristea, G., Dima, C., Georgescu, C., Dima, D., Deleanu, L. & Solea, L. 2018a. Evaluating Lubrication capability of soybean oil with nano carbon additive. *Tribology in Industry*, 40, 66–72.

Cristea, G. C., Radulescu, A., Georgescu, C., Radulescu, I. & Deleanu, L. 2018b. Influence of additive concentration in soybean oil on rheological and tribological behavior. *INCAS Bulletin*, 10, 35–43.

Darminesh, S. P., Sidik, N. A. C., Najafi, G., Mamat, R., Ken, T. L. & Asako, Y. 2017. Recent development on biodegradable nanolubricant: A review. *International Communications in Heat and Mass Transfer*, 86, 159–165.

Dossat, V., Combes, D. & Marty, A. 2002. Efficient lipase catalysed production of a lubricant and surfactant formulation using a continuous solvent-free process. *Journal of Biotechnology*, 97, 117–124.

Gerbig, Y., Ahmed, S. U., Gerbig, F. & Haefke, H. 2004. Suitability of vegetable oils as industrial lubricants. *Journal of Synthetic Lubrication*, 21, 177–191.

Guezmil, M., Bensalah, W. & Mezlini, S. 2016. Effect of bio-lubrication on the tribological behavior of UHMWPE against M30NW stainless steel. *Tribology International*, 94, 550–559.

Habibullah, M., Masjuki, H. H., Kalam, M., Ashraful, A., Habib, M. A. & Mobarak, H. 2014. Effect of bio-lubricant on tribological characteristics of steel. *Procedia Engineering*, 90, 740–745.

Jain, A. K. & Suhane, A. 2013. Biotechnology: A way to control environmental pollution by alternative lubricants. *Research in Biotechnology*, 4, 38–42.

Kalam, M., Masjuki, H., Cho, H. M., Mosarof, M., Mahmud, M. I., Chowdhury, M. A. & Zulkifli, N. 2017. Influences of thermal stability, and lubrication performance of biodegradable oil as an engine oil for improving the efficiency of heavy duty diesel engine. *Fuel*, 196, 36–46.

Kanagasabapathi, N., Balamurugan, K. & Mayilsamy, K. 2012. Wear and thermal conductivity studies on nano copper particle suspended soya bean lubricant. *Journal of Scientific and Industrial Research*, 71, 492–495.

Kiriliauskaitė, V., Bendikienė, V. & Juodka, B. 2011. Synthesis of trimethylolpropane esters of oleic acid by Lipoprime 50T. *Journal of Industrial Microbiology & Biotechnology*, 38, 1561–1566.

Koshy, C. P., Rajendrakumar, P. K. & Thottackkad, M. V. 2015. Evaluation of the tribological and thermo-physical properties of coconut oil added with MoS_2 nanoparticles at elevated temperatures. *Wear*, 330, 288–308.

Kozma, M. 1997. Investigation into the scuffing load capacity of environmentally-friendly lubricating oils. *Journal of Synthetic Lubrication*, 14, 249–258.

Li, K., Zhang, X., Du, C., Yang, J., Wu, B., Guo, Z., Dong, C., Lin, N. & Yuan, C. 2019. Friction reduction and viscosity modification of cellulose nanocrystals as biolubricant additives in polyalphaolefin oil. *Carbohydrate Polymers*, 220, 228–235.

Linko, Y.-Y., Tervakangas, T., Lämsä, M. & Linko, P. 1997. Production of trimethylolpropane esters of rapeseed oil fatty acids by immobilized lipase. *Biotechnology Techniques*, 11, 889–892.

Mcnutt, J. 2016. Development of biolubricants from vegetable oils via chemical modification. *Journal of industrial and Engineering Chemistry*, 36, 1–12.

Mohanraj, T., Shankar, S., Rajasekar, R., Deivasigamani, R. & Arunkumar, P. M. 2019. Tool condition monitoring in the milling process with vegetable based cutting fluids using vibration signatures. *Materials Testing*, 61, 282–288.

Panchal, T. M., Patel, A., Chauhan, D., Thomas, M. & Patel, J. V. 2017. A methodological review on bio-lubricants from vegetable oil based resources. *Renewable and Sustainable Energy Reviews*, 70, 65–70.

Quinchia, L., Delgado, M., Reddyhoff, T., Gallegos, C. & Spikes, H. 2014a. Tribological studies of potential vegetable oil-based lubricants containing environmentally friendly viscosity modifiers. *Tribology International*, 69, 110–117.

Rani, S. 2017. The evaluation of lubricant properties and environmental effect of bio-lubricant developed from rice bran oil. *International Journal of Surface Science and Engineering*, 11, 403–417.

Rani, S., Joy, M. & Nair, K. P. 2015. Evaluation of physiochemical and tribological properties of rice bran oil–biodegradable and potential base stoke for industrial lubricants. *Industrial Crops and Products*, 65, 328–333.

Reeves, C. J., Menezes, P. L., Jen, T.-C. & Lovell, M. R. 2015. The influence of fatty acids on tribological and thermal properties of natural oils as sustainable biolubricants. *Tribology International*, 90, 123–134.

Reeves, C. J., Menezes, P. L., Lovell, M. R. & Jen, T.-C. 2013. The size effect of boron nitride particles on the tribological performance of biolubricants for energy conservation and sustainability. *Tribology Letters*, 51, 437–452.

Ruggiero, A., D'amato, R., Merola, M., Valašek, P. & Müller, M. 2017. Tribological characterization of vegetal lubricants: Comparative experimental investigation on Jatropha curcas L. oil, Rapeseed Methyl Ester oil, Hydrotreated Rapeseed oil. *Tribology International*, 109, 529–540.

Salimon, J. & Salih, N. 2009a. Oleic acid diesters: Synthesis, characterization and low temperature properties. *European Journal of Scientific Research*, 32, 216–222.

Salimon, J. & Salih, N. 2009b. Preparation and characteristic of 9, 10-epoxyoleic acid α-hydroxy ester derivatives as biolubricant base oil. *European Journal of Scientific Research*, 31, 265–272.

Sapawe, N., Samion, S., Ibrahim, M. I., Daud, M. R., Yahya, A. & Hanafi, M. F. 2017. Tribological testing of hemispherical titanium pin lubricated by novel palm oil: Evaluating anti-wear and anti-friction properties. *Chinese Journal of Mechanical Engineering*, 30, 644–651.

Shafi, W. K., Raina, A. & Haq, M. I. U. 2018. Friction and wear characteristics of vegetable oils using nanoparticles for sustainable lubrication. *Tribology - Materials, Surfaces & Interfaces*, 12, 27–43.

Shahabuddin, M., Masjuki, H. H. & Kalam, M. A. 2013a. Experimental investigation into tribological characteristics of bio-lubricant formulated from jatropha oil. *Procedia Engineering*, 56, 597–606.

Shahabuddin, M., Masjuki, H. H., Kalam, M. A., Bhuiya, M. M. K. & Mehat, H. 2013b. Comparative tribological investigation of bio-lubricant formulated from a non-edible oil source (jatropha oil). *Industrial Crops and Products*, 47, 323–330.

Shankar, S., Mohanraj, T. & Ponappa, K. 2017. Influence of vegetable based cutting fluids on cutting force and vibration signature during milling of aluminium metal matrix composites. *Jurnal Tribologi*, 12, 1–17.

Shankar, S., Mohanraj, T. & Pramanik, A. 2019a. Tool condition monitoring while using vegetable based cutting fluids during milling of inconel 625. *Journal of Advanced Manufacturing Systems*, 18, 563–581.

Shankar, S., Mohanraj, T. & Rajasekar, R. 2019b. Prediction of cutting tool wear during milling process using artificial intelligence techniques. *International Journal of Computer Integrated Manufacturing*, 32, 174–182.

Sharma, B. K., Adhvaryu, A., Liu, Z. & Erhan, S. Z. 2006. Chemical modification of vegetable oils for lubricant applications. *Journal of the American Oil Chemists' Society*, 83, 129–136.

Sharma, U. C. & Sachan, S. 2019. Friction and wear behavior of karanja oil derived biolubricant base oil. *SN Applied Sciences*, 1, 668.

Sharma, R. V., Somidi, A. K. R. & Dalai, A. K. 2015. Preparation and properties evaluation of biolubricants derived from canola oil and canola biodiesel. *Journal of Agricultural and Food Chemistry*, 63, 3235–3242.

Shashidhara, Y. & Jayaram, S. 2012. Tribological studies on AISI 1040 with raw and modified versions of Pongam and Jatropha vegetable oils as lubricants. *Advances in Tribology*, 2012, 1–6.

Silva, M. S., Foletto, E. L., Alves, S. M., De Castro Dantas, T. N. & Neto, A. A. D. 2015. New hydraulic biolubricants based on passion fruit and moringa oils and their epoxy. *Industrial Crops and Products*, 69, 362–370.

Siniawski, M. T., Saniei, N., Adhikari, B. & Doezema, L. A. 2007. Influence of fatty acid composition on the tribological performance of two vegetable-based lubricants. *Journal of Synthetic Lubrication*, 24, 101–110.

Siniawski, M. T., Saniei, N. & Stoyanov, P. 2011. Influence of humidity on the tribological performance of unmodified soybean and sunflower oils. *Lubrication Science*, 23, 301–311.

Sneha, E., Rani, S. & Arif, M. 2019. Evaluation of lubricant properties of vegetable oils as base oil for industrial lubricant. *IOP Conference Series: Materials Science and Engineering*, IOP Publishing, England, 012022.

Suganeswaran, K., Parameshwaran, R., Mohanraj, T. & Radhika, N. 2020. Influence of secondary phase particles Al2O3/SiC on the microstructure and tribological characteristics of AA7075-based surface hybrid composites tailored using friction stir processing. *Proceedings of the Institution of Mechanical Engineers, Part C: Journal of Mechanical Engineering Science*, 235, 161–178.

Suresh, S., Kumar, R. M., Arun, G. & Siva, M. 2020. An experimental investigation on tribological behaviour of jatropha biolubricant with pyrolysed cardanol biolubricant at varying loads. *MS&E*, 764, 012024.

Suresha, B., Hemanth, G., Rakesh, A. & Adarsh, K. 2020. Tribological behaviour of neem oil with and without graphene nanoplatelets using four-ball tester. *Advances in Tribology*. 2020, 1–11.

Tobón, A. E. D. & Chaparro, W. A. A. 2019. Evaluation of polyethylene wax and paraffin as anti wear and extreme pressure additives in virgin sesame base stock. *Tecciencia*, 14, 23–32.

Trajano, M. F., Moura, E. I. F., Ribeiro, K. S. B. & Alves, S. M. 2014. Study of oxide nanoparticles as additives for vegetable lubricants. *Materials Research*, 17, 1124–1128.

Uosukainen, E., Linko, Y.-Y., Lämsä, M., Tervakangas, T. & Linko, P. 1998. Transesterification of trimethylolpropane and rapeseed oil methyl ester to environmentally acceptable lubricants. *Journal of the American Oil Chemists' Society*, 75, 1557–1563.

Yadav, G. D. & Lathi, P. S. 2003. Kinetics and mechanism of synthesis of butyl isobutyrate over immobilised lipases. *Biochemical Engineering Journal*, 16, 245–252.

Yuan, S., Hou, X., Wang, L. & Chen, B. 2018. Experimental investigation on the compatibility of nanoparticles with vegetable oils for nanofluid minimum quantity lubrication machining. *Tribology Letters*, 66, 1–10.

Zaidi, A., Gainer, J., Carta, G., Mrani, A., Kadiri, T., Belarbi, Y. & Mir, A. 2002. Esterification of fatty acids using nylon-immobilized lipase in n-hexane: Kinetic parameters and chain-length effects. *Journal of Biotechnology*, 93, 209–216.

Zulkifli, N., Azman, S., Kalam, M., Masjuki, H. H., Yunus, R. & Gulzar, M. 2016. Lubricity of bio-based lubricant derived from different chemically modified fatty acid methyl ester. *Tribology International*, 93, 555–562.

Zulkifli, N. W. M., Kalam, M., Masjuki, H. H., Shahabuddin, M. & Yunus, R. 2013. Wear prevention characteristics of a palm oil-based TMP (trimethylolpropane) ester as an engine lubricant. *Energy*, 54, 167–173.

11 Group IV Base Stock: *Polyalphaolefin – A High-Performance Base Oil for Tribological Applications*

Homender Kumar and A.P. Harsha
Indian Institute of Technology (Banaras Hindu University)

CONTENTS

11.1 INTRODUCTION

Friction, which acts as a resistance to motion, has always been and still an interesting physical phenomenon for researchers and scientists. It has gained the attention of mankind throughout history due to its many impacts (favourable or unfavourable) in our life. It is well known that friction and wear between the moving components of the automotive engine directly influence its fuel economy, durability and emission characteristics. It has been stated that approximately one-third of fuel energy consumed by light- and heavy-duty vehicles is dissipated due to friction and other mechanical losses in engines, transmission, tyres and brakes (Holmberg,

Andersson and Erdemir, 2012; Holmberg et al., 2014). The recent statistical information shows that the sum of motorcycles on the planet has reached more than one billion, which is expected to rise much more in future, primarily due to the acute growth of transport markets in developing countries such as China and India (Erdemir and Holmberg, 2015). Since the industrial reformation began, several scientists and engineers have investigated the source of friction and wear to attenuate their antagonistic influences on energy, component durability and the environment. Multiple scientific pieces of knowledge on wear and friction mechanisms have been developed, and consequently, the development of novel solid and liquid lubricants to mitigate their detrimental impact on all mobile mechanical devices. In the last few decades, various materials and techniques have been deployed to minimize the friction, such as ZDDP/organic molybdenum compounds and other friction modifiers, ionic liquids, nanolubricants, fullerene, graphene, DLC and surface texturing (Erdemir and Holmberg, 2015).

11.2 LIQUID LUBRICANTS

The fundamental role of lubricants is to control friction and wear. However, the accompanying features of lubricants can be depicted as a coolant, corrosion resistant, wear particle remover, etc. A typical lubricant comprises 95% base oil and 5% additives. Lubricants have been characterized into three different categories, i.e. solid, liquid and gaseous. A variety of approaches can be used to classify the liquid lubricants, but one of the best methods to categorize the liquid lubricant is based on the type of base oil used. The most common types of base oil are mineral oil (derived from crude oil), synthetic oil and biological oils (derived from plants and animals). The physical properties of lubricants primarily rely on their base stocks (Pownraj and Valan Arasu, 2020). The lubricating oils available in the market are mainly originated from three sources, as shown in Figure 11.1.

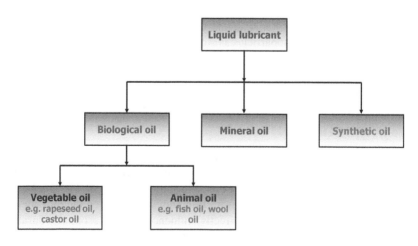

FIGURE 11.1 Classification of oils.

11.2.1 SYNTHETIC LUBRICANTS

Synthetic lubricants are different from conventional (mineral) lubricants in terms of components adopted in the formulation. The synthetic base stock is the main component in synthetic lubricants. There are several definitions to describe synthetic lubricants. According to the American Society for Testing and Materials (ASTM), the synthetic lubricant is a blend that contains the stocks produced by chemical synthesis and constituting essential functional additives (Rao, Srivastava and Mehta, 1987). The final base stocks are designed through molecular rebuilding by a specific chemical reaction to obtain predetermined chemical and physical properties.

11.2.2 WHY DO WE USE SYNTHETIC LUBRICANTS?

There are two significant rationales to use synthetic lubricants as a replacement for petroleum-based lubricants.

1. When equipment requires distinct performance characteristics such as exceptionally high or low operating temperature, stability under extreme conditions and long service life that cannot be fulfilled by conventional petroleum-based lubricants.
2. When synthetic lubricants can provide economic advantages such as lower energy dissipation, machine reliability, expanded oil change interval and increased power output.

11.2.3 OVERVIEW OF SYNTHETIC BASE OILS

The structural formulas of major types of synthetic base oil are displayed in Table 11.1. Nowadays, for lubricating purposes, the following five base oils can be commercially considered as the most significant representatives (Mang, 2014):

1. Polyalphaolefins (PAOs)
2. Neopentyl polyol esters
3. Polyalkylene glycols
4. Perfluorinated polyethers
5. Silicone oils (polysiloxanes).

Other important synthetic base stocks are alkylated aromatics, polybutenes, phosphate ester, polyphenylene ethers, etc. but their applications are limited because of high costs or performance restrictions. It is estimated that around 80% of the global market for synthetic lubricants is corroborated by three groups of synthetic base oils (Wu, Ho and Forbus, 2006):

1. PAOs (45%)
2. Esters (25%)
3. Polyalkylene glycols (10%).

TABLE 11.1

Typical Chemical Structures of the Most Common Synthetic Base Oils (Stachowiak and Batchelor, 2000)

S.No.	Name of Lubricants	Typical Structural Formula
1	Alkylbenzenes	
2	Polybutene	
3	Polyalphaolefin	
4	Diester	
5	Polyol esters	
6	Phosphate ester	
7	Neopentyl polyol esters	
8	Polyalkylene glycol	

1. Alkylbenzenes

R (top), R (bottom), benzene ring with $-R$

2. Polybutene

$$\left(-CH_2-CH_2-CH_2-CH_2-\right)_n$$

3. Polyalphaolefin

$$CH_3-CH-CH_2-CH-CH_2-CH_2$$
with C_8H_{17}, C_8H_{17}, C_8H_{17}

4. Diester

$$R-O-\overset{\overset{O}{\|}}{C}-(CH_2)_n-\overset{\overset{O}{\|}}{C}-O-R$$

5. Polyol esters

$$C-\left(CH_2-O-\overset{\overset{O}{\|}}{C}-R\right)_4$$

6. Phosphate ester

$$R-O-\overset{\overset{O}{\|}}{\underset{\underset{OR'}{|}}{P}}-O-R''$$

7. Neopentyl polyol esters

$$CH_3-CH_2-\overset{\overset{CH_2-OOC-C_8H_{17}}{|}}{\underset{\underset{CH_2-OOC-C_8H_{17}}{|}}{C}}-OOC-C_8H_{17}$$

8. Polyalkylene glycol

$$RO\left[-CH_2-\overset{\overset{R'}{|}}{CH}-O-\right]_n R''$$

(*Continued*)

TABLE 11.1 (*Continued*)
Typical Chemical Structures of the Most Common Synthetic Base Oils (Stachowiak and Batchelor, 2000)

S.No.	Name of Lubricants	Typical Structural Formula
9	Silicon oils (dimethyl siloxane)	
10	Perfluoropolyalkylether	

The applications of other synthetic lubricants are limited when compared to PAO-, ester- and polyalkylene glycol-based lubricants. For all synthetic base stocks except phosphate esters and silicones, starting ingredients are originated from basic petrochemicals such as ethylene, propylene, butene, higher olefins, toluene, benzene, xylene and naphthalene. Generally, synthetic lubricants have a high price that impacts the market expansion. The cost of synthetic lubricants can be 2–5 times higher than the cost of mineral oil-based lubricants. The value of synthetic silicone lubricant can reach up to 20 times the value of mineral oil-based lubricants. To opt synthetic type of lubricants in main developing markets, e.g. China, India, Africa and Brazil, the price of lubricants plays a crucial role.

11.3 POLYALPHAOLEFIN (PAO)

The term 'polyalphaolefin' (PAO) is derived from the lubricant base stock, i.e. saturated α-olefin oligomers. It is projected that during 2018–2023, PAO-based lubricants would dominate the synthetic lubricant market. PAOs are hydrogenated olefin polymers that are produced by catalytic oligomerization of linear-α-olefin. An α-olefin is a member of an olefin family containing carbon–carbon double bond at first and second carbons of the molecular chain. Since the oligomers are saturated, therefore, PAOs, usually termed synthetic hydrocarbons, are the primary class of synthetic base oils. The American Petroleum Institute (API) has categorized the character of PAO into group IV (Table 11.2) and applied them as lubricant specification worldwide. Currently, PAO products are engineered with a broad viscosity range from 2 to 100 cSt at 100°C and traded commercially (Goze et al., 2010). PAO having viscosities 2, 4 and 6 cSt at 100°C referred to as low-viscosity PAOs and 8, 10, 40, 100 cSt are considered as a medium to high-viscosity PAOs.

TABLE 11.2
API Classification of Base Stock (Mang, 2014)

Group	Sulphur (wt.%)	Saturates (wt. %)	Viscosity Index
Group I	>0.03	<90	80–119
Group II	<0.03	>90	80–119
Group III	<0.03	>90	>120
Group IV	Polyalphaolefin (PAO)		
Group V	All other base stocks not comprised in Group I–IV		

11.3.1 SYNTHESIS PROCESS OF PAO

PAOs are produced by two-step reaction process from linear α-olefin (1-decene) which is the basic olefin (or fundamental building block) for the production of PAO. The linear α-olefin is derived from ethylene. In the first step, the ethylene is polymerized to 1-decene. The linear 1-decene is further oligomerized in the presence of a catalyst, yielding low molecular weight unsaturated oligomers, such as a dimer, trimer, tetramer and some higher oligomers (Moore et al., 2003). For the development of low-viscosity PAOs (i.e. 2–10 cSt at 100°C), Lewis acid catalysts such as boron trifluoride and alcohol or water as cocatalyst are used. In contrast, Ziegler–Natta catalysts such as trialkylaluminium or alkylaluminium halide with second moiety, namely a halogen or halide source, are preferred for the production of high-viscosity PAOs (40 and 100 cSt) (Shubkin, Corporation and Rouge, 1991).

In the final step of the synthesis process, these unsaturated oligomers are hydrogenated by using a metal catalyst as palladium or nickel to develop a fully saturated hydrocarbon mixture. These saturated hydrocarbon mixtures are fractionated through the distillation to obtain desired viscosity of PAO. The hydrogenation may be performed before or after distillation. The molecular structures of dimers, trimers and tetramers of 1- decene are displayed in Figure 11.2. All these unsaturated oligomers exhibit various isomeric forms. The word PAO is only adopted when the fluid is saturated in downstream chemical hydrogenation. A graphical representation of the manufacturing process for PAO products is depicted in Figure 11.3. The physical properties of final products (PAOs) are greatly influenced by the various process parameters (reaction parameters), i.e., the physical properties of PAO products required for end-use applications can be tailored by controlling the process parameters (Rudnick, 2005). Some critical reaction parameters are as follows:

1. Catalysts and their concentration
2. Cocatalysts
3. The chain length of olefin feedstock
4. Temperature
5. Time
6. Pressure

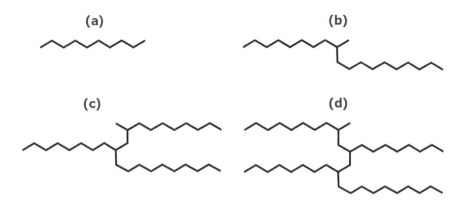

FIGURE 11.2 Molecular structure of 1-decene (PAO) components (a) monomer (b) dimer (c) trimer and (d) tetramer.

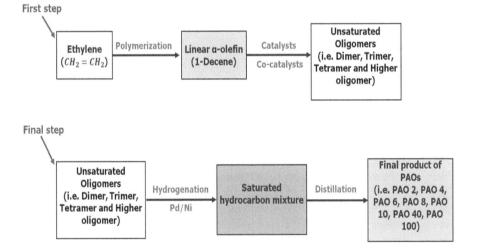

FIGURE 11.3 Manufacturing process of PAOs.

7. Catalysts for hydrogenation
8. Distillation.

11.3.2 PHYSICAL PROPERTIES OF PAO

The typical physical characteristics of some commercially available 1-decene-based PAOs are listed in Table 11.3. For designating the different grades, the most common convention is to use the kinematic viscosity in cSt (mm²/s) at 100°C. Therefore, a PAO oil having kinematic viscosity 2 cSt can be denoted as PAO 2. For PAO 2, the viscosity index is not mentioned in Table 11.3, since the viscosity index for any base oil acquiring a kinematic viscosity less than 2.0 cSt at 100°C is undefined (Rudnick, 2005).

TABLE 11.3

Physical Properties of Different Grades of PAOs (Rudnick, 2005)

Property	ASTM Standard	PAO 2	PAO 4	PAO 6	PAO 8	PAO 10	PAO 40	PAO 100
Kinematic viscosity (cSt @100°C)	D445	1.8	4.1	5.93	7.74	9.87	39	100
Kinematic viscosity (cSt @40°C)	D445	5.5	19	31	46.3	64.5	396	1150
Viscosity index	D2270	–	126	138	136	137	147	179
Density (g/ml)	D1298	0.798	0.821	0.827	0.833	0.835	0.850	0.853
Pour point (°C)	D97	−66	−66	−63	−57	−53	−36	−30
Flash point (°C)	D92	165	220	246	252	264	281	300
Noack volatility (wt. %)	D5800	99	14	6.8	3	2	0.7	0.6

11.3.3 Comparison of PAOs with Petroleum-Based Mineral Oils

PAOs show a distinct chemical composition as compared to conventional mineral based oils. A blob of synthetic oil under a microscope reveals countless molecules having approximately the same size and shape. Conversely, a blob of traditional oil under a microscope features myriad molecules in diverse sizes, shapes and structures. Today, most of the vehicles are suitable for employing either PAO based oils or natural mineral oil that meets the API and the International Lubricant Standardization and Advisory Committee (ILSAC) requirements. A particular PAO oil may acquire certain advantages, but no individual lubricant is superior in all respects. However, PAOs are often offered excellent performance as compared to petroleum-based oil. This higher performance may only be linked to specific attributes but not others. It may be possible that some traditional mineral oil-based formulations may demonstrate superior performance on specific properties. Based on the test results conducted on various important characteristics such as physical, chemical and performance properties specifically low-temperature pumpability, shear stability, oxidation resistance, deposit formation, volatility and rheological (viscosity) changes caused by oxidation, American Automation Association (AAA) illustrated that PAOs outperformed mineral oil by approximately 47% (Trout and Fitch, 2020). Nevertheless, in the past few years, PAOs have faced a tough competition in the market comparably from low-cost group III lubricants. Although, with the increasingly growing demand for engine efficiency and strict emission standards, the market for PAOs will continue to be high in the following years.

11.3.3.1 Advantages of PAO Oils

- **Higher viscosity index** – This implies that the viscosity varies less (more stable) as temperature alters during normal engine's start-up and working conditions. Viscosity is an important characteristic of lubricants, which produces fluid film or clearance between sliding or rotating metal surfaces. Intense friction and wear will take place without this fluid film.

- **Better oxidative stability** – Conventional petroleum-based mineral oils are normally more likely to be chemically degraded (oxidation) than PAOs when subjected to certain conditions such as combustion by-products, fuel dirtiness, acids, pro-oxidants, water adulteration, metal particles and extreme heat (i.e. from combustion). Such conditions are normally exposed in engines. Oil decay can lead to sludge formation, corrosion, varnish or deposits, change in viscosity and decrease in engine efficiency.
- **Lower temperature fluidity** – Mineral oils (in comparison with PAOs) can become so thicker (highly viscous) at extremely low temperatures that the oil is not pumpable or is not able to flow adequately in the engine. The scarcity of oil flow can lead to lubricant starvation conditions and malfunctioning of engine operations.
- **Lower volatility** – PAO based engine oils usually exhibit low volatility as compared to mineral oils. This indicates that the oil is less wasted through the exhaust flux of the engine, which causes environmental harm. Consequently, the necessity of make-up oil between oil change interval is reduced.
- **Greater thermal stability** – PAOs withstand higher temperature with minimum decomposition as compared to conventional mineral oils.
- **Longer life span** – The biggest advantage of PAO oils are that they have a longer service life as compared to traditional petroleum base mineral oils. For PAO oils, the suggested change period is about every 5,000–7,000 miles, whereas for some brands, the change interval is approximately 15,-000–25,000 miles.

PAO also possess other key features depending on the application over mineral oils are as follows:

- Low toxicity
- Hydrolytically stable
- Compatible with mineral oils, esters and alkylated naphthalene
- Lower traction force
- Low corrosion.

11.3.3.2 Disadvantages of PAO Oils

PAO base oils do have some negative effects, and these are as follows:

- **High cost** – The cost is the most noticeable drawback of PAOs. The value of PAOs is around two to four times higher than that of the value of mineral oils.
- **Limited additive stability** – Due to composition purity and very low polarity, PAOs provide poor solvency for common additives essential for fully formulated lubricants. Also, additive stability in PAOs under cold temperature conditions over a long period may be a great challenge.

- **Seal shrinkage risk** – Highly non-polar characteristics of PAOs may affect the seal performance under some situations. Therefore, PAO-based lubricants are generally mixed with ester oil to counteract the shrinking effect of PAO molecules.

The comparisons of typical features of PAOs such as viscosity range, oxidation stability, thermal stability and corrosion stability with mineral oil and other synthetic oils are presented in Table 11.4.

11.3.4 RECENT DEVELOPMENTS

After the successful acceptability of PAOs in various applications, ExxonMobil Chemical Co. launched a new class of PAOs, i.e. SpectraSyn Ultra™ PAO. The SpectraSyn Ultra™ PAOs are manufactured from 1-decene, the same basic stock that is used to produce traditional PAO. The physical attributes of commercial SpectraSyn Ultra™ PAOs are summarized in Table 11.5. The SpectraSyn Ultra™ PAOs exhibit higher VI, lower pour point and higher viscosity ranges compared to conventional PAOs. These novel lubricants can be employed in automotive engine oil and industrial oil formulations to offer benefits in terms of viscosity characteristics, shear stability and augmented film thickness of lubricants.

11.3.5 APPLICATIONS OF PAOS

The PAOs are acquiring rapid admissibility in many niche applications because of their inherent and attractive characteristics. The application areas of PAO base oils based on their viscosity grades are shown in Table 11.6.

11.4 LUBRICANT ADDITIVES

Lubricant additives are chemicals nearly organic or organometallic incorporated into base oil in little weight percentage to enhance the lubricating performance and durability of the base oils. Over the past decades, conventional additives like antiwear (AW) and extreme pressure (EP) agent, corrosion and rust inhibitor, detergent, dispersant, friction modifier were used as lubricant additives, but they have demonstrated peak performance. To increase the performance of lubricants, different kinds of nanomaterials were used as additives and are currently being examined. Nanoparticles are comparably a new category of lubricant additives in the development history of lubricant additives (Spikes, 2015) as shown in Figure 11.4. Nanoparticles as lubricant additives have many potential advantages over traditional lubricant additives.

1. Insolubleness in non-polar base oils
2. Low reactivity with other additives in the lubricant
3. Higher probability to form a lubricant film between several types of mating surfaces
4. Great sustainability
5. Very low volatility to withstand high temperatures.

TABLE 11.4

Comparison of Typical Properties of PAOs with Mineral and Other Synthetic Oils (Murphy, Blain and Galiano-Roth, 2002)

Properties	Polyalphaolefins (PAOs)	Mineral Oil	Diester	Polyol Ester	Phosphate Ester	Polyalkylene Glycol	Silicon Oil	Perfluoropolyether
Cost	M	L	M	M	VH	M	H	VH
Viscosity range	VW	W	L	L	L	V	W	W
Viscosity temperature	VG	F	G	G	P	E	E	NA
Temperature range	VG	BA	G	V	G	G	E	E
Oxidative stability	VG	BA	G	VG	VG	VG	E	E
Thermal stability	G	F	G	G	F	F	G	NA
Low volatility	E	F	E	E	G	G	G	NA
Corrosion stability	E	E	BA	BA	VG	G	G	BA
Hydrolytic stability	E	E	F	F	F	VG	G	NA
Compatibility with mineral oil	E	E	G	G	BA	BA	BA	BA
Pour point	L	G	L	L	VG	M	VL	L
Flash point	VG	M	G	VG	E	G	E	E
Additive solubility	G	E	VG	VG	G	F	P	NA
Seal compatibility	VG	VG	BA	BA	BA	G	VG	VG

where L, Low; VL, Very low; M, Medium; H, High; VH, Very high; W, Wide; VW, Very wide; F, Fair; G, Good; VG, Very good; E, Excellent; P, Poor; BA, Below average or less than ideal; NA, Not available

TABLE 11.5

Physical Attributes of a New Class of PAO – SpectraSyn Ultra™ (ExxonMobil Chemical Advanced Synthetic Base Stocks)

Property	ASTM Standard	SpectraSyn Ultra™		
		PAO 150	PAO 300	PAO 1000
Kinematic viscosity (cSt @100°C)	D445	150	300	1,000
Kinematic viscosity (cSt @40°C)	D445	1,500	3,100	10,000
Viscosity index	D2270	218	241	307
Density (g/ml)	D1298	0.850	0.852	0.855
Pour point (°C)	D97	−33	−27	−18
Flash point (°C)	D92	>265	>265	>265

TABLE 11.6

Application of Different Grades of PAOs in Various Areas (ExxonMobil Chemical)

Applications	Conventional PAOs							SpectraSyn Ultra™		
	PAO 2	PAO 4	PAO 6	PAO 8	PAO 10	PAO 40	PAO 100	PAO 150	PAO 300	PAO 1000
Gasoline and Diesel engines	–	XX	XX	XX	–	XX	–	XX	XX	–
Automatic transmissions	XX	XX	–	–	–	–	–	XX	–	–
Industrial/automotive gear and transmissions	X	XX	XX	X	X	XX	XX	XX	–	–
Hydraulic systems	XX	XX	XX	X	X	XX	XX	XX	–	–
Industrial bearings	–	–	XX	X	X	XX	XX	XX	XX	–
Rotary air and gas compressor	–	X	XX	XX	XX	XX	XX	–	–	–
Hydrocarbon refrigeration compressor	X	X	X	X	X	X	–	–	–	–
Greases	X	X	XX	X	X	XX	XX	XX	XX	XX
Turbines	–	XX	XX	X	X	XX	XX	–	–	–
Heat transfer fluids	XX	X	–	–	–	–	–	–	–	–
Automotive hydraulic fluids	XX	–	–	–	–	–	–	–	–	–
Mist lubricants	–	XX	XX	X	X	X	X	XX	–	–

where XX- most commonly used application, X- Less commonly used applications

The nanoparticles can be classified into various categories according to their applications. In this investigation, the nanoparticles have been categorized based on tribological applications. Figure 11.5 demonstrates the three fundamental types of nanoparticles.

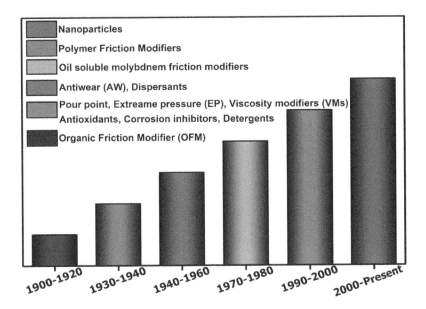

FIGURE 11.4 Chronology of the evolution of lubricant additives (Spikes, 2015).

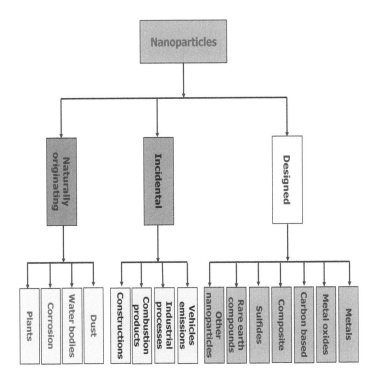

FIGURE 11.5 Classification of nanoparticles on the basis of their origin (Gulzar et al., 2016).

TABLE 11.7

Summary of Investigated Nanoparticles as Lubricant Additives

S.N.	Type of Nanoparticles	Name of Nanoparticles
1.	Metal based	Cu (Choi et al., 2009; Padgurskas et al., 2013), Ni (Chou et al., 2010), Al (Peng et al., 2010), Fe (Padgurskas et al., 2013) and Pd (Abad and Sánchez-López, 2013)
2.	Metal oxide based	Al_2O_3 (Peña-Parás et al., 2015), CuO (Kashyap and Harsha, 2016; Gupta and Harsha, 2018b; Rawat et al., 2019a), ZnO (Rawat et al., 2019a), ZrO_2 (Hernández Battez et al., 2008), SiO_2 (Rawat, Harsha and Deepak, 2019) and Fe_2O_3 (Gao et al., 2013)
3.	Carbon and its derivative based	Fullerene (Upadhyay and Kumar, 2019), SWCNTs (Joly-Pottuz et al., 2004), MWCNTs (Kumar and Harsha, 2020), graphene (Upadhyay and Kumar, 2019), graphene oxide (Rawat et al., 2020), nanodiamond (Nunn et al., 2015) and graphite (Kogovšek and Kalin, 2014)
4.	Composite nanoparticles	Carbon-coated copper (Viesca et al., 2011), Al_2O_3/SiO_2 (Jiao et al., 2011), ZrO_2/SiO_2 (Li et al., 2011) and $ZnO/GO/MoS_2$ (Chouhan et al., 2020)
5.	Rare earth compound based	LaF_3 (Li et al., 2014) and CeO_2 (Gupta and Harsha, 2018a)
6.	Sulphide based	MoS_2 (Rawat et al., 2019b) and WS_2 (Kogovšek and Kalin, 2014)
7.	Others	$CaCO_3$ (Zhang et al., 2009), $ZnAl_2O_4$ (Song et al., 2012), h-BN (Bondarev et al., 2020) and PTFE (Saini et al., 2020)

The nanoparticles used by researchers and scientists as additives in various lubricating oils in recent decades are listed in Table 11.7.

11.4.1 ROLE OF NANOPARTICLES AS ADDITIVES

The examination of lubrication mechanisms of nanoparticles is regarded as a critical parameter to comprehend the role of nanoparticles in lubricants. However, in various research studies dealt with nanoparticle-based lubrication systems, identifying the active mechanisms remains a topic of discussion. The researchers have suggested multiple mechanisms using surface analysis techniques to illustrate the advancement in lubrication performance by the inclusion of nanoparticles in lubricating oil. These mechanisms can be divided into two primary categories. The first is the direct impact of the nanoparticles that comprise the ball bearing effect and protective/tribofilm formation. Another is the secondary impact that leads to surface improvement by mending effect and polishing effect (Gulzar et al., 2016).

11.5 TRIBOLOGICAL PERFORMANCE OF PAOs-BASED NANOLUBRICANTS

In the past few years, various types of nanoparticles such as metal-, metal oxide-, carbon-based nanomaterials and rare earth compound have been adopted as an additive in PAO base oils to enhance their tribological properties. Many research probes highlighted the enhancement of tribological performances of PAO based nanolubricants. It was demonstrated that the improvement is governed by several parameters such as type of nanoparticles, the concentration of nanoparticles, temperature, type of base oil and the stability of additive in the base oil. The various types of lubrication mechanisms are proposed for the improvement in performance, and these are sliding, rolling, mending and tribofilm formation by the nanoparticles on friction surfaces (Hernández Battez et al., 2008; Zhang et al., 2009; Chou et al., 2010; Viesca et al., 2011; Zhang, Wang and Liu, 2013; Kogovšek and Kalin, 2014; Nunn et al., 2015; Peña-Parás et al., 2015; Sui et al., 2016; Alves et al., 2016; Azman et al., 2016; Huang et al., 2017; Raina and Anand, 2017; Kumara, Meyer and Qu, 2019; Liñeira del Río et al., 2019; Qi et al., 2020; Zhang et al., 2020; Tang et al., 2020; Wang et al., 2020; Wen et al., 2020; Bondarev et al., 2020; Kassim et al., 2020).

The tribological experimentation of PAO-based nanolubricants has been performed by using a variety of instruments such as pin-on-disc, four-ball tribotester, and high-frequency reciprocating rig (HFRR). However, the most used tribological instruments are four-ball tribometers and a pin-on-disk. The pin-on-disc tribometer facilitates to measure the coefficient of friction and the wear of the liquid based nanolubricants. In contrast, the four-ball tribotester is used both liquid based nanolubricants and semisolid lubricants such as grease to measure the friction coefficient and wear. Furthermore, four-ball tribotester is employed to estimate the EP properties of lubricants.

The lubrication mechanisms of nanoparticles on the worn surfaces are characterized using transmission electron microscopy (TEM), scanning electron microscopy (SEM), energy-dispersive X-ray spectroscopy (EDS), X-ray photoelectron microscopy (XPS), atomic force microscopy (AFM), optical microscope, Raman spectroscopy and 3D-profilometer. Generally, SEM, TEM, optical microscope, AFM and profilometer instruments are adopted to recognize the surface morphology and wear scar diameter (WSD) of worn surfaces of tested specimens. Simultaneously, XPS, EDX and Raman spectrum are employed to analyse the existence of several chemical elements, bonds and their binding energy and development of tribofilm on the friction surfaces. The impacts of various nanoparticles on the tribological performance of different grades of PAO-based nanolubricants are listed in Table 11.8.

Bondarev et al. (2020) demonstrated the role of spherical tungsten (W), hexagonal boron nitride (h-BN) nanoparticles and their mixture for enhancing the tribological performance of PAO 6 oil by molecular dynamics (MD) simulation and in situ TEM mechanical test (Figure 11.6). The results revealed that the positive influence of inclusion of spherical W nanoparticles in PAO 6 could be ascribed to rolling and sliding action at the interfaces of mating pairs, whereas exfoliation and sliding of BN layers under tribological contact can credit to the diminution of friction and wear. Furthermore, the synergistic impact of the cumulative introduction of both types of nanoparticles in PAO 6 produced the superior lubricity due to the development of W/BN core/shell structures by covering W nanoparticles by h-BN sheets.

TABLE 11.8

Tribological Performance of Different Grades of PAO–Base Oils with Various Nanoparticles as an Additive

Nanoparticles	Base Oil	Nanoparticle Source	Average Particle Size (APS)	Shape	Optimum Concentration	Coefficient of Friction (COF)	Wear (WSD)/Wear Volume (WV)	Extreme Pressure (EP)	Morphological Analyses and Techniques	References
CuO, ZnO, ZrO$_2$	PAO 6	Commercial	30–50 nm, 20 nm, 20–30 nm	Spherical	0.5 wt.% 2 wt.% 1 wt.%	–	–	EP ↑	SEM, EDS	Hernández Battez et al. (2007)
CaCO$_3$	PAO 10	Fabricated	40 nm	Nearly spherical	1 wt.%	COF ↓	WV ↓ 83.3%	EP ↑ 200 N to 600 N	XPS, Surface profilometer	Zhang et al. (2005)
Ni	PAO 6	Commercial	20 nm	Spherical	0.5 wt.%	COF ↓ 30%	WV ↓ 54 %	EP ↑ LWI–201.81 N to 257.59 N	SEM, EDS	Chou et al. (2010)
Carbon-coated copper (Cu25C), uncoated copper (Cu25)	PAO 6	Commercial	25 nm	Nearly spherical	0.5 wt.%	–	WV ↓ Cu25C–31% Cu25–50%	EP ↑ Cu25C–21.2% Cu25–8.8 %	SEM, EDS	Viesca et al. (2011)
BN, MoS$_2$	PAO 10	Commercial	70 nm 50 nm	Spherical	3 wt.% +1 wt.% S S–Surfactant (benzethonium chloride)	COF ↓ But COFN↓ of MoS$_2$> BN	WV ↓	–	SEM, Raman spectroscopy	Demas et al. (2012)

(Continued)

TABLE 11.8 (Continued)

Tribological Performance of Different Grades of PAO–Base Oils with Various Nanoparticles as an Additive

Nanoparticles	Base Oil	Nanoparticle Source	Average Particle Size (APS)	Shape	Optimum Concentration	Coefficient of Friction (COF)	Wear (WSD)/Wear Volume (WV)	Extreme Pressure (EP)	Morphological Analyses and Techniques	References
Triazine derivative, NNBO, NN	PAO	Fabricated	–	Sticky liquid	1.2 wt. % (COF),0.6 wt.% (WSD)	COF↓ NN exhibited better friction reduction than NNBO	WSD↓ NNMO–41% NN–36%	EP↑ NNBO–380N to 550 N NN–380N to 520N	SEM, FTIR, TGA, XPS	Yang et al. (2013)
LaF$_3$	PAO 2	Fabricated	10 nm	Nearly Spherical	0.5 wt.%	COF↓	WV↓	EP↑ 100–600 N	XPS, SEM	Zhang, Wang and Liu (2013)
NGP, SWCNTs, MWCNTs, OLC–L, OLC–S, DND	PAO 6	Commercial	10–20 nm, 1–2 nm, 10–20 nm, 200 nm, 40 nm, 4–5 nm	Platelets, tubular, onion–like, nearly spherical	0.015 wt. %+0.15 MoDDP (DND) 0.5 wt. %, for all remaining nanoparticles	COF↓	WV↓	–	XPS, SEM	Nunn et al. (2015)
Hairy silica nanoparticles (HSNs)	PAO 100	Fabricated	100 nm	Spherical	1 wt. %	COF↓ (40%)	WSD↓ (60%)	–	SEM, EDS	Sui et al. (2016)

(Continued)

TABLE 11.8 (Continued)
Tribological Performance of Different Grades of PAO–Base Oils with Various Nanoparticles as an Additive

Nanoparticles	Base Oil	Nanoparticle Source	Average Particle Size (APS)	Shape	Optimum Concentration	Coefficient of Friction (COF)	Wear (WSD)/Wear Volume (WV)	Extreme Pressure (EP)	Morphological Analyses and Techniques	References
Graphene platelets	PAO 10–based ester blend (95% PAO 10 +5 % TMP)	Commercial	100 nm	Lamellar	0.05 wt. %	COF↓ (5%)	WSD↓ (15%)	–	SEM, EDS	Azmar et al. (2016)
Mo/B oleic diethanolamide derivatives, i.e. YXM and YXB	PAO 6	Fabricated	–	–	1 wt.% (COF) 2.5 wt. % and 0.5 wt. % (WSD) 3wt. % (EP)	COF↓ (YXB>YXM)	WSD↓ (YXM>YXB)	EP↑ (YXM>YXB)	SEM, XPS	Huang et al. (2017)
Lipophilic magnetite nanoparticles (MagNPs)	PAO 8	fabricated	5–10 nm	Nearly spherical	6.7 wt. %	COF↓	WV↓ (68%)	–	White light interferometry, SEM	Zuin et al. (2017)
Pd, Ag	PAO 4	fabricated	2–4 nm 3–6 nm	Nearly spherical	1.0 wt. %	COF↓ (Pd >Ag)	WV↓ (Pd >Ag)	–	SEM, XPS	Kumara et al. (2018)
TiO$_2$	PAO 2	Commercial	100 nm, 165 nm, 0.3 µm, 0.5 µm	Nearly spherical	0.05 wt. %	COF↓	WV↓	–	SEM, EDS	Peña–Parás et al. (2018)
ND, MoS, WS$_2$, ND+ MoS$_2$, ND+ WS$_2$	PAO 4	Commercial	60–90 nm	Nearly spherical	0.5 wt. % (MoS$_2$, WS$_2$) 0.2 wt. % (ND)	COF↓	WV↓	–	SEM, EDS	Raina and Anand (2018; (Continued)

TABLE 11.8 (Continued)
Tribological Performance of Different Grades of PAO–Base Oils with Various Nanoparticles as an Additive

Nanoparticles	Base Oil	Nanoparticle Source	Average Particle Size (APS)	Shape	Optimum Concentration	Coefficient of Friction (COF)	Wear (WSD)/Wear Volume (WV)	Extreme Pressure (EP)	Morphological Analyses and Techniques	References
Pd, Ag, Pd+Ag	PAO 4	Fabricated	2–4 nm 3–6 nm	Nearly spherical	0.05 wt. % (Pd) 0.05 wt. % (Ag)	COF ↓ (30%–40%)	MWV↓ (80%)	—	SEM, EDS, XPS	Kumara, Meyer and Qu (2019)
rGO	PAO 40	Fabricated	60 nm	Lamellar	0.25 wt. %	COF ↓ (24.3%)	WSD↓ (16.7%)	—	3 D profilometer, Raman spectroscopy	Liñeira del Río et al. (2019)
COOH–functionalized MWCNTs	PAO 4, PAO 6, PAO 40, PAO 100	Commercial	20–30 nm	Tubular	0.075 wt. % (PAO 4, PAO 6, PAO 100) 0.025 wt. % (PAO 40)	COF ↓ (20%, 31%, 27.6%, 22.5%)	MWV↓ (57%, 36%, 58%, 61%)	EP ↑ (in case of PAO 40 and PAO 100 only)	SEM, EDS, 2 D profilometer	Kumar and Harsha (2020)
Amine additive (Octa, Ar18D, Ar HTD, MeOc, Ar DM18D, Duo CD, Duo O, Tria T, Etho O/12, Stearyl etho), ZDDP	PAO 4	Commercial	—	—	1 wt.%	COF ↓	WSD↓	—	XPS	Massoud et al. (2020)

(Continued)

TABLE 11.8 (Continued)

Tribological Performance of Different Grades of PAO–Base Oils with Various Nanoparticles as an Additive

Nanoparticles	Base Oil	Nanoparticle Source	Average Particle Size (APS)	Shape	Optimum Concentration	Coefficient of Friction (COF)	Wear (WSD)/Wear Volume (WV)	Extreme Pressure (EP)	Morphological Analyses and Techniques	References
rGO, Fe$_3$O$_4$,rGO/ Fe$_3$O$_4$	PAO 6	Fabricated	5–10 μm 5–10 nm	Lamellar, spherical	0.1 wt. %	COF ↓ (7%)	MWV ↓ (52.27%)	–	SEM, Raman spectroscopy, XPS	Zhang et al. (2020)
Modified MoS$_2$ Modified rGO	PAO 6	Fabricated	0.5–5 μm .3–5 μm	Lamellar	0.3 wt. % (5:5)	COF ↓ (48.3%)	WSD ↓ (31.71%)	–	SEM, XPS	Wang et al. (2020)
Graphene–like covalent organic framework (GCF)	PAO 10	Fabricated	Thickness =3.1–3.3 nm	Lamellar	0.008 wt. %	COF ↓ (53.5%)	MWV ↓ (95.4%)	–	SEM, XPS	Wen et al. (2020)
3D Graphene (Gr), 2D h–BN, Gr+h–BN	PAO 4	Commercial	–	Lamellar	3 wt.%	COF ↓ (Gr+h–BN> Gr, h–BN)	Wear rate ↓ (Gr+h–BN> Gr, h–BN)	–	SEM, EDS, 3D profilometer	Qi et al. (2022)
W, h–BN, W+h–BN	PAO 6	Fabricated	20—70 nm 10–80 nm	Spherical, lamellar	0.01 wt.%, (W) 0.1 wt. %, (h–BN)	COF ↓ (20%) (30%) (50%)	Wear rate ↓ (93.9%) (95.4%) (99.3%)	–	SEM, EDS	Bondarev et al. (2020)

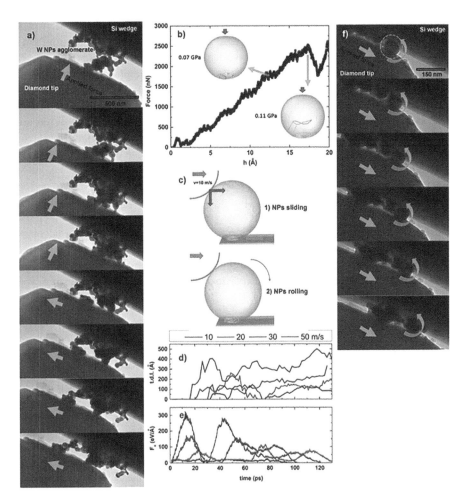

FIGURE 11.6 Images of MD simulation (a, f) displaying the response of W nanoparticles under compressive and shear loads; (b) force-displacement curve for W nanoparticles under compression; (c) illustration of W nanoparticles' deformation and motion; (d) total dislocation length in W nanoparticles with respect to time for several speeds; (e) force-time dependences during movement of W nanoparticles (Reproduced with the permission of Bondarev et al., 2020).

11.6 CONCLUSIONS

The use of synthetic lubricants in the modern age is becoming essential. This would not only conserve scarce petroleum products but also increase the life and performance of equipment, i.e. reduced maintenance and waste, lower emissions and pollution, higher reliability and efficiency. Among the presently used synthetic base oil for lubrication, PAOs maintain a prominent position over their closest challenger such as organic esters and PAGs. As per the report of MarketsandMarkets™ 2019, the market size of synthetic lubricants is considered to be USD 32.2 billion in 2018 and is expected to achieve USD 37.3 billion by 2023 at a compound annual growth rate (CAGR) of 3.0% between 2018 and 2023 and during the forecast duration, PAO is predicted to dominate the market of synthetic lubricants (MarketsandMarkets™, 2019). Multiple techniques may be employed to modify the physical properties of PAOs to meet the requirements of specific end-use applications. In the majority of industrial applications, the availability of high-viscosity PAOs offers a special benefit over conventional mineral oil that is usually restricted to low grade. Also, PAOs are miscible with petroleum base mineral oils. In the last few years, many researchers and scientists have explored various nanoadditives to enhance the tribological properties of PAOs positively, although challenges remain to be solved. The most important challenge is to prepare homogenous nanolubricants for an extended period because of the propensity of most of the nanoparticles to aggregate in PAO base oils and deposit at the bottom of PAOs according to the density of agglomerated cluster. The literature highlighted many lubricating mechanisms that are accountable for enhanced tribological performance, but a theoretical understanding of these mechanisms is still insufficient. To fully understand the lubricating mechanism of nanolubricants, more theoretical and experimental investigations are requisite.

REFERENCES

Abad, M. D. and Sánchez-López, J. C. (2013) 'Tribological properties of surface-modified Pd nanoparticles for electrical contacts', *Wear*, 297(1–2), pp. 943–951. Doi: 10.1016/j.wear.2012.11.009.

Alves, S. M. et al. (2016) 'Nanolubricants developed from tiny CuO nanoparticles', *Tribology International*, 100, pp. 263–271. Doi: 10.1016/j.triboint.2016.01.050.

Azman, S. S. N. et al. (2016) 'Study of tribological properties of lubricating oil blend added with graphene nanoplatelets', *Journal of Materials Research*, 31(13), pp. 1932–1938. Doi: 10.1557/jmr.2016.24.

Bondarev, A. V. et al. (2020) 'Mechanisms of friction and wear reduction by h-BN nanosheet and spherical W nanoparticle additives to base oil: Experimental study and molecular dynamics simulation', *Tribology International*, 151(November), p. 106493. Doi: 10.1016/j.triboint.2020.106493.

Choi, Y. et al. (2009) 'Tribological behavior of copper nanoparticles as additives in oil', *Current Applied Physics*. Elsevier B.V., 9(2), pp. e124–e127. Doi: 10.1016/j.cap.2008.12.050.

Chou, R. et al. (2010) 'Tribological behavior of polyalphaolefin with the addition of nickel nanoparticles', *Tribology International*, 43(12), pp. 2327–2332. doi: 10.1016/j.triboint.2010.08.006.

Chouhan, A. et al. (2020) 'Synergistic lubrication performance by incommensurately stacked ZnO-decorated reduced graphene oxide/MoS2 heterostructure', *Journal of Colloid and Interface Science*. Elsevier Inc., 580(July), pp. 730–739. Doi: 10.1016/j.jcis.2020.07.033.

Demas, N. G. et al. (2012) 'Tribological effects of BN and MOS$_2$ nanoparticles added to polyalphaolefin oil in piston skirt/cylinder liner tests', *Tribology Letters*, 47(1), pp. 91–102. Doi: 10.1115/IJTC2012–61062.

Erdemir, A., and Holmberg, K. (2015) 'Energy consumption due to friction in motored vehicles and low-friction coatings to reduce it.', in Cha, S. C. and Erdemir, A. (eds) *Coating Technology for Vehicle Applications*. Cham: Springer International Publishing, pp. 1–23. Doi: 10.1007/978-3-319-14771-0.

ExxonMobil Chemical (n.d.) Polyalphaolefin synthetic basestocks Group IV basestocks, *Datasheet*.

ExxonMobil Chemical Advanced synthetic base stocks (n.d.). Available at: https://www.exxonmobilchemical.com/en/products/synthetic-base-stocks.

Gao, C. et al. (2013) 'Tribological properties of magnetite nanoparticles with various morphologies as lubricating additives', *Journal of Nanoparticle Research*, 15(3), p. 1502. Doi: 10.1007/s11051-013-1502-z.

Goze, M. C. et al. (2010) *Method of Making Low Viscosity PAO'*. United States Patent.

Gulzar, M. et al. (2016) 'Tribological performance of nanoparticles as lubricating oil additives', *Journal of Nanoparticle Research*. Springer Netherlands, 18(8), p. 223. Doi: 10.1007/s11051-016-3537-4.

Gupta, R. N. and Harsha, A. (2018a) 'Antiwear and extreme pressure performance of castor oil with nano-additives', *Proceedings of the Institution of Mechanical Engineers, Part J: Journal of Engineering Tribology*, 232(9), pp. 1055–1067. Doi: 10.1177/1350650117739159.

Gupta, R. N. and Harsha, A. P. (2018b) 'Tribological study of castor oil with surface-modified CuO nanoparticles in boundary lubrication', *Industrial Lubrication and Tribology*, 70 (4), pp. 700–710. Doi: 10.1108/ILT-02-2017-0030.

Hernández Battez, A. et al. (2007) 'Wear prevention behaviour of nanoparticle suspension under extreme pressure conditions', *Wear*, 263(7–12 SPEC. ISS.), pp. 1568–1574. Doi: 10.1016/j.wear.2007.01.093.

Hernández Battez, A. et al. (2008) 'CuO, ZrO_2 and ZnO nanoparticles as antiwear additive in oil lubricants', *Wear*, 265(3–4), pp. 422–428. Doi: 10.1016/j.wear.2007.11.013.

Holmberg, K., Andersson, P. and Erdemir, A. (2012) 'Global energy consumption due to friction in passenger cars', *Tribology International*. Elsevier, 47(March), pp. 221–234. Doi: 10.1016/j.triboint.2011.11.022.

Holmberg, K. et al. (2014) 'Global energy consumption due to friction in trucks and buses', *Tribology International*. Elsevier, 78(October), pp. 94–114. Doi: 10.1016/j.triboint.2014.05.004.

Huang, J.-M. et al. (2017) 'Tribological properties of 2 novel Mo/B-based lubricant additives in polyalphaolefin', *Lubrication Science*, 29(7), pp. 475–484. Doi: 10.1002/ls.1381.

Jiao, D. et al. (2011) 'The tribology properties of alumina/silica composite nanoparticles as lubricant additives', *Applied Surface Science*. Elsevier B.V., 257(13), pp. 5720–5725. Doi: 10.1016/j.apsusc.2011.01.084.

Joly-Pottuz, L. et al. (2004) 'Ultralow friction and wear behaviour of Ni/Y-based single wall carbon nanotubes (SWNTs)', *Tribology International*, 37(11–12), pp. 1013–1018. Doi: 10.1016/j.triboint.2004.07.019.

Kashyap, A. and Harsha, A. (2016) 'Tribological studies on chemically modified rapeseed oil with CuO and CeO_2 nanoparticles', *Proceedings of the Institution of Mechanical Engineers, Part J: Journal of Engineering Tribology*, 230(12), pp. 1562–1571. Doi: 10.1177/1350650116641328.

Kassim, K. A. M. et al. (2020) 'The wear classification of MoDTC-derived particles on silicon and hydrogenated diamond-like carbon at room temperature', *Tribology International*. Elsevier Ltd, p. 106176. Doi: 10.1016/j.triboint.2020.106176.

Kogovšek, J. and Kalin, M. (2014) 'Various MoS_2-, WS_2- and C-based micro- and nanoparticles in boundary lubrication', *Tribology Letters*, 53(3), pp. 585–597. Doi: 10.1007/s11249-014-0296-1.

Kumar, H. and Harsha, A. P. (2020) 'Investigation on friction, anti-wear, and extreme pressure properties of different grades of polyalphaolefins with functionalized multi-walled carbon nanotubes as an additive', *Journal of Tribology*, 142(8), pp. 1702–1714. doi: 10.1115/1.4046571.

Kumara, C. et al. (2018) 'Palladium Nanoparticle-Enabled Ultrathick Tribofilm with Unique Composition', *ACS Applied Materials and Interfaces*. American Chemical Society, 10 (37), pp. 31804–31812. Doi: 10.1021/acsami.8b11213.

Kumara, C., Meyer, H. M. and Qu, J. (2019) 'Synergistic interactions between silver and palladium nanoparticles in lubrication', *ACS Applied Nano Materials*, 2(8), pp. 5302–5309. Doi: 10.1021/acsanm.9b01248.

Li, W. et al. (2011) 'Friction and wear properties of ZrO$_2$/SiO$_2$ composite nanoparticles', *Journal of Nanoparticle Research*, 13(5), pp. 2129–2137. Doi: 10.1007/s11051-010-9970-x.

Li, Z. et al. (2014) 'Preparation of lanthanum trifluoride nanoparticles surface-capped by tributyl phosphate and evaluation of their tribological properties as lubricant additive in liquid paraffin', *Applied Surface Science*. Elsevier B.V., 292, pp. 971–977. Doi: 10.1016/j.apsusc.2013.12.089.

Liñeira del Río, J. M. et al. (2019) 'Tribological properties of dispersions based on reduced graphene oxide sheets and trimethylolpropane trioleate or PAO 40 oils', *Journal of Molecular Liquids*. Elsevier B.V., 274, pp. 568–576. Doi: 10.1016/j.molliq.2018.10.107.

Mang, T. (2014) *Encyclopedia of Lubricants and Lubrication, Encyclopedia of Lubricants and Lubrication*. Edited by T. Mang. New York: Springer. Doi: 10.1007/978-3-642-22647-2.

MarketsandMarkets™ (2019) *Synthetic Lubricants Market by Type (PAO, PAG, Esters, Group III), Application (Engine Oil, Hydraulic Fluids, Metalworking Fluids, Compressor Oil, Gear Oil, Refrigeration Oil, Transmission Fluids, Turbine Oil), Region - Global Forecast to 2023, MarketsandMarkets™*. Available at: https://www.marketsandmarkets.com/Market-Reports/synthetic-lubricant-market-141429702.html.

Massoud, T. et al. (2020) 'Effect of ZDDP on lubrication mechanisms of linear fatty amines under boundary lubrication conditions', *Tribology International*. Elsevier Ltd, 141, p. 105954. Doi: 10.1016/j.triboint.2019.105954.

Moore, L. D. et al. (2003) 'PAO-based synthetic lubricants in industrial applications', *Tribology & Lubrication Technology*, 59(1), pp. 23–30.

Murphy, W. R., Blain, D. A. and Galiano-Roth, A. S. (2002) 'Synthetics basics benefits of synthetic lubricants in industrial', *Journal of Synthetic Lubrication*, 18(4), pp. 301–325. Doi: 10.1002/jsl.3000180406.

Nunn, N. et al. (2015) 'Tribological properties of polyalphaolefin oil modified with nanocarbon additives', *Diamond and Related Materials*, 54, pp. 97–102. Doi: 10.1016/j.diamond.2014.09.003.

Padgurskas, J. et al. (2013) 'Tribological properties of lubricant additives of Fe, Cu and Co nanoparticles', *Tribology International*, 60, pp. 224–232. Doi: 10.1016/j.triboint.2012.10.024.

Peña-Parás, L. et al. (2015) 'Effect of CuO and Al$_2$O$_3$ nanoparticle additives on the tribological behavior of fully formulated oils', *Wear*, 332–333, pp. 1256–1261. Doi: 10.1016/j.wear.2015.02.038.

Peña-Parás, L. et al. (2018) 'Effects of substrate surface roughness and nano/micro particle additive size on friction and wear in lubricated sliding', *Tribology International*, 119(November), pp. 88–98. Doi: 10.1016/j.triboint.2017.09.009.

Peng, D. X. et al. (2010) 'Dispersion and tribological properties of liquid paraffin with added aluminum nanoparticles', *Industrial Lubrication and Tribology*, 62(6), pp. 341–348. Doi: 10.1108/00368791011076236.

Pownraj, C. and Valan Arasu, A. (2020) 'Effect of dispersing single and hybrid nanoparticles on tribological, thermo-physical, and stability characteristics of lubricants: A review', *Journal of Thermal Analysis and Calorimetry*. Springer International Publishing, (0123456789). Doi: 10.1007/s10973-020-09837-y.

Qi, S. et al. (2020) 'Synergistic lubricating behaviors of 3D graphene and 2D hexagonal boron nitride dispersed in PAO$_4$ for steel/steel contact', *Advanced Materials Interfaces*, 7(8), pp. 1–9. Doi: 10.1002/admi.201901893.

Raina, A. and Anand, A. (2017) 'Tribological investigation of diamond nanoparticles for steel/steel contacts in boundary lubrication regime', *Applied Nanoscience*. Springer Berlin Heidelberg, 7(7), pp. 371–388. Doi: 10.1007/s13204-017-0590-y.

Raina, A. and Anand, A. (2018) 'Effect of nanodiamond on friction and wear behavior of metal dichalcogenides in synthetic oil', *Applied Nanoscience (Switzerland)*. Springer Berlin Heidelberg, 8(4), pp. 581–591. Doi: 10.1007/s13204-018-0695-y.

Rao, A. M., Srivastava, S. P. and Mehta, K. C. (1987) 'Synthetic lubricants in India — an overview', *Journal of Synthetic Lubrication*, 4(2), pp. 137–145. Doi: 10.1002/jsl.3000040204.

Rawat, S. S., Harsha, A. P. and Deepak, A. P. (2019) 'Tribological performance of paraffin grease with silica nanoparticles as an additive', *Applied Nanoscience (Switzerland)*. Springer International Publishing, 9(3), pp. 305–315. Doi: 10.1007/s13204-018-0911-9.

Rawat, S. S. et al. (2019a) 'Effect of CuO and ZnO nano-additives on the tribological performance of paraffin oil–based lithium grease', *Tribology Transactions*, Taylor & Francis, pp. 1–11. Doi: 10.1080/10402004.2019.1664684.

Rawat, S. S. et al. (2019b) 'Pristine and alkylated MoS$_2$ nanosheets for enhancement of tribological performance of paraffin grease under boundary lubrication regime', *Journal of Tribology*, 141(7), pp. 072102–12. Doi: 10.1115/1.4043606.

Rawat, S. S. et al. (2020) 'Effect of graphene-based nanoadditives on the tribological and rheological performance of paraffin grease', *Journal of Materials Engineering and Performance*. Springer US, 63(1), pp. 90–100. Doi: 10.1007/s11665-020-04789-8.

Rudnick, L. R. (2005) *Synthetics, Mineral Oils, and Bio-Based Lubricants: Chemistry and Technology*. Edited by L. R. Rudnick. New York: CRC Press, Taylor and Francis Group. Doi: 10.1201/9781420027181.

Saini, V. et al. (2020) 'Role of base oils in developing extreme pressure lubricants by exploring nano-PTFE particles', *Tribology International*. Elsevier Ltd, 143(November), p. 106071. Doi: 10.1016/j.triboint.2019.106071.

Shubkin, R. L., Corporation, M. E. K. E. and Rouge, B. (1991) 'Tailor-making Polyalphaolefins', *Journal of Synthetic Lubrication*, 8(2), pp. 115–134.

Song, X. et al. (2012) 'Synthesis of monodispersed ZnAl$_2$O$_4$ nanoparticles and their tribology properties as lubricant additives', *Materials Research Bulletin*. Elsevier Ltd, 47(12), pp. 4305–4310. Doi: 10.1016/j.materresbull.2012.09.013.

Spikes, H. (2015) 'Friction modifier additives', *Tribology Letters*. Springer US, 60(1), pp. 1–26. Doi: 10.1007/s11249-015-0589-z.

Stachowiak, G. W. and Batchelor, A. W. (2000) *Engineering Tribology*. Butterworth-Heinemann.

Sui, T. et al. (2016) 'Bifunctional hairy silica nanoparticles as high-performance additives for lubricant', *Scientific Reports*. Nature Publishing Group, 6, p. 22696. Doi: 10.1038/srep22696.

Tang, G. et al. (2020) '2D black phosphorus dotted with silver nanoparticles: An excellent lubricant additive for tribological applications', *Chemical Engineering Journal*. Elsevier, 392(November), p. 123631. Doi: 10.1016/j.cej.2019.123631.

Trout, J. and Fitch, J. (2020) *Synthetic Oil: What Consumers Need to Know*. Noria Corporation, pp. 1–10. Available at: https://www.machinerylubrication.com/Articles/Print/31800.

Upadhyay, R. K. and Kumar, A. (2019) 'Boundary lubrication properties and contact mechanism of carbon/MoS$_2$ based nanolubricants under steel/steel contact', *Colloid and Interface Science Communications*. Elsevier, 31(May), p. 100186. Doi: 10.1016/j.colcom.2019.100186.

Viesca, J. L. et al. (2011) 'Antiwear properties of carbon-coated copper nanoparticles used as an additive to a polyalphaolefin', *Tribology International*, 44(7–8), pp. 829–833. Doi: 10.1016/j.triboint.2011.02.006.

Wang, S. et al. (2020) 'Dispersion stability and tribological properties of additives introduced by ultrasonic and microwave assisted ball milling in oil', *RSC Advances*. Royal Society of Chemistry, 10(42), pp. 25177–25185. Doi: 10.1039/d0ra03414b.

Wen, P. et al. (2020) 'Two-dimension layered nanomaterial as lubricant additives: Covalent organic frameworks beyond oxide graphene and reduced oxide graphene', *Tribology International*. Elsevier Ltd, 143, p. 106051. Doi: 10.1016/j.triboint.2019.106051.

Wu, M. M., Ho, S. C. and Forbus, T. R. (2006) 'Synthetic lubricant base stock processes and products', in *In Practical advances in petroleum processing*. New York: Springer, pp. 553–577. Doi: 10.1007/978-0-387-25789-1.

Yang, G. et al. (2013) 'Preparation of triazine derivatives and evaluation of their tribological properties as lubricant additives in poly-alpha olefin', *Tribology International*. Elsevier, 62, pp. 163–170. Doi: 10.1016/j.triboint.2013.02.024.

Zhang, M., Wang, X. and Liu, W. (2013) 'Tribological behavior of LaF$_3$ nanoparticles as additives in poly-alpha-olefin', *Industrial Lubrication and Tribology*, 65(4), pp. 226–235. Doi: 10.1108/00368791311331202.

Zhang, M. et al. (2009) 'Performance and anti-wear mechanism of CaCO$_3$ nanoparticles as a green additive in poly-alpha-olefin', *Tribology International*, 42(7), pp. 1029–1039. Doi: 10.1016/j.triboint.2009.02.012.

Zhang, Q. et al. (2020) 'Preparation, characterization and tribological properties of poly-alphaolefin with magnetic reduced graphene oxide/Fe$_3$O$_4$', *Tribology International*, 141(September 2019). Doi: 10.1016/j.triboint.2019.105952.

Zuin, A. et al. (2017) 'Lipophilic magnetite nanoparticles coated with stearic acid: A potential agent for friction and wear reduction', *Tribology International*. Elsevier Ltd, 112(August), pp. 10–19. Doi: 10.1016/j.triboint.2017.03.028.

12 Role of Surfactants and Their Concentrations on the Tribological Characteristics of MWCNT-in-Oil Lubricants for Hybrid AMMC–Steel Sliding Contact

Hiralal Bhowmick and Harpreet Singh
Thapar Institute of Engineering and Technology

CONTENTS

12.1 INTRODUCTION

With an increased demand of globalization, energy efficiency and miniaturization
of machine elements, the usage of solid lubricants as the potential particle additives in enhancing the dry and wet lubrication characteristics of the composites or
coated/cladded surfaces has gained a huge attention from the tribologists more than
before. Many experimental shreds of evidence using different solid lubricants such
as graphite, metal dichalcogenides (MoS_2/WS_2) and h-BN have proved their ability
to enhance the tribological behaviour of loaded components [1–10]. The purpose of
solid additive-based liquid lubrication has been more manifested owing to minimize
the friction-generating points and subsequent wearing out in the various tribological components of automobiles. These aspects help in facilitating the life of mating
components [11,12]. Recently, carbon nanotubes (CNTs) have gained unusual attention owing to their superior properties attributed by their shape, high aspect ratio
and flexibility [13]. These salient features help in reducing friction and wear of composites when used as reinforcements [14–19], shielding (or protective) films [20–23]
and solid lubricants [21–24]. However, limited numbers of studies are available in
the literature depicting the use of CNTs as potential oil additive, most of which are
reported for the steel–steel tribocontacts or aqueous lubrication [25–32].

Particulate-reinforced metal matrix composites (PRMMCs) have established their
visibility as a fine alternative to the conventional metal alloys or monolithic materials, due to their high stiffness and strength with a bonus feature of low-density
applications in various sectors like industries, automobiles and aerospace [33–35].
Aluminium metal matrix composites (AMMCs) are found to be the front runner for
various lightweight applications, including the automobile industries. These reinforced composites can be formed by stir casting method that is a less expensive and
agreeable program for mass production [33,36]. The specific properties of strength,

modulus, and low friction and wear make them superior candidate for many engineering conditions where sliding contact is anticipated. Interestingly, dry sliding friction–wear behaviour of PRMMCs using different types of reinforcements has been reported by many authors [37–39]. The study of AMMC–steel or AMMC–cast iron tribopair is an important aspect for the improved engine tribology. The potential application of composite piston using Al-SiC had been explored by a few manufacturers including Duralcan and Martin Marietta [36]. Taguchi's design of experiments (DOEs) is a powerful technique to pursue the experiments and perform the data analysis to study the effect of processing parameters on particular responses [40,41]. This approach, along with the regression analysis and contour plotting can be used to examine the wear performance and mapping of the wear transition mechanism in the case of MMCs [42].

On the contrary, wet tribological behaviour of PRMMC is a less discovered area. Despite its huge potential, very limited work has been reported on the wet tribological behaviour of AMMC. Some of the studies related to single-particle-reinforced AMMC–steel contacts under oil with micro-/nanosized particle additive or without any additive have been reported by a few authors [43,44].

The underlying mechanism of friction and wear in composite contacts under the lubricated sliding is very complicated. The effectiveness of lubricant formulation using the solid particle additives is influenced by the particle's dispersion capability and stability in the suspension [17,26,28]. The stability, dispersibility of nanoparticles, as well as their effectiveness, can be enhanced by functionalizing the nanoparticles or using surfactant and dispersants. However, such studies involving the functionalized nanodispersions for the improved tribological behaviour of composite surfaces are available in the literature for the contacts other than the MMCs [17,25,27,31,32]. Hwang et al. [17] in their study used alkyl-aryl sulfonate to functionalize the multiwalled carbon nanotubes (MWCNTs). Although they reported a low value of friction coefficient of 0.03, however, they also observed that the agglomerated CNT-based particles do not effectively play the role of ball bearings in comparison with other carbon-based particles, such as graphite and carbon black.

Chebattina et al. [31] in their work used Span 80 to coat the surface of the MWCNTs using a mechano-chemical method. From their study, it was revealed that the stability, as well as the triboperformance of lubricant, can be enhanced significantly with the dispersion of shortened MWCNTs. Muzakkir et al. [32] used Triton X-100 surfactants to functionalize the MWCNT particles. They reported that the lubricant containing surfactant-functionalized (SF) MWCNT is able to minimize the coefficient of friction (COF) as compared to the base oil without MWCNT, and oil sample containing MWCNT but with no surfactant.

In addition to the above facts, the frictional response under boundary or mixed lubricated regimes may be greatly influenced by oxide and other forms of layers, which can result in stern changes in the surface wear phenomena. Also, the interaction of particle-based liquid lubricants with hybrid composite materials enhances the level of complexity of interpretation.

In the pursuit of the deeper insights into some of the above-mentioned challenges, the authors of this chapter had carried out a systematic investigation [45–48], and hence, the important results and findings of the research carried out by the authors will be referred to in this chapter while discussing the various aspects of the particulate lubrication of the composite contact. In this chapter, the authors present a systemic attempt made to investigate the lubrication mechanism and tribological characteristics of the functionalized MWCNT-oil dispersion under the hybrid AMMC–steel sliding contact. The role of surfactant in enhancing the lubricity properties is illustrated through the experimental findings. The lubricated tribology of the SF nanodispersions is performed as per an L9-orthogonal array using the binary composite as the pin and EN31 as the disc. Taguchi-ANOVA was also conducted, further details of which can be found in the later sections of this chapter. The various influencing factors, as well as the possible interactions between the oil-particles-surfactant, are explored in detail to understand the underlying lubrication mechanism. To identify the regime of lubrication, surface roughness and film thicknesses are also estimated. Finally, the wear tracks are examined to understand the underlying friction and wear mechanisms that prevailed for the selected lubricated contact.

12.2 MATERIALS AND METHODOLOGY

12.2.1 SELECTION OF RAW MATERIALS, ADDITIVES AND THEIR PROCESSING

12.2.1.1 Selection of Matrix, Reinforcements and Fabrication Technique for Composite

In the present study, commercially available Al6061 alloy was selected as the base matrix, whereas SiC (average size: 30–50 μm, density 3.2 g/cm³) and graphite (average size: 40 μm, density 2.2 g/cm³) have been used as the reinforcements of the hybrid composites (Al6061/10%SiC/3%Gr), fabricated by a low-cost stir casting method. The process of selection materials (i.e. the matrix and reinforcements and their compositions) and the choice of a fabrication technique for the alloy composite fabrication were based on the cost, ease of fabrication and the desired functional properties of the fabricated composites, such as adequate hardness and antiseizure properties [45–48]. Earlier reported literature supports the improvement in tribological properties by the addition of these reinforcements and this fabrication method [49–52].

12.2.1.2 Selection of Oil Additive and Surfactant

It is worth mentioning here that the selection of oil additive is significantly influenced by the nature of oil-particle-tribomaterial surface interactions. Grade-I mineral oil (SN500) is selected as the base fluid. Looking at the potential benefits that can be achieved by the addition of the solid lubricant in developing the commercial lubricant, the present work will focus on MWCNTs as the potential oil additive. For this purpose, MWCNTs of 10–20 nm diameter (length of 3–8 μm) were purchased from Nanoshell, Inc., India. The rationale of using the optimum concentration of the selected particle additive was developed based on the pilot tribology studies involving lubricants with various concentrations of MWCNT in oil [45–48]. The concentration of MWCNT (0.1 vol. %) is finalized following a series of preliminary investigation on the rheological and tribological characteristics. In the absence of any surfactant or dispersant, it was observed that a high fraction of the MWCNTs turned into clustered lumps. The surface modification/functionalization of the particles has been done using a nonionic sorbitan monooleate surfactant, Span 80. Following a CMC study of surfactant, a concentration of approximately 0.04% Span 80 is found to be the most preferable to use and chosen to utilize the potential benefit of MWCNT-laden oil, for the detailed tribological investigation. The performance of the lubricants with the addition of various concentrations of surfactant and the determination of the optimum concentration of the surfactant is detailed in the later section. The decent choice of the surfactant was based on its low HLB, ease of facilitating the adequate solubility characteristics with mineral oil [53–55]. It is sensible to mention that there are a few nonionic surfactants, such as Span 85, that have lower HLB than Span 80. However, the head group of Span 85 has only one -OH group, which, in turn, causes very weak adsorption affinity to the interacting surface in contrast to Span 80. Furthermore, the length of the tail group of Span 85 is much larger than Span 80 that induces substantial interference to the adsorption of the other close-bound particles [56]. It is also observed that the surfactants with shorter branches can penetrate more easily into the gaps of the nanoparticle clusters [57].

12.2.2   FRICTION–WEAR TESTS

For tribotesting, a pin-on-disc tribometer (DUCOM Pvt Ltd) is used. The dimensions of pin-disc specimens, initial roughness, sliding condition, tribological parameters such as load, speed, sliding distance, ambient conditions, the repeatability and reproducibility criteria were outlined in the previous works [46–48]. Before each friction–wear test, pin and disc were cleaned with acetone. For the lubricated tribology runs, the lubricant was supplied at the contact zone at a flow rate of 1.5 ml/100 s. All the tribotesting were conducted at room temperature ~ 30°C. Each of the friction–wear tests was repeated at least thrice to ensure the reproducibility of the results and the mean value is reported. The maximum standard deviation for each set of readings is found to be less than 10%.

12.2.3   CHARACTERIZATIONS

12.2.3.1   Characterization of Lubricant (Oil-Particle-Surfactant)

The process of ultrasonication is used to prepare the stable dispersions of nanoparticles. The detailed dispersion stability characteristics of the formulated concentrations were examined by a UV-Vis spectroscopy and sedimentation process, previously reported in the recent works [45,48]. For the stability studies, the authors carried out the sedimentation tests to ensure that any agglomerates have not appeared in the prepared samples. It was observed from the dispersion of MWCNTs and SF MWCNT in oil that in the absence of surfactant the MWCNT particles are agglomerated in the suspension and settled to the bottom after a span of 4 weeks. However, functionalization of MWCNTs with the use of surfactants enhanced the dispersion stability, as reported by Singh et al. [45]. In the UV results, it was observed that in the case of MWCNT-based lubricant, there is an insignificant change in the absorbency over the span of more than 3 weeks and the settling time of dispersed particles extended much beyond 1 month. The grafting of surfactant MWCNT was confirmed through FTIR analysis (FTIR peaks near 1744.84, 1167.97 and 3433.26 cm^{-1}) [48].

12.2.3.2 Microstructural, Morphological and Chemical Characterization of Wear Tracks

Microstructural analysis was carried out based on the images obtained using a scanning electron microscope (SEM) (Make: JSM-6510LV, JEOL Ltd., Tokyo, Japan) at an acceleration voltage of 15 kV or more, equipped with an EDS detector. Roughness measurements were done with SJ-400 apparatus for better insight into topographic changes. The chemical compositions of samples and wear tracks were analysed using EDS spectra.

12.2.4 STATISTICAL ANALYSIS: TAGUCHI-ANOVA

Tribological experiments for the surfactant-assisted MWCNT-lubricated trails were carried out following a DOE based on Taguchi's L-9 orthogonal array. The experimental results for various operating conditions such as normal load, sliding speed and sliding distance were examined. The prime objective of the experimental design was to identify the key parameters and arrangement of these parameters based on their influence on the wearing out and lubrication phenomenon, as well as to gain the insight into the optimum formulation of additive-based oil for the application of sliding composites with an improved tribology. The statistical analysis involved the construction of SN-ratio plots, ANOVA and regression analysis.

12.3 RESULTS AND DISCUSSION

12.3.1 TRIBOLOGICAL CHARACTERISTICS, RHEOLOGICAL AND ELECTRICAL PROPERTIES

In the present study, different surfactant concentrations ranging from 0.025 to 0.08 vol. % were used in 0.1% MWCNT-oil dispersions and subsequently friction and wear responses were compared with that of MWCNT-oil dispersion and pure oil lubrication in the absence of any additives. It is observed that the time-dependent response of COF follows the same trend for the different surfactant concentrations (Figure 12.1). However, it is clear from these plots that at a certain surfactant concentration (0.04%) COF reduced to a minimal, which is coincidentally the critical micelle concentration (CMC) of the surfactant. This significant reduction as compared to fresh oil and MWCNT-dispersed oil without any surfactant depicts the importance and usage of surfactant-modified MWCNT in oils for the purpose of low-friction applications. The stability of the friction response at this specified surfactant concentration is also found to be remarkably enhanced. Thus, the formulation of MWCNT-oil with a surfactant concentration of nearly 0.04 vol. % (measured by electrical conductivity (Figure 12.2)) is proven to be quite helpful for the low frictional response.

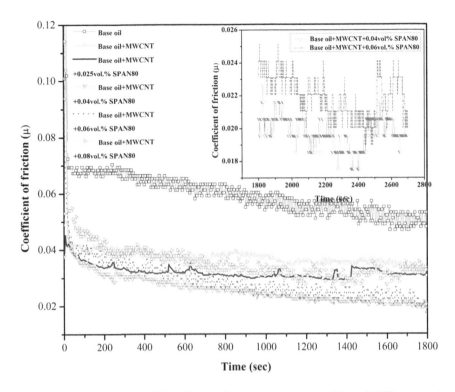

FIGURE 12.1 Variation of COF with time (inset: steady-state condition of COF response.)

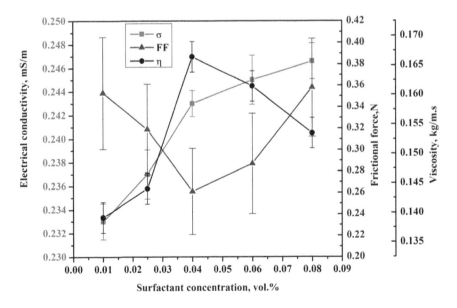

FIGURE 12.2 Typical variation of electrical conductivity, frictional force and viscosity with surfactant concentration (σ = electrical conductivity, FF = friction force, η = viscosity).

The increase in viscosity, as shown in Figure 12.2, is attributed to the exposed surface areas of solid additives, the filament-like morphology of MWCNTs and their aggregation and the large asymmetric geometric distributions of these particles in the fresh oil that lead to the confined movement of the oil layers [11,12]. Moreover, a highly viscous molecule of Span 80 (approximately viscosity of 2,000 cP) also affects the viscosity of dispersions [58]. The surfactant molecules encapsulate MWCNT particles by surface interaction and their formed aggregation influences the movement of base fluid layers. Besides, the addition of surfactant, by steric effect helps in the uniform dispersion of MWCNT, leading to a stable suspension [54,59].

Thus, the role of surfactant addition in particle dispersions is not only to tailor the tribocharacteristics but also to modify the rheology via redesigning the shear forces and also to provide the possibility of formation of micellar structures. These micellar structures might be helpful in reducing the asperity contacts under the shear loading, which, in turn, promotes the lubricity of the oil. A noticeable high value of measured viscosity observed near the CMC as shown in Figure 12.2 supports this fact.

Although this is one of the possible causes for the improved antifriction and antiwear characteristics, a further investigation on the relevant aspects of the lubrication regime and effect of the sliding parameters as discussed in the later section will definitely provides the deeper insight into the underlying lubrication mechanism under the given operating conditions.

Following the realization of the low friction capability of the SF MWCNT in oil, further investigation is made with this formulated oil to evaluate the antiwear performance under similar operating conditions. For this purpose, volumetric wear losses were calculated for different lubricated sliding contacts. Thereafter, corresponding wear coefficients or specific wear rates were calculated (Table 12.1), based on Archard's wear model (Eq. 12.1).

$$V = K_W \cdot W \cdot L \tag{12.1}$$

K_w, W, V and L are termed as dimensional wear coefficient, load, wear volume and sliding distance, respectively.

It is clear from Table 12.1 that the surfactant-modified MWCNT in oils have a significant influence on the wear performance of composite–steel contacts. The significant

TABLE 12.1

Wear Rates for Composite Pins under Different Lubricated Conditions (Load = 9.81, Speed = 0.5 m/s)

Lubricated Condition	Wear Coefficient, K_w (mm³/Nm × 10⁻⁷)
SN500	4.077
SN500+MWCNT	0.815
SN500+MWCNT+ 0.025 vol. % Span 80	0.764
SN500+MWCNT+ 0.04 vol. % Span 80	0.693
SN500+MWCNT+ 0.06vol. % Span 80	0.744
SN500+MWCNT+ 0.08vol. % Span 80	0.795

enhancement of friction and wear characteristics for such a contact in the presence of surfactant and particle additive may be attributed to a number of interfacial phenomena. This definitely requires a multifaceted investigation to get insight of the underlying lubrication mechanism. Such an attempt is also made in the subsequent sections.

12.3.2 PROPHECY OF LUBRICATION REGIME

For the purpose of the detailed investigation on the underlying lubrication mechanism, the relationship between the frictional forces and customized Stribeck number (Eqs. 12.2 and 12.3) is developed and plotted for different wet lubricated conditions such as base oil and base oil with SF MWCNTs.

The Stribeck curve is a general relationship of COF *vs* Stribeck number.

$$\frac{\text{Frictional force}}{\text{Applied force}} \quad \text{versus} \quad \frac{Z.V}{\text{Applied force}\Big/\text{Slidind distance}} \qquad (12.2)$$

Multiply the relation (12.2) by the 'applied force' on both sides, we get

$$\text{Frictional force versus } Z.V.SD \, (\text{customized relationship}) \qquad (12.3)$$

where Z, V and SD denote viscosity, sliding velocity and sliding distance, respectively.

This is a pathway for the prediction of lubrication regime under the generalized operating conditions. Accordingly, the relationship is developed between sliding velocities (here, 0.1–1.25 m/s) and the applied loading (here, normal load = 9.81 N) and is shown in Figure 12.3.

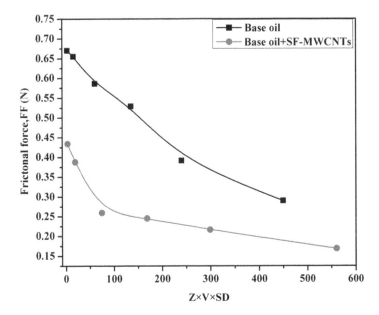

FIGURE 12.3 Variation of friction force for (a) base oil and (b) base oil with SF MWCNTs.

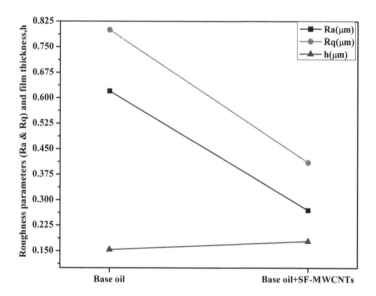

FIGURE 12.4 Variation of roughness and film thickness, for two different wet sliding conditions: (a) base oil and (b) base oil with SF MWCNTs.

The procedure for the estimation of empirical film thickness is discussed elsewhere [48]. It was observed that The induced film thickness is influenced by the surface morphology, fluid rheology and operating conditions. In the case of thin-film lubrication, such as boundary or mixed lubrication, it is a well-known fact that the regime is influenced by the coupling factors of boundary lubrication (asperity contact effect and material properties) and oil film properties [60].

Following a detailed investigation, the decrease in average roughness as shown in Figure 12.4 is attributed to the entrapment of MWCNT leading to the formation of mixed layers. It was observed that the presence and penetration of nano-sized MWCNT found to be certainly fruitful in smoothening the softer surface by shielding the asperity contacts. Subsequently, it reduces the friction between the surfaces that is evident from the frictional force plots (Figure 12.1). The increase in film thickness and reduction of roughness parameter and their ranges in the case of surfactant-modified MWCNT lubrication confirm the prevailing of mixed lubrication regime [45,48,61].

Moreover, it is observed that with the addition of the SF MWCNT in the base oil, the lubrication regime gradually shifts from the unified boundary-mixed to the mixed lubrication regime. Hence, there is a reduction in the friction, as well as the wear under the various operating conditions. Bhowmick et al. [45,48] performed a detailed analysis of this regime shifting for their work at an operating speed of 0.5 m/s. Their study reveals the fact that a unified boundary-mixed regime of lubrication exists under the SF MWCNT-in-oil-lubricated contact. In the later section of this chapter, a mapping of lubrication regime will also be presented based on the statistical analysis which will further validate the author's claim of this operating lubrication regime, at this specified operating and lubrication condition.

12.3.3 ROLE OF THE OPERATING PARAMETERS ON THE TRIBOPERFORMANCE OF SF MWCNT IN OIL UNDER THE AMMC–STEEL CONTACT: A STATISTICAL ANALYSIS USING THE TAGUCHI-ANOVA AND MULTIPLE LINEAR REGRESSION (MLR) MODELLING APPROACH

12.3.3.1 Trend Analysis of Dominating Factors and Estimation of the Optimal Operating Condition for the Response Variables (COF and WR)

For the purpose of statistical analysis, a set of tribological experiments were conducted using the selected tribopair (h-AMMC–EN31) in the presence of particle additive (0.1 vol. % MWCNT) and surfactant (0.04 vol. % Span 80) in mineral oil. These results will provide a better insight into the utility, efficacy and best suited operating regime of the surfactant-based oil dispersions. The brief overview and description of these experimental results are discussed in the subsequent paragraphs.

The experimental results were analysed for various set of operating conditions, viz. normal load, sliding speed and sliding distances. The experiments were carried out for the surfactant-assisted MWCNT-lubricated trails according to an L-9 orthogonal array. The COF and WR data were analysed using Minitab 17. Taguchi proposed the analysis of signal-to-noise (SN) ratio of response (for instance, WR or COF) that entails plotting of main effects and recognizing the noteworthy parameters visually [62]. The experimental results for WR and COF of SF MWCNT oil-dispersed lubricated contacts are listed in Table 12.2.

The effect of the influencing parameters on the friction–wear performance were analysed using SN-ratio response (smaller is better). The ranking of the operating parameters using SN ratio indicates the extent of their influence on the dependent variables (Table 12.3). From Table 12.3, it is evident that the normal load is the most dominating factor (followed by sliding velocity and sliding distance) on the WR and COF.

The influence of the operating parameters on the response variable is illustrated graphically in Figures 12.5 and 12.6. The investigation of these results using SN ratio

---

TABLE 12.2
Orthogonal Array and COF as Well as Wear Rate Results of AMMC after SF MWCNT-Oil-Lubricated Sliding

Load (N)	Sliding Velocity (m/s)	Sliding Distance (m)	WR (mm^3/m) $\times 10^{-4}$	SN Ratio (db)	COF (μ)	SN Ratio (db)
9.81	0.1	180	0.00827	41.65	0.0442	27.09
9.81	0.5	900	0.00680	43.35	0.0264	31.57
9.81	1.25	2250	0.00583	44.68	0.0172	35.31
29.43	0.1	900	0.00921	40.71	0.0544	25.35
29.43	0.5	2250	0.00783	42.12	0.0367	28.70
29.43	1.25	180	0.00776	42.20	0.0359	28.91
49.05	0.1	2250	0.00986	40.12	0.0639	23.89
49.05	0.5	180	0.00962	40.33	0.0578	24.76
49.05	1.25	900	0.00849	41.42	0.0393	28.12

TABLE 12.3

Response Table for SN Ratios of WR and COF

	Wear Rate (WR)				Coefficient of Friction (COF)		
Level	Load (L)	Speed (S)	Distance (D)	Level	Load (L)	Speed (S)	Distance (D)
1	43.23	40.83	41.39	1	31.32	25.44	26.92
2	41.67	41.93	41.83	2	27.65	28.34	28.34
3	40.63	42.77	42.31	3	25.59	30.78	29.30
Delta	2.60	1.94	0.91	Delta	5.73	5.33	2.38
Rank	1	2	3	Rank	1	2	3

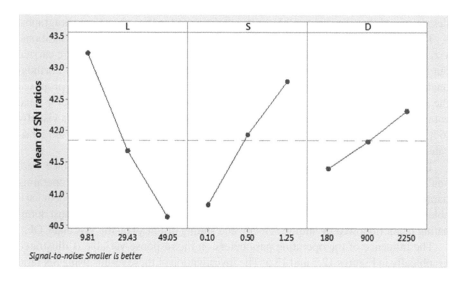

FIGURE 12.5   Main effect plots for SN ratios (wear rate).

provides the insight into the optimal conditions resulting in minimum WR as well as COF under the selected tribopair and lubricating condition.

From Figures 12.5 and 12.6, the optimal condition for resulting WR and COF can be identified as L_1, S_3 and D_3. Hence, this optimum condition of operating factors governs for better wear resistance and antifriction property of hybrid AMMC sliding under functionalized MWCNT-dispersed contacts. This implies that formulated oils can be applied in a better way for low-load and high-speed applications.

12.3.3.2   Analysis of Variance (ANOVA) for the SF MWCNT-in-Oil Tribological Tests

12.3.3.2.1   Significance and the Contribution of the Operating Parameters

The percentage-wise contribution of each sliding parameter and their level of significance based on the probability measures are discussed in this section. This

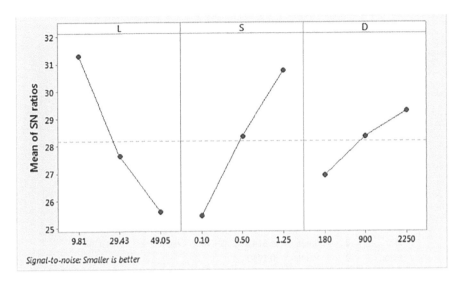

FIGURE 12.6 Main effect plots for SN ratios (COF).

analysis is crucial for the adequacy and preparation of the predictive modelling of WR and COF for the SF MWCNT-based oil dispersions. In order to examine the significant influence of operating variables on the friction–wear performance measures, ANOVA approach is applied to the experimental results so that a decision can be made regarding which independent variable dominates over the other variable.

The ANOVA findings for SN ratios of WR and COF for three independent predictors assorted at three different levels are listed in Table 12.4. A significance level of $\alpha = 0.05$ (using a confidence level of 95%) is taken for carrying out the statistical analysis. This implies that P-value <0.05 were regarded as a statistically significant effect of parameter on the performance of the system. It is clear from Table 12.4 that the normal load has a significant effect on the response variable (WR or COF), as compared to the sliding distance and speed, for the selected composition of the lubricant and the tribopair.

12.3.3.3 Development of a Relationship between the Response Variable and the Operating Parameters: Multiple Linear Regression (MLR) Modelling

Multiple linear regression (MLR) analysis is used to develop a model that gives a relationship between multiple independent predictors and a response variable. This is done by fitting a linear equation to the experimental dataset [63,64].

Provided L, D and S are nonzero values, the regression model formed for WR is given below (Eq. 12.4):

$$WR = 0.007984 + 0.000057\,L - 0.001086\,S - 0.000001\,D$$
$$- 0.000018\,L*S + 0.000000\,L*D + 0.000000\,S*D \qquad (12.4)$$

TABLE 12.4

ANOVA for SN Ratios of WR and COF

			Wear Rate (WR)			
Source	DF	Adj SS	Adj MS	F-value	P-value	Contribution (%)
L	2	10.2665	5.13327	180.17	0.006	59.491
S	2	5.6895	2.84476	99.85	0.010	32.969
D	2	1.2440	0.62199	21.83	0.044	7.208
Error	2	0.0570	0.02849			0.330
Total	8	17.2570				100
			Coefficient of Friction (COF)			
L	2	50.627	25.3137	127.71	0.008	49.4245
S	2	42.780	21.3902	107.91	0.009	41.7638
D	2	8.628	4.3142	21.77	0.044	8.4230
Error	2	0.396	0.1982			0.3865
Total	8	102.433				100

The regression model formed for COF is given as follows (Eq. 12.5):

$$COF = 0.04054 + 0.000717\,L - 0.00613\,S - 0.000020\,D$$
$$- 0.000535\,L*S + 0.000000\,L*D + 0.000007\,S*D \qquad (12.5)$$

The adequacy of the model is verified by using the normal probability distribution of the residuals for COF (Figure 12.7).

The direct influence of the applied load on the response variables is discussed in this section. When the load increases and induces stress beyond the rupture strength of the load-bearing asperities, SiCs are fragmented out of the matrix. These fractured SiC materials may accelerate the transfer of materials from the subsurface during sliding action [63,64]. Due to the more and more asperity interlocking followed by the fragmentation of the hard and soft reinforcements thereof and increase in the metallic contacts, the frictional heating of the contacting surfaces goes up. However, the presence of graphite inside the alloy matrix [65–68] and MWCNT from the lubricant helps in imparting the self-lubricity of the contact. These solid lubricant additives flush onto the sliding pin surface to form a friction-controlling layer during sliding.

The inverse relationship of speed and distance with friction and wear is attributed to the formation of stable oxide tribofilm enriched with tribofavourable product(s) [65,66], that is validated by the presence of the moderate amount of oxygen in the EDS analysis and is subsequently discussed in the later section. The decreasing trend of COF with sliding distance is attributed to the formation of a discrete antifriction layer under the contact in the case of lubricated sliding surfaces. However, this factor is not significant and hence, the slope of SN-ratio plot is quite low.

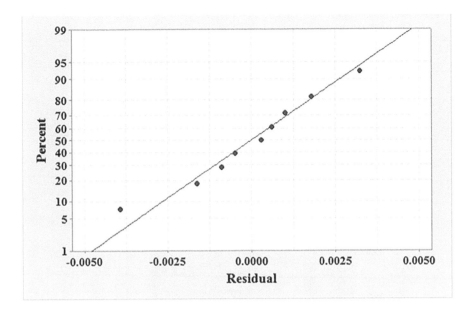

FIGURE 12.7   Normplot of residuals for COF.

12.3.3.4   Mapping of Lubrication Regime

This section highlights the key role of the surfactant in the formulation of lubricant and demonstrates the prediction of the safe-working or best suited operating regime as per the tribocontact and the sliding parameters. Contour plots of COF (Figure 12.8) under different lubrication regimes are created using Minitab 17 software based on three independent factors as predictor variables; load (L), speed (S) and sliding distance (D), and COF as the response variable. Distance-power interpolation method has been used for this purpose. In these maps, contour lines of constant COF, under different operating conditions are plotted. The boundaries for COF transition are identified using the relevant parameters such as COFs for particular load–speed condition, film thickness, roughness, etc., along with the information extracted from the worn-out surface analysis. The rationale of developing the regime mapping is in agreement with the reported literature [69–72]. Based on the aforementioned facts and the observed results of COF, lubrication regimes are classified as follows:

- <0.020 (Transition between mixed and EHD regimes)
- 0.020–0.035(Mixed regime)
- 0.035–0.055(Unified boundary-mixed regime)
- >0.055 (Boundary regime).

12.3.4   A Final Look into the Underlying Friction–Wear–Lubrication Mechanism Prevailed under the SF MWCNT-in-Oil Lubrication

As evident from the contour mapping, it is sensible to mention that the basic role of surfactant is to confront the prevailing of boundary regime characteristics.

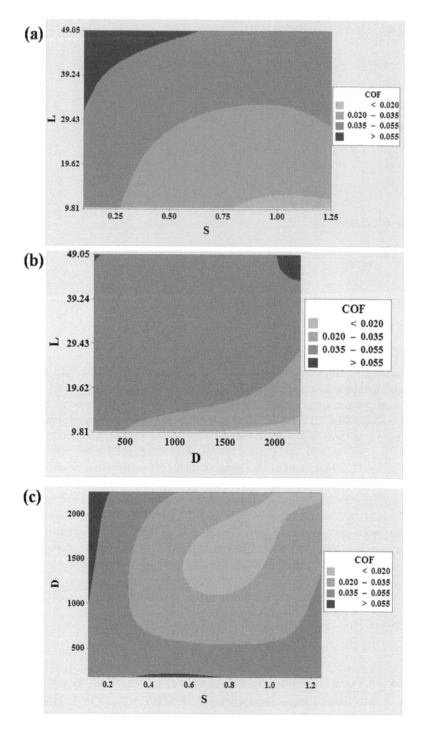

FIGURE 12.8 Contour plot of COF *versus* (a) L, S; (b) L, D; and (c) D, S.

Figure 12.8 clearly shows that a large fraction of area is occupied by the mixed and unified boundary-mixed lubrication regime in a varying range of load and speed, whereas a small faction is concealed for the boundary lubrication regime (represented by dark green colour). It is also observed that the lubrication regime mapping and MLR modelling are in-line with each other. This also validates the author's claim of the predicted lubrication regime based on the film thickness analysis.

Based on the present study, along with the findings reported earlier [45–48] it has been identified that a few interfacial phenomena play a predominant role in enhancing the underlying lubrication mechanism as compared to those of base oil lubrication. Apart from the unified boundary-mixed regime characteristics, a few other associated factors such as improved rheology and improved MWCNT-oil-surfactant interaction favouring the formation and shear sliding of the micellar structures for SF MWCNT-oil dispersion are believed to be responsible for the enhanced tribological characteristics of the prepared lubricant under the selected tribopair. It is important to mention here that due to its bilayered structure, the formed micelles involve less amount of energy to break and shear under the traction forces and releases oil under the loaded contact, promoting the lubricity in the process [48,73,74]. All these factors pose a favourable effect on the friction and wear performance of sliding pairs. The detailed characterization of formed tribolayer on the wear track may enlighten some other aspects, too. For this purpose, a detailed analysis of the worn-out surfaces is provided in the following section.

Figure 12.9 shows the SEM micrographs and corresponding EDS analysis of wear tracks generated at two test conditions (29.43 N, 0.5 m/s) and (49.05, 0.5 m/s),

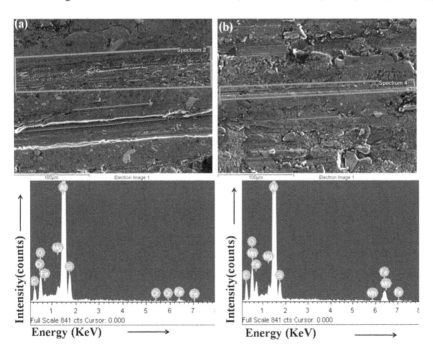

FIGURE 12.9 SEM–EDS of self-lubricating composite sliding under (a) 29.43 N, 0.5 m/s and (b) 49.05, 0.5 m/s.

TABLE 12.5

Elemental Analysis (at.%) on the Worn-Out Hybrid AMMC Pins Sliding against Steel Surfaces under SF MWCNT-Based Oil

Element	Atomic %	
	29.43 N, 0.5 m/s	49.05 N, 0.5 m/s
C K	33.61	37.26
O K	13.48	15.24
Mg K	0.46	0.57
Al K	48.99	41.89
Si K	2.95	1.86
Mn K	0.01	0.15
Fe K	0.50	3.03
Totals	100	100

sliding under the SF MWCNT-in-oil lubrication. In the case of the functionalized MWCNT-oil dispersions, the aligned abrasive scars are seen in SEM micrographs (Figure 12.9). Metallic wear or formation of MML is greatly reduced as obvious from the reduced content of Fe, as shown in EDS spectra (Table 12.5).

It is worth mentioning that a high amount of carbon in the EDS spectra is detected, i.e., approximately more than 33%–37% (wt.). This supports the fact that when the pin is sliding under the SF MWCNTs in oil, there is the formation of stable carbonaceous layers of MWCNTs, graphene sheets or various forms of carbides along with the possible presence of SiC from the matrix. The deposition of carbonaceous layer on the wear track may happen by the 'mending effect' or 'exfoliation' of the MWCNTs [48,75–77]. The exfoliation of the MWCNT into individual graphene sheets under shear stress and the breaking of the nanotubes have been demonstrated by the authors in their earlier work [48]. However, from Figure 12.9, it is also observed that the wear surface has scattered crevices and pits, indicating that the mending effect of nanoparticles may also play a predominant role in controlling the friction and compensating the wear loss.

The presence of oxygen in a moderate amount, i.e. ~13%–15%, might be due to the formation of oxide layers of iron and aluminium. These oxide layers influence the tribological properties of sliding pairs. In a nutshell, dominance of abrasive wear is significantly reduced by using functionalized MWCNTs with the simultaneous formation of continuous carbonaceous layers and composite layers of oxides and carbides. The synergic effects of these tribolayers improve the friction–wear performance of the sliding pairs.

Hence, based on the findings of the present study, it can be rationalized that self- and forced lubrication technologies equipped with carbon-based materials may lead to long-lasting, sustainable and green solutions for automotive, transport and aerospace industries to provide an edge towards the composite tribology and machining sciences.

12.4 CONCLUSIONS

The prime objective of the present investigation is to study the role of nonionic SF MWCNT-in-oil lubricant in the tribological characteristics of the selected hybrid AMMC–EN31 pair. For better insight and comparative assessment, some of the experiments were conducted with nonadditized base oil under similar contact conditions. The study reveals the following interesting facts.

- It has been observed that antifriction and antiwear performance of lubricant is remarkably enhanced by the addition of SF MWCNT. A reduction of wear rate by 83% and COF by 56% has been recorded from the friction–wear test using the specified composition of lubricant and operating condition (9.81 N load and 0.5 m/s), as shown in Figure 12.1 and Table 12.1.
- Operating lubrication regime can be shifted from the boundary to the unified boundary-mixed regime or mixed regime for the SF MWCNT-based oils, by suitably selecting the operating parameters and lubricant compositions.
- Based on the response table, main effect plots (for SN ratios for COF and WR), and ANOVA analysis, it is illustrated that the normal load is the most dominating influencing factor (~59% in WR and ~49% in COF), followed by sliding velocity and sliding distance.
- The SN-ratio analysis indicates that the optimum operating condition using this formulated oil can be realized for low-load and high-speed applications (L_1, S_3 and D_3), to depict the better wear resistance and low-friction characteristics of hybrid AMMC.
- A well-developed lubrication regime mapping of the tribosystem (Figure 12.8), with the help of contour plots of COF as the function of operating parameters, is undoubtedly an important diagnostic tool for the identification of operating lubrication regime.
- Based on the present study, along with the findings reported earlier [45–48] it has been identified that the formations of continuous carbonaceous and composite tribolayers, as well as the grafting of the surfactant on the MWCNT, are responsible for the enhanced tribological behaviour of h-AMMC–steel tribopair under surfactant-modified MWCNT-oil dispersion.

REFERENCES

1. Dwivedi DK. Surface modification by developing coating and cladding. In: *Surface Engineering*. 2018; Springer, New Delhi. Doi: 10.1007/978-81-322-3779-2_6.
2. Omrani E, Moghadam AD, Menezes PL, Rohatgi PK. Influences of graphite reinforcement on the tribological properties of self-lubricating aluminum matrix composites for green tribology, sustainability, and energy efficiency—a review. *The International Journal of Advanced Manufacturing Technology*. 2016;83(1–4):325–46.
3. Omrani E, Moghadam AD, Algazzar M, Menezes PL, Rohatgi PK. Effect of graphite particles on improving tribological properties Al-16Si-5Ni-5Graphite self-lubricating composite under fully flooded and starved lubrication conditions for transportation applications. *The International Journal of Advanced Manufacturing Technology*. 2016;87(1–4):929–39.

4. Gupta A, Mohan S, Anand A, Haq MI, Raina A, Kumar R, Singh RA, Jayalakshmi S, Kamal M. Tribological behaviour of Fe–C–Ni self-lubricating composites with WS_2 solid lubricant. *Materials Research Express*. 2019;6(12):126507.

5. Rajaganapathy C, Vasudevan D, Bruno AD, Rajkumar T. Tribological and mechanical characteristics of AA6082/TiC/WC-based aluminium composites in dry and wet conditions. *Materials Today: Proceedings*. 2020.

6. Srivyas PD, Charoo MS. Tribological behavior of aluminum silicon eutectic alloy based composites under dry and wet sliding for variable load and sliding distance. *SN Applied Sciences*. 2020;2(10):1–21.

7. Kuang W, Zhao B, Yang C, Ding W. Effects of h-BN particles on the microstructure and tribological property of self-lubrication CBN abrasive composites. *Ceramics International*. 2020;46(2):2457–64.

8. Mahathanabodee S, Palathai T, Raadnui S, Tongsri R, Sombatsompop N. Dry sliding wear behavior of SS316L composites containing h-BN and MoS_2 solid lubricants. *Wear*. 2014;316(1–2):37–48.

9. Panda JN, Bijwe J, Pandey RK. Role of micro and nano-particles of hBN as a secondary solid lubricant for improving tribo-potential of PAEK composite. *Tribology International*. 2019;130:400–12.

10. Prasad SV, McConnell BD. Tribology of aluminum metal-matrix composites: Lubrication by graphite. *Wear*. 1991;149(1–2):241–53.

11. Martin JM, Ohmae N. *Nanolubricants*. 2008 (Wiley, West Sussex).

12. Neville A, Morina A, Haque T, et al. Compatibility between tribological surfaces and lubricant additives-how friction and wear reduction can be controlled by surface/lube synergies. *Tribology International*. 2007;40:1680–1695.

13. Ruoff RS, Lorents DC. Mechanical and thermal properties of carbon nanotubes. *Carbon*. 1995;33(7):925–30.

14. Tan, J., Yu, T., Xu, B, et al. Microstructure and wear resistance of nickel–carbon nanotube composite coating from brush plating technique. *Tribology Letters*. 2006;2:107–111.

15. Scharf T, Neira A, Hwang J, et al. Self-lubricating carbon nanotube reinforced nickel matrix composites. *Journal of Applied Physics*. 2009;106: 013508.

16. Kim KT, Cha SI, Hong SH. Hardness and wear resistance of carbon nanotube reinforced Cu matrix nanocomposites. *Materials Science and Engineering: A Journal*. 2007;449:46–50.

17. Hwang Y, Lee C, Choi Y, et al. Effect of the size and morphology of particles dispersed in nano-oil on friction performance between rotating discs. *Journal of Mechanical Science and Technology*. 2011; 25 (11), 2853–2857.

18. Chen W, Tu J, Wang L, et al. Tribological application of carbon nanotubes in a metal-based composite coating and composites. *Carbon*. 2003;41:215–222.

19. Suarez S, Rosenkranz A, Gachot C, et al. Enhanced tribological properties of MWCNT/Ni bulk composites–Influence of processing on friction and wear behaviour. *Carbon*. 2014;66:164–171.

20. Hu J, Jo S, Ren Z, et al. Tribological behavior and graphitization of carbon nanotubes grown on 440C stainless steel. *Tribology Letters*. 2005;19:119–125.

21. Hirata A, Yoshioka N. Sliding friction properties of carbon nanotube coatings deposited by microwave plasma chemical vapor deposition. *Tribology Letters*. 2004;37:893–898.

22. Miyoshi K, Street JK, Vander Wal R, et al. Solid lubrication by multiwalled carbon nanotubes in air and in vacuum. *Tribology Letters*. 2005;19:191–201.

23. Dickrell PL, Pal SK, Bourne GR, et al. Tunable friction behavior of oriented carbon nanotube films. *Tribology Letters*. 2006;24:85–90.

24. Zhang X, Luster B, Church A, et al. Carbon nanotube–MoS_2 composites as solid lubricants. *ACS Applied Materials & Interfaces*. 2009;1:735–744.

25. Peng Y, Hu Y, Wang H. Tribological behaviors of surfactant-functionalized carbon nanotubes as lubricant additive in water. *Tribology Letters*. 2007; 25:247–253.

26. Lu HF, Fei B, Xin JH, et al. Synthesis and lubricating performance of a carbon nanotube seeded miniemulsion. *Carbon* 2007;45:936–942.

27. Chen C, Chen X, Xu L, et al. Modification of multi-walled carbon nanotubes with fatty acid and their tribological properties as lubricant additive. *Carbon*. 2005;43:1660–1666.

28. Kristiansen K, Zeng H, Wang P, et al. Microtribology of aqueous carbon nanotube dispersions. *Advanced Functional Materials*. 2011;21:4555–4564.

29. Singh KGK, Suresh R. Behavior of composite nanofluids under extreme pressure condition. *International Journal of Engineering Research & Technology*. 2012;1, 1–7.

30. Nunn N, Mahbooba, Ivanov ZMG, Ivanov DM, Brenner DW, Shenderova O. Tribological properties of polyalphaolefin oil modified with nanocarbon additives. *Diamond and Related Materials*, 2015;54:97–102.

31. Chebattina KRR, Srinivas V, and Rao NM. Effect of size of multiwalled carbon nanotubes dispersed in gear oils for improvement of tribological properties. *Advances in Tribology*, 2018, Article ID 2328108:1–13.

32. Muzakkir SM, Lijesh KP & Hirani H. Influence of surfactants on tribological behaviors of MWCNTs (multi-walled carbon nano-tubes). *Tribology - Materials, Surfaces & Interfaces*, 2016. Doi: 10.1080/17515831.2016.1138636.

33. Clyne T. Metal matrix composites: Matrices and processing. *Encyclopedia of Materials: Science and Technology*. 2001; 8.

34. Pawar P, Utpat AA. Development of aluminium based silicon carbide particulate metal matrix composite for spur gear. *Procedia Materials Science*. 2014;6:1150–1156.

35. Lindroos V, Talvitie M. Recent advances in metal matrix composites. *Journal of Materials Processing Technology*. 1995;53:273–278.

36. Bisane VP, Sable YS, Dhobe MM, Sonawane PM. Recent development and challenges in processing of ceramics reinforced Al matrix composite through stir casting process: A review. *International Journal of Engineering and Applied Sciences* 2015; 2:11–16.

37. Dixit G, Khan MM. Sliding wear response of an aluminium metal matrix composite: Effect of solid lubricant particle size. *Jordan Journal of Mechanical and Industrial Engineering*. 2014;8:351–358.

38. Bindumadhavan P, Chia T, Chandrasekaran M, Wah HK, Lam LN, and Prabhakar O. Effect of particle-porosity clusters on tribological behavior of cast aluminum alloy A356-SiCp metal matrix composites. *Materials Science and Engineering: A - Journal*. 2001;315(1):217–222.

39. Dharmalingam S, Subramanian R, and Vinoth KS. Analysis of dry sliding friction and wear behavior of aluminum-alumina composites using Taguchi's techniques. *Journal of Composite Materials*. 2010;44(18):2161–2177.

40. Kaushik N, Singhal S. Hybrid combination of Taguchi GRA-PCA for optimization of wear behavior in AA6063/SiCp matrix composite. *Production and Manufacturing Research*. 2018;6:171–89.

41. Satyanarayana T, Rao PS, Krishna MG. Influence of wear parameters on friction performance of A356 aluminum–graphite/granite particles reinforced metal matrix hybrid composites. *Heliyon*, 2019;5:e01770.

42. Kaushik N, Singhal S. Wear conduct of aluminum matrix composites: A parametric strategy using Taguchi based GRA integrated with weight method. *Cogent Engineering* 2018;5:1.

43. Akhlaghi F, Zare-Bidaki A. Influence of graphite content on the dry sliding and oil impregnated sliding wear behavior of Al 2024–graphite composites produced by in situ powder metallurgy method. *Wear*. 2009;266:37–45.

44. Walker J, Rainforth W, Jones H. Lubricated sliding wear behaviour of aluminium alloy composites. *Wear*. 2005;259:577–589.

45. Singh H, Bhowmick H. Lubricated tribology of hybrid AMMC–steel sliding contact: A comparative investigation between fully formulated commercial engine oils and surfactant functionalized MWCNT–base oil formulation. *Proceedings of the Institution of Mechanical Engineers. Part J: Journal of Engineering Tribology.* 2020:1350650119901221.

46. Singh H, Bhowmick H. Tribological behaviour of hybrid AMMC sliding against steel and cast iron under MWCNT-Oil lubrication. *Tribology International.* 2018;127:509–519.

47. Singh H, Singh PP, Bhowmick H. Influence of MoS_2, H_3BO_3, and MWCNT additives on the dry and lubricated sliding tribology of AMMC–Steel contacts. *The Journal of Tribology.* 2018;140: 041801.

48. Singh H, Bhowmick H. Lubrication characteristics and wear mechanism mapping for hybrid aluminium metal matrix composite sliding under surfactant functionalized MWCNT-oil. *Tribology International.* 2020;145:106152

49. Krishna MV, Xavior AM. An investigation on the mechanical properties of hybrid metal matrix composites. *Procedia Engineering.* 2014;97:918–924.

50. Velmurugan C, Subramanian R, Thirugnanam S, et al. Investigation of friction and wear behavior of hybrid aluminium composites. *Industrial Lubrication and Tribology.* 2012;64:152–163.

51. Karthikeyan A, Nallusamy S. Investigation on mechanical properties and wear behavior of Al-Si-SiC-Graphite composite using SEM and EDAX. *IOP Conference Series: Materials Science and Engineering.* 2017;225:012281.

52. Singh J. Fabrication characteristics and tribological behavior of Al/SiC/Gr hybrid aluminum matrix composites: A review. *Friction.* 2016;4:1–17.

53. Lee HJ, Chin BD, Yang SM, et al. Surfactant effect on the stability and electrorheological properties of polyaniline particle suspension. *Journal of Colloid and Interface Science* 1998;206:424–438.

54. Cui H, Yan X, Monasterio M, and Xing F. Effects of various surfactants on the dispersion of MWCNTs–OH in aqueous solution. *Nanomaterials.* 2017;7:262;1–14.

55. Li H, Qiu Y. Dispersion, sedimentation and aggregation of multi-walled carbon nanotubes as affected by single and binary mixed surfactants. *Royal Society Open Science.* 2019;6:190241; 1–9.

56. Tan Y, Resasco DE. Dispersion of single-walled carbon nanotubes of narrow diameter distribution. *The Journal of Physical Chemistry B.* 2005;109:14454–14460.

57. Karimian H, Moghbeli MR. Influence of single-wall carbon nanotubes and polypyrrole thin layer coating on the electrical conductivity of PolyHIPE foams. *Polymer-Plastics Technology and Engineering.* 2014;53(4):344–52.

58. Goodarzi F and Zendehboudi S. A comprehensive review on emulsions and emulsion stability in chemical and energy industries. *The Canadian Journal of Chemical Engineering.* 2019;97:281–309.

59. Teng TP, Fang YB, Hsu YC, and Lin L. Evaluating stability of aqueous multiwalled carbon nanotube nanofluids by using different stabilizers. *Journal of Nanomaterials,* 2014; Article ID 693459, 15 p.

60. Menezes PL, Kailas SV. Influence of roughness parameters on coefficient of friction under lubricated conditions. *Sadhana* 2008;33(3):181–90.

61. Guangteng G, and Spikes HA. An experimental study of film thickness in the mixed lubrication regime. *Tribology Series,* 1997;32:159–166.

62. Montgomery DC, Runger GC. *Applied Statistics and Probability for Engineers.* 1999 (John Wiley & Sons, New York).

63. Basavarajappa S, Chandramohan G. Wear studies on metal matrix composites: A Taguchi approach. *Journal of Materials Science and Technology.* 2005;21:845–850.

64. Alphas AT, Zhang J. Effect of SiC particulate reinforcement on the dry sliding wear of aluminium-silicon alloys (A356). *Wear.* 1992;155:83–104.

65. Basavarajappa S, Chandra Mohan G, Mukund K, Ashwin M, Prabu M. Dry sliding wear behaviour of Al 2219/SiC/Gr hybrid metal matrix composites. *Journal of Materials Engineering and Performance.* 2006;15:668–674.
66. Sahin Y. Wear behaviour of aluminium alloy and its composites reinforced by SiC particles using statistical analysis. *The Journal of Materials: Design.* 2003;24:95–103.
67. Banerji A, Prasad SY, Surappa MK, Rohatgi PK. Abrasive wears of cast aluminium alloy-zircon particle composites. *Wear.* 1982;82:141–151.
68. Mahmoud T S. Tribological behaviour of A390/Grp metal–matrix composites fabricated using a combination of rheocasting and squeeze casting techniques. *Journal of Mechanical Engineering Science* 2008;222:257–265.
69. Srinivasan V, Maheshkumar KV, Karthikeyan R. Application of probabilistic neural network for the development of wear mechanism map for glass fiber reinforced plastics. *Journal of Reinforced Plastics and Composites.* 2007;26:1893–914.
70. Prajapati DK, Tiwari M. Effect of topography parameter, load, and surface roughness on friction coefficient in mixed lubrication regime. *Lubrication Science.* 2019;31(5):218–28.
71. Yan Y. Tribology and tribo-corrosion testing and analysis of metallic biomaterials. *Metals. in Metals for Biomedical Devices,* 2010: 178–201.
72. Guegan J, Kadiric A, Gabelli A, Spikes H. The relationship between friction and film thickness in EHD point contacts in the presence of longitudinal roughness, *Tribology Letters.* 2016;64:33. Doi 10.1007/s11249-016-0768-6.
73. Danov KD, Kralchevsky PA, Stoyanov SD, Cook JL, Stott JP, Pelan EG. Growth of wormlike micelles in nonionic surfactant solutions: Quantitative theory vs. experiment. *Advances in Colloid and Interface Science.* 2018; 256:1–22.
74. Wang RK, Chen WC, Campos DK, Ziegler KJ. Swelling the micelle core surrounding single-walled carbon nanotubes with water-immiscible organic solvents. *Journal of the American Chemical Society.* 2008;13048:16330–16337.
75. Liu G, Li X, Qin B, Xing D, Guo Y, Fan R. Investigation of the mending effect and mechanism of copper nano-particles on a tribologically stressed surface. *Tribology Letters.* 2004 Nov 1;17(4):961–966.
76. Sharma A, Sharma VM, Sahoo B, Joseph J, Paul J. Study of nano-mechanical, electro-chemical and Raman spectroscopic behavior of Al6061-SiC-Graphite hybrid surface composite fabricated through friction stir processing. *Journal of Composites Science.* 2018; 2:1–17.
77. Kalin M, Kogovšek J, Remškar M. Mechanisms and improvements in the friction and wear behavior using MoS_2 nanotubes as potential oil additives. *Wear.* 2012 Mar;20(280):36–45.

13 A Nexus of Tribology and Rheology to Study Thin-Film Mechanics of Asphalt–Aggregate Interaction during Mixing and Compaction

Saqib Gulzar, Cassie Castorena,
and Shane Underwood
North Carolina State University

CONTENTS

13.1   INTRODUCTION

Asphalt mixture compaction is an extremely complex process. It is governed by various factors such as material, field conditions, boundary condition properties and gradation. In order to control the process of mixing and compaction, the gradation of the aggregates cannot be changed due to obvious economic and mix design considerations; however, two main factors are controlled to obtain the desired asphalt mixture compaction: temperature and compactive effort. The pavement density, controlled by the percentage of air voids in the compacted mix, is used as a measure to achieve a desired pavement performance. The scope of this chapter is limited to provide an introduction to the mixing and compaction temperature selection and summarize the underlying discerning mechanisms occurring during the mixing and compaction process. The advent of modified binders has further complicated the mixing and compaction phenomena, and the traditional estimators of compaction temperature have been proven to be insufficient (Hanz et al. 2010, Puchalski 2012, Hanz and Bahia 2013). Traditionally, rheological measure of viscosity has been used as an estimator for mixing and compaction temperatures. But many times, mixtures of the same gradation and binder viscosity yield different in-place densities for the same compactive effort. This phenomenon has led researchers to study the underlying mechanism at a deeper level (Puchalski 2012). It has come to the forefront that tribology (the study of friction, lubrication and wear) offers an additional insight into the thin-film mechanics of asphalt–aggregate interactions during mixing and compaction.

In this review, the background of asphalt mixing and compaction is given first. Next, the fundamental concepts of tribology in general and vis-à-vis asphalt mixtures and binders are discussed in detail, highlighting the important underlying mechanisms. The tests employed for asphalt mixing and compaction based on rheology and tribology are briefly presented and summarized. Finally, the results on tribological characterization of asphalt binders from the literature are reviewed and critical gaps and potential future areas of work are highlighted.

13.2   ASPHALT MIXING AND COMPACTION

13.2.1   MIXING AND COMPACTION PROCESS

At an asphalt mixture batch production plant, once the aggregates are dried and treated, they are emptied in a pugmill to be mixed with each other and with the asphalt binder. A pugmill has two mixing shafts with paddles, which rotate in opposite directions. There are two stages of mixing in the pugmill such as dry mixing and wet mixing. In dry mixing, the aggregates are mixed with each other. In wet mixing, asphalt binder is introduced in the mixing process, either using gravity or pressurized spray. Among the controlling parameters, the critical ones are the mixing time and temperature which deeply affect the quality of the asphalt mixture produced as well as the rate of production.

Once transported to the project site, asphalt mix is placed down and compacted to a certain degree with a paver. A paver is a rubber- or steel-track mounted and consists of a hopper to receive the material from the truck, an assembly of conveyors, augers, screeds, etc., which ensures that a desired depth of asphalt concrete is laid down to a preliminary degree of compaction. Next, rollers are used to achieve the compaction of layers by reducing the volume of air in the mix as well as by rearrangement and reorientation of aggregate particles in the mix. Rollers are typically very heavy and may cause the wear and breakdown of the aggregate particles too during the compaction process (Mallick 2013).

It can be seen that both mixing and compaction involve the asphalt binder, aggregates and their interaction with each other. While the binder properties have been used to determine the mixing and compaction parameters for the bulk mixtures, it is essential to investigate the mechanisms occurring as part of asphalt–aggregate interactions. A rigorous characterization of the underlying mechanisms will enable better tuning of the mixing and compaction process, which may not only produce better quality mixes but also have tremendous environmental and technical benefits (Zhai et al. 2000, Sefidmazgi et al. 2013).

13.2.2 Determination of Mixing and Compaction Temperatures

One of the important considerations in the production and compaction of asphalt mixtures in the field is the selection of mixing and compaction temperatures. Conventionally, these are determined from viscosity considerations and viscosity is determined using the rotational viscometer AASHTO T316. In Section 13.4, detailed tests used to determine mixing and compaction temperatures and their advantages and disadvantages are given. In this section, the range of shear rates an asphalt binder experiences in the lab only is tabulated, as shown in Table 13.1. Based on the literature (Gulzar and Underwood 2019a, 2019b, 2020a, 2020b, 2020c, 2020d, Daryaee et al. 2020), asphalt binders, especially the modified asphalt binders, have shown non-Newtonian behaviour, such as shear thinning or shear thickening, as such there is a need to characterize the behaviour of asphalt binders for a range of shear rates. At the same time, it causes a different film thickness of asphalt binder around the aggregates in an asphalt mixture. As will be shown in Sections 13.3.2 and 13.3.3, the film thickness has a huge impact of the lubricating regime in which asphalt–aggregate interaction occurs during mixing

---

TABLE 13.1

Summary of Range of Shear Rates for Some Field and Lab Equipment (Wes et al., 2010)

Laboratory Device	Model	RPM	Radius(in)	Tangential Velocity(min/s)	Shear Rate(1/s)
Bucket mixer	KOL M-60	65	5.6	961	9,6100
Pugmill mixer	7590-H	128	3.9	1,341	134,100
Workability device	InstroTek	20	6	319	31,919
Bowl mixer	Hobart A200	48	4	425	42,558
Rotational viscometer	Brookfield DV-ll+	20	8.4mm	17.5	6.8
Dynamic shear rheometer	MCR series	10rad/s	25mm	–	125
			8mm		20

and compaction. Hence, there is a need to study asphalt binders through the lens of both tribology and rheology, in order to have a rigorous characterization of underlying mechanisms occurring during mixing and compaction.

13.2.3 WMA Additives and the Need for Tribology

Warm mix asphalt (WMA) additives refer to the technology that produces asphalt mixtures that are produced and placed at temperatures 20°C–55°C lower than typical hot mix asphalt (HMA). WMA offers several environmental and technical benefits and have been successfully deployed in the field. However, the mechanisms that make this temperature reduction possible are not completely understood. Initially, it was believed that WMA additives cause a viscosity reduction, which causes a reduction in mixing and compaction temperatures; however, the viscosity measurements at similar temperatures have shown little effect of WMA additives (Puchalski 2012). Although some developers of surfactant additives reported that these additives improve workability without any viscosity reduction, the cause of increased workability and ease of compaction so consistently reported in field experiments is unknown. This mismatch between the effect of WMA additives on viscosity and asphalt mixture densification over a similar range of temperatures indicates that viscosity is not the only mechanism by which WMA additives allow for compaction at reduced temperatures. It has been shown that the reduction of internal friction or the improvement of lubricating properties of asphalt binder can be a possible mechanism taking place (Hanz et al. 2010, Puchalski 2012, Hanz and Bahia 2013). This prompted the researchers to investigate the tribological properties of asphalt binders. Thus, there is a need to study the thin-film lubricating behaviour of asphalt binder in an asphalt mixture, characterize it and then use it to study the effectiveness of WMA additives. The internal friction coefficient of asphalt binders is referred to as asphalt lubricity as it allows the assessment of the lubricating properties of asphalt binders. In the subsequent section, the fundamentals of tribology will be presented, followed by the tribological tests used or devised by researchers to characterize the lubricating properties of asphalt binders.

13.3 FUNDAMENTALS OF TRIBOLOGY

13.3.1 Background

Tribology deals with the study of friction, lubrication and wear of surfaces that are in contact or in relative motion. Primarily, tribology has a wide application in the field of mechanical engineering due to the kinematic nature of machines and their components. Tribology is also used in the field of lubrication, where lubricants are used to reduce wear and tear of mechanical parts as well as to reduce the friction between the parts moving relative to one another. According to Rabinowicz (1995), the primary interaction phenomena occurring between the materials when in contact or in relative motion have been described as follows:

- **Friction** –'The friction effects are those that arise from the tangential forces transmitted across the interface of contact when solid surfaces are pressed together by a normal force' (Rabinowicz 1995).

- **Wear** –'The wear phenomena consist of the removal of material from the surfaces of one of the contacting bodies, as a result of interaction with the other contacting body' (Rabinowicz 1995).
- **Adhesion** – '(…) the ability of contacting bodies to withstand tensile forces after being pressed together (…) It seldom occurs to any marked extent and has been much less investigated than the others' (Rabinowicz 1995).
- **Lubricant** –'Lubricant is a substance capable of altering the nature of surface interactions between contacting solids' (Rabinowicz 1995). Proper use of a lubricant may significantly reduce the friction and wear between two contacting surfaces (Puchalski 2012).

13.3.2   TRIBOLOGY VIS-À-VIS ASPHALT MIXTURE

As discussed above, tribology and its use in the area of asphalt technology have become most evident and relevant during the compaction process of asphalt mixtures. When asphalt concrete is placed on the road, compaction happens due to particle rearrangement and reorientation under loads. Asphalt concrete consists of aggregate particles of different particle sizes. In order to achieve a particular asphalt concrete consistency in the field, a very high compactive effort is required to achieve the set density/air void targets. This compactive effort is used to overcome the friction between the aggregate particles to compact them, and in the finishing stages of compaction process, substantial wear may occur between aggregate particles. The frictional force, F_f, between two objects is given as follows:

$$F_f = \mu.F_N \tag{13.1}$$

where μ is the coefficient of friction and F_N is the normal reaction force. The coefficient of friction is a function of the type of the solid materials in contact.

Asphalt binder plays a very vital role in the entire compaction process. It acts as a lubricant, reducing friction between aggregate particles, protects against surface wear and aids the overall reorientation of aggregate particles. The tribology of asphalt binder revolves around the concept of the thin film it forms around the aggregates and has been discussed in detail in the next section.

13.3.3   TRIBOLOGY VIS-À-VIS ASPHALT BINDER

As explained in the previous section, asphalt binder acts as a lubricant between the aggregate particles during the process of mixing and compaction. The properties of asphalt binder as a lubricant should be related to the compactive effort required to achieve a particular degree of compaction in the asphalt concrete. Thus, the better asphalt binder acts a lubricant, the less the required compactive will be (Puchalski 2012).

The behaviour of asphalt binder as a lubricant during the mixing and compaction of asphalt concrete is characterized by the thin-film mechanics of lubrication. As previously described, the aggregate particles produce high friction when in constant movement during the compact process and may be accompanied by the wear

of surfaces due to rubbing during rearrangement and reorientation under heavy compactive loads. The presence of asphalt binder creates a thin film between the two
aggregate surfaces, acting as a lubricant and thereby reducing friction as well as
the wear. The friction behaviour and regimes between two solids in relative motion
separated by a lubricating film can be studied using the Stribeck curve. This curve
describes the changes in the measures of friction (e.g. the coefficient of friction) as
a function of the properties of the lubricant film system such as absolute viscosity
of film, relative speed of the two moving bodies and load per unit area. Generally,
the evolution of the coefficient of friction with relative speed is observed. A function
known as Sommerfeld number (SN) is also used to characterize the change in the
control variables (Rabinowicz 1995). It is given as follows:

$$SN = \frac{\eta \times N}{P} \tag{13.2}$$

where η is the absolute viscosity of lubricant, N is the relative speed between the two
moving solid bodies and P is the load per unit area.

In a typical Stribeck curve, there are four distinct lubricating regimes. Each of
the regimes is governed by a different physical mechanism. These four regimes are
mentioned below:

a. Hydrodynamic regime
b. Elasto-hydrodynamic
c. Mixed regime
d. Boundary regime

These four regimes in a typical Stribeck curve are schematically shown in Figure 13.1.

These four regimes are briefly explained in the context of asphalt compaction,
thin-film mechanics and the governing physical mechanism as shown in Figure 13.2.

13.4   TESTS FOR ASPHALT BINDERS

13.4.1   Tests Based on Rheology

Traditionally, just like asphalt mixture production, asphalt mixing and compaction have been required to have a specific viscosity for efficient aggregate coating.
So numerous tests have been used to measure viscosity and subsequently estimate
mixing and compaction temperatures. The test for conventional binders is based on
AASHTO T 316, while for modified binders (AASHTO 2011), the low shear viscosity and phase angle methods outlined in NCHRP reports 459 and 648 are more commonly used (Bahia et al. 2001, West et al. 2010). All of these tests rely on rheology
to characterize the bulk properties of the mixture. A summary of research methods
used for determining mixing and compaction temperatures using rotational viscometer is given in Table 13.2. Similarly, a summary of test methods employing dynamic
shear rheometer and mixture-based methods is given in Table 13.3. It can be seen that
most of these methods rely on some measures of viscosity to determine mixing and
compaction temperatures.

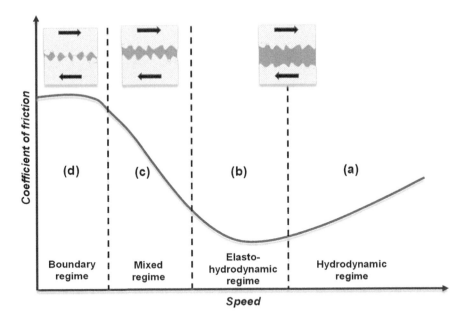

FIGURE 13.1   A typical Stribeck curve with four lubricating regimes.

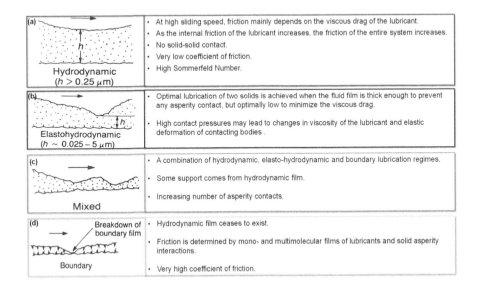

FIGURE 13.2   Four lubricating regimes and their mechanisms (Bhushan 2002).

TABLE 13.2

Summary of Test Methods Using Rotational Viscometer (West et al., 2010)

Method	Equipment	Description	Advantages	Disadvantages
Equiviscous temperatures	Rotational viscometer	• Viscosity at 2 temperatures and 1 shear rate is determined • Viscosity vs. temperature plotted • Temperature range corresponding to 0.17 ± 0.02 Pa-s is chosen for mixing • Temperature range corresponding to 0.28 ± 0.03 Pa-s is chosen for compaction	• Simple to obtain and analyse results • Can be completed in less than 1 h	• Assumes linear relationship between viscosity and temperature • Assumes that all asphalt binders are Newtonian liquids • Does not account for shear rate dependency • Can result in unnecessarily high mixing and compacting temperatures
High shear rate viscosity	Rotational viscometer	• Determines the shear rate dependency of an asphalt binder at 2 temperatures (135°C,165°C) • For each temp., the data are fit to an inverse power curve and extrapolated to estimate the viscosity at a shear rate of 490 1/s • High shear viscosities are plotted versus temperature • Same mixing and compaction temp as above	• Considers the shear rate dependency of modified asphalt binders • Testing is simple to perform • Does not require complicated modelling	• Requires extrapolation of results to a high shear rate
Zero (low) shear viscosity	Rotational viscometer	• Determines the shear rate dependency of an asphalt binder at 3 temperatures (120,135,165°C) • The Cross-Williams model is used to fit a curve to the data at each temperature from which the viscosity at a shear rate of 0.001 1/s is estimated • The low shear viscosities are plotted versus temperature • Mixing and compaction temperature ranges are determined at target value of 3.0 Pa-s and 6.0 Pa-s, respectively	• Considers the shear rate dependency of modified asphalt binders • Testing is simple to perform • Results in lower mixing and compaction temperatures for modified asphalt binders	• Requires extrapolation • Not accurate • Cross-Williams model is complicated • No clear definition of ZSV • May yield unrealistically low mixing and compaction temperatures

TABLE 13.3

Summary of Test Methods Using DSR and Mixture-Based Methods (West et al. 2010)

Method	Equipment	Description	Advantages	Disadvantages
Steady Shear Flow	Dynamic shear rheometer	• Uses a DSR steady-state flow test at 76°C, 82°C, 88°C and 94°C • Measurements of steady-state viscosity made over a range of 0.16 to 500 Pa are plotted versus temperatures • Mixing and compaction temperature ranges are determined at target values of 0.17±0.02 Pa-s and 0.35±0.03 Pa-s, respectively	• Simple to perform • Uses standard DSR equipment and testing procedures	• Can be consuming for modified asphalts • Not all modified asphalts reach a state of steady shear by 500 Pa • Requires extrapolation of viscosity to much higher temperatures
Mixture Workability	Stirring device	• Uses a large stirring device to measure the torque required to stir a mix as it cools • Torque is inversely proportional to workability and • The relationship between workability and temperature range where a mix is easiest to work	• Considers the effects of aggregate particle shapes and size on compactability of mixtures	• New equipment • Time-consuming procedure • Not practical for routine use • Aggregate characteristics and graduation may overwhelm binder effects
Compaction Test	Superpave Gyratory compactor	• A standard mix is compacted with an unmodified 'control' binder to establish a baseline density • The modified binder is then added to the standard mix and samples are compacted at temperature intervals • The temperature that provides the same density as the control binder is the compaction temperature for the modified binder	• Easy to analyse based on density and volumetric properties	• Time-consuming procedure • SGC is insensitive to binder consistency • Only provides results for compaction temperature • Results are dependent on the 'standard' mixture. Other mixes may provide different results

13.4.2 Tests Based on Tribology

In order to determine the lubrication property of asphalt binders, three main testing geometries or set-ups have been used in the literature. All three tests use a special type of fixture in the DSR to determine the lubricity of asphalt binders.

13.4.2.1 Ball-on-Three-Plate Configuration

In this tribological set-up, a ball attached to the DSR shaft slides against three plates as shown in Figure 13.3. When the DSR shaft rotates, it produces a sliding speed for the ball relative to the three base plates at the contact points. In order to obtain coefficient of friction, the frictional force and normal force need to be determined. Using the geometry of the set-up as shown in Figure 13.3, the frictional force can be calculated from the resulting torque as follows (Shahrivar et al. 2014):

$$F_{f,\mathrm{TOT}} = 3 \times \left(\frac{T}{3 \times r_{\mathrm{ball}} \times \sin\alpha} \right) = \frac{T}{r_{\mathrm{ball}} \sin\alpha} \tag{13.3}$$

where T is the total torque, $F_{f,\mathrm{TOT}}$ is the total friction experienced by the sample, r_{ball} is the radius of the ball and α is the angle between the plates and the horizontal plane.

Similarly, the axial force is divided equally into the three contact points and is transmitted as normal force to the normal plates, which is given as follows [13]:

$$F_{f,\mathrm{triboTOT}} = 3 \times \left(\frac{F_N}{3 \times \cos\alpha} \right) = \frac{F_N}{\cos\alpha} \tag{13.4}$$

where F_N is the axial force of the DSR. Thus, the tribological properties are calculated as follows:

$$\mu = \frac{F_{f,\mathrm{TOT}}}{F_{f,\mathrm{triboTOT}}} = \frac{T}{r_{\mathrm{ball}} F_N \tan\alpha} \tag{13.5}$$

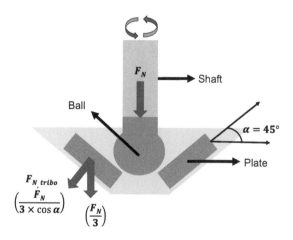

FIGURE 13.3 Ball-on-three-plate tribology set-up (Canestrari et al. 2017).

$$v_s = \sin \alpha \times \left(\frac{2\pi}{60}\right) \times v \times r_{ball} \tag{13.6}$$

where μ is the coefficient of friction, v is the rotational speed in rpm and v_s is the sliding speed in m/s. In order to test different combinations of materials, both the ball and the plates can be exchanged. Many researchers have used this readily available measuring cell to study the lubrication properties of asphalt binders. A summary of these is provided in Section 13.4.3.4.

13.4.2.2 Four-Ball Configuration

The four-ball tribology set-up was developed at the University of Wisconsin-Madison and is popularly referred to as asphalt lubricity test (Hanz et al. 2010, Hanz and Bahia 2013). It is similar to the ball-on-three-plate tribology set-up, except that instead of three plates, this test uses three balls. The force distribution is similar to the previous test and the governing principle is the same. This test was selected based on the literature review of available test methods for lubricating fluids in the engine industry, considering ASTM D5183-05. However, in order to use this set-up in the DSR, it was scaled down appropriately. Figure 13.4 shows the picture of the asphalt lubricity testing set-up.

The tribological property was calculated as follows:

$$\mu = C \times \frac{T}{F_N \times d} \tag{13.7}$$

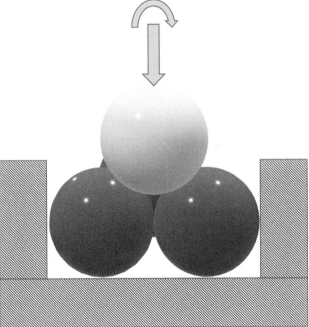

FIGURE 13.4 Four-ball tribology set-up.

where $C = 2.842$ is a constant for the asphalt lubricity testing fixture, T is the torque in Nm, F_N is the normal force in N and d is the diameter of the balls in m.

13.4.2.3 Pin-on-Flat Geometry Configuration

In continuation of tribology studies of asphalt binders, researchers at the University of Wisconsin-Madison proposed another tribological test to address the limitations of the previous two tests (Puchalski 2012). It was hypothesized that at high temperatures, a boundary regime of lubrication exists between the aggregates and the type of aggregates, significantly affected the effectiveness of asphalt binder lubricity. In order to test this hypothesis, a pin-on-flat geometry was chosen due to two reasons. Firstly, aggregate disks could be easily prepared. Secondly, this geometry employs a circular motion, which was compatible with DSR. This test is popularly referred to as asphalt boundary lubrication test.

The tribological property in asphalt binder lubrication test is given as follows:

$$\mu = \frac{T}{F_N \times d'} \tag{13.8}$$

where T is the torque in Nm, F_N is the normal force in N and d' is the distance between the centroids of the top surfaces of the two pins in m. Figure 13.5 shows the actual picture of the testing set-up and a schematic of the testing geometry parts.

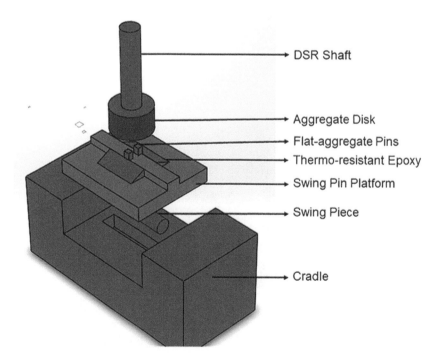

FIGURE 13.5   Pin-on-flat tribology set-up.

Pin-on-flat setup (Asphalt Boundary Lubrication Test)

Year/Ref	Testing Temperature (°C)	Speed	Normal Force/Stress	Substrate	Type
2012 [3]	90, 115, 145	0.05 rad/s	100 kPa, 500kPa, 1Mpa	Limestone; Granite	Rotatory

Stribeck curve is not obtained, hence, lubrication regimes are not identified. Multiple measurements needed.

Ball-on-three plates setup (A DSR Fixture)

Year/Ref	Testing Temperature (°C)	Speed	Normal Force/Stress	Substrate	Type
2012 [3]	85, 115, 145	0.01-3000 rpm	21 N	Steel; Teflon	Rotatory
2016 [15]	25, 50, 135	0.001-1000 rpm	10 N	Steel	Rotatory
2013 [16]	115, 130, 145, 160	0-1433 rpm	10 N	Steel	Oscillatory
2012 [17]	115, 130, 145, 160	0-1433 rpm	10 N	Steel	Rotatory

Entire Stribeck curve is obtained with a single measurement as a range of rotation speed if considered.

Four-ball setup (Asphalt Lubricity Test)

Year/Ref	Testing Temperature (°C)	Speed	Normal Force/Stress	Substrate	Type
2010 [1]	80, 90, 100	50 rpm	20, 30 N	Steel	Rotatory
2013 [2]	85, 115, 145	50 rpm	10 N	Steel	Rotatory
2012 [1]	85, 115, 145	50 rpm	10 N	Steel	Rotatory

Stribeck curve is not obtained, hence, lubrication regimes are not identified. Multiple measurements needed.

FIGURE 13.6 Summary of testing conditions for tribological tests of asphalt binders (Canestrari et al. 2017).

13.4.2.4 Summary of Testing Conditions

Various researchers have conducted tribological tests on asphalt binders under different testing conditions. These testing conditions vary for the type of testing set-up, objective and scope of the study, as well as limitations of each geometry. A summary of all the testing conditions is given in Figure 13.6.

13.5 RESULTS AND DISCUSSION

In this section, some of the results from tribological characterization of asphalt binders will be briefly presented and discussed. One typical result from each type of tribological test listed in Section 13.4.2 is presented and briefly discussed.

Figure 13.7 shows the typical result from the ball-on-three-plate tribology set-up. The Stribeck curve is shown for a neat asphalt binder at a testing temperature of 115°C and a normal force of 10 N. It can be clearly seen that the binder shows all four lubrication regimes as the speed increases. This result shows that the hypothesis about the lubrication characteristics of asphalt binder has validity, but it is limited since the test used metal balls.

Figure 13.8 shows the coefficient of friction as a function of temperature for asphalt binders with and without WMA from a typical four-ball tribology testing set-up. It can be seen that this set-up is able to capture two important properties. Firstly, the difference in high-temperature PG grade is captured. Secondly, the decrease in the coefficient of friction due to the addition of WMA additive (Revix) is also captured.

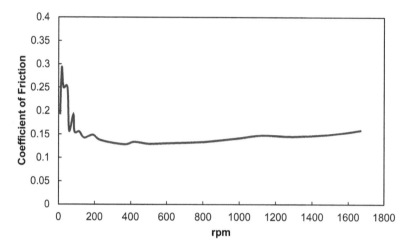

FIGURE 13.7 Typical schematic of the Stribeck curve for a neat asphalt binder using ball-on-three-plate tribology set-up (Baumgardner and Reinke 2013).

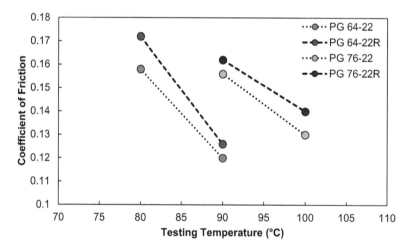

FIGURE 13.8 Coefficient of friction as a function of temperature, for asphalt binder with and without WMA additive (Hanz et al. 2010).

Figure 13.9 shows the results from asphalt boundary lubrication test for neat and WMA-modified asphalt binders. It can be seen that the boundary regime of the lubrication has been captured, and the effect of WMA additive in varying dosages is evident in the boundary coefficient of friction values. This shows that WMA additives do affect the lubricity of asphalt binders, and thus, lubrication does have a role in the

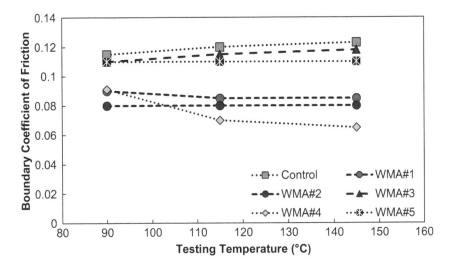

FIGURE 13.9 Typical schematic result from pin-on-flat set-up test for neat and WMA-modified asphalt binders (Puchalski 2012).

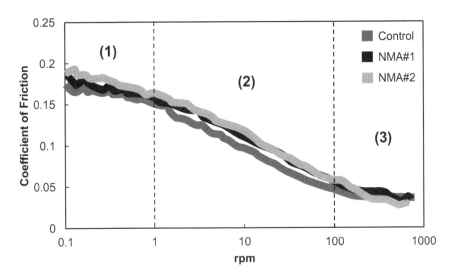

FIGURE 13.10 Schematic of the Stribeck curve for a neat and nano-modified asphalt binder (Ingrassia et al. 2019).

reduction of mixing and compaction temperatures due to the addition of WMA additives to the asphalt binders.

Figure 13.10 shows the Stribeck curve of neat and nano-modified asphalt binders in dosages of 3% and 6%. It can be seen that the addition of nano-platelets increases the coefficient of friction, especially in the boundary regime.

13.6   CRITICAL GAPS AND FUTURE WORK

Some of the critical gaps in the tribo-rheological characterization of asphalt binders (Ingrassia et al. n.d., Geisler et al. 2016, Bairgi et al. 2019) are given below:

- The effect of WMA binders produced with foaming process has not been adequately captured.
- The type of test, oscillatory or rotational, to be used for testing, is not clear.
- The current methods capture only some effects of WMA additives. More rigorous tests are required.
- The influence of wear on the substrate was not taken care of.
- Testing conditions and procedures seem to affect the results. Hence, the true material characterization has not been done yet.
- Representative samples and set-up of asphalt mixture has not been studied in particular.
- Lot of interaction effects might be occurring which shadow the true tribological characterization of asphalt binders.

In order to properly use the theory of tribology and its principles, it is essential to conduct further research in the following areas:

- A standard test that captures the effect of asphalt binder on the tribological behaviour only. The effect of substrate, wear, etc., should be accountable for separately.
- DSR fixtures that allow testing over a wide range in a single test should be developed.
- Better control of sample preparation and testing set-up is required to create the ideal and representative conditions of the field.
- The reliability and repeatability of each test much be ascertained before using it for routine use.
- There is a need to study the correlation of asphalt binder viscosity and lubricity and may be come up with a parameter that takes into account both.
- There is a need to correlate tribological properties of asphalt binders with workability of asphalt mixtures.
- To standardize a testing set-up that enables to study the effect of WMA additives in a more discerning and pronounced way.

13.7   CONCLUSION

In this review, a background on the asphalt mixture and compaction was given and the need to study thin-film mechanics as part of asphalt–aggregate interaction was discussed. The fundamentals of tribology have been presented, and the importance of tribology vis-à-vis asphalt mixtures and asphalt binders was illustrated. The tests based on rheology and tribology to identify and characterize the underlying mechanism occurring during asphalt mixing and compaction have been summarized. The results from various tribological studies have been reviewed. Some of the main observations include the importance of testing geometry, procedure and conditions in determining the tribological behaviour of asphalt binders; the lubricating behaviour is a function of speed, load per unit area and temperature as illustrated by Stribeck curve; the current tribological tests are sensitive to the type of asphalt binder, its chemistry, as well as aggregate substrate; and finally, it was difficult to ascertain a distinctive effect of WMA additives on the tribological behaviour of asphalt binders. The limitations and critical gaps of the studies reported in the literature have been highlighted, and the future areas of work are proposed to obtain a true tribological characterization of asphalt binders, with or without additives.

REFERENCES

AASHTO, 2011. *AASHTO: 316-11 Viscosity Determination of Asphalt Binder Using Rotational Viscometer.* AASHTO.

Bahia, H., Hanson, D., Zeng, M., Zhai, H., and Khatri, M., 2001. Characterization of modified asphalt binders in superpave mix design. No. Project 9-10 FY'96.

Bairgi, B.K., Manna, U.A., and Tarefder, R.A., 2019. Tribological evaluation of asphalt binder with chemical warm-mix additives. *In: Airfield and Highway Pavements 2019: Testing and Characterization of Pavement Materials - Selected Papers from the International Airfield and Highway Pavements Conference 2019.* American Society of Civil Engineers (ASCE), USA, 266–273.

Baumgardner, G.L. and Reinke, G., 2013. Binder additives for warm mix asphalt technology. asphalt paving technology. *In: Association of Asphalt Paving Technologists (AAPT) – Proceedings of the Technical Sessions, 82,* 685–709. 697.

Bhushan, B., 2002. *Introduction to Tribology,* 2nd ed., Wiley, John Wiley & Sons, Inc., USA.

Canestrari, F., Ingrassia, L.P., Ferrotti, G., and Lu, X., 2017. State of the art of tribological tests for bituminous binders. *Construction and Building Materials,* 157, 718–728.

Daryaee, D., Habibpour, M., Gulzar, S., and Underwood, B.S., 2020. Combined effect of waste polymer and rejuvenator on performance properties of reclaimed asphalt binder. *Construction and Building Materials,* 268, 121059.

Geisler, F., Kapsa, P., and Lapalu, L., 2016. Tribological and wettability study of non foaming warm mix asphalt additives at mixing and compaction temperatures. In *6th Eurasphalt & Eurobitume Congress, Prague, Czech Republic.*

Gulzar, S. and Underwood, S., 2019a. Use of polymer nanocomposites in asphalt binder modification. In: *Advanced Functional Textiles and Polymers.* Wiley, USA, 405–432.

Gulzar, S. and Underwood, S., 2019b. Nonlinear rheological behavior of asphalt binders. In: *91st Annual Meeting of The Society of Rheology.* Raleigh, NC: The Society of Rheology.

Gulzar, S. and Underwood, B.S., 2020a. Nonlinear viscoelastic response of crumb rubber modified asphalt binder under large strains. *Transportation Research Record: Journal of the Transportation Research Board,* 2674 (3), 139–149.

Gulzar, S. and Underwood, S., 2020b. Stress decomposition of nonlinear response of modified asphalt binder under large strains. In: *Advances in Materials and Pavement Performance Prediction (AM3P 2020)* (p. 436). San Antonio, TX: CRC Press.

Gulzar, S. and Underwood, S., 2020c. Fourier transform rheology of asphalt binders. In: *Advances in Materials and Pavement Performance Prediction II: Contributions to the 2nd International Conference on Advances in Materials and Pavement Performance Prediction (AM3P 2020)* (p. 436). San Antonio, TX: CRC Press.

Gulzar, S. and Underwood, S., 2020d. Large amplitude oscillatory shear of modified asphalt binder. In: *Advances in Materials and Pavement Performance Prediction II: Contributions to the 2nd International Conference on Advances in Materials and Pavement Performance Prediction (AM3P 2020)* (p. 436). San Antonio, TX: CRC Press.

Hanz, A.J. and Bahia, H.U., 2013. Asphalt binder contribution to mixture workability and application of asphalt lubricity test to estimate compactability temperatures for warm-mix asphalt. *Transportation Research Record: Journal of the Transportation Research Board*, 2371 (1), 87–95.

Hanz, A.J., Faheem, A., Mahmoud, E., and Bahia, H.U., 2010. Measuring effects of warm-mix additives. *Transportation Research Record: Journal of the Transportation Research Board*, 2180 (1), 85–92.

Ingrassia, L.P., Lu, X., Canestrari, F., and Ferrotti, G., n.d. Tribological characterization of bituminous binders with warm mix asphalt additives. *Construction and Building Materials*, 172, 309–318

Ingrassia, L.P., Lu, X., Marasteanu, M., and Canestrari, F., 2019. Tribological characterization of graphene nano-platelet (GNP) bituminous binders. *In: Airfield and Highway Pavements 2019: Innovation and Sustainability in Highway and Airfield Pavement Technology - Selected Papers from the International Airfield and Highway Pavements Conference 2019*. American Society of Civil Engineers (ASCE), 96–105.

Mallick, R.B. and El-Korchi, T., 2013. *Pavement Engineering: Principles and Practice.* CRC Press. USA.

Puchalski, S., 2012. *Investigation of warm mix asphalt additives using the science of tribology to explain improvements in mixture compaction* (Doctoral dissertation). University of Wisconsin-Madison.

Rabinowicz, E., 1995. *Friction and Wear of Materials,* 2nd ed. New York: Wiler, John Wiley & Sons, Inc.

Sefidmazgi, N.R., Teymourpour, P., and Bahia, H.U., 2013. Effect of particle mobility on aggregate structure formation in asphalt mixtures. *Road Materials and Pavement Design*, 14, 16–34.

Shahrivar, K., Ortiz, A.L., and De Vicente, J., 2014. A comparative study of the tribological performance of ferrofluids and magnetorheological fluids within steel-steel point contacts. *Tribology International*, 78, 125–133.

West, R., Watson, D., Turner, P., and Casola, J., 2010. Mixing and compaction temperatures of asphalt binders in hot-mix asphalt (No. Project 9-39), USA.

Zhai, H., Bahia, H.U., and Erickson, S., 2000. Effect of film thickness on rheological behavior of asphalt binders. *Transportation Research Record: Journal of the Transportation Research Board*, 1728 (1), 7–14.

14 Effect of Fatty Acid Composition on the Lubricating Properties of Bio-Based Green Lubricants

Wani Khalid Shafi, M.S. Charoo, and M. Hanief
National Institute of Technology,
Hazratbal , Srinagar, J & K, India.

CONTENTS

14.1 INTRODUCTION

The replacement of non-renewable sources (petroleum-based oils) has been a case study for almost three decades now. This is mainly due to the increase in demand for petroleum-based oils and the need for decreasing the pollutants released by petroleum-based emissions (Baba et al., 2019). The use of petroleum-based lubricants is associated with environmental problems like soil pollution, water pollution and other health issues. The USA has introduced many initiatives like the Great Lakes water quality initiative for the protection, maintenance and restoration of great lake resources. Executive order 12,873 was also introduced encouraging the use of environment-friendly oils (Woma et al., 2019). In Poland, the forest authorities have deemed necessary the use of sorbents and sorbent mats with all the machines and equipment for absorbing the spilled oil and fuel (Nowak et al., 2019). The initiative like the dangerous substances directive introduced by the European community

231

establishes the criteria for the possible impact of a product on the aquatic environment (Woma et al., 2019). Consequently, the use of sustainable lubricants becomes inevitable. Vegetable oils are forthcoming as a capable replacement of mineral oils. Vegetable oils possess the following (Salimon et al., 2010; Kodali, 2002; Agrawal et al., 2014; Shafi et al., 2019):

- Higher lubricity, resulting in minimizing friction and power losses.
- Excellent biodegradability, ensuring minimum environmental problems.
- Lower volatility, thereby decreasing exhaust emissions.
- High viscosity index.
- Higher dispersancy.
- High flashpoints.
- Lower evaporation.
- Better compatibility with human skin, resulting in fewer dermatological problems.

Vegetable oils are mostly composed of triglycerides in which glycerol molecules and fatty acids are attached via ester linkages. Vegetable oils contain polar and non-polar groups. Polar groups include a functional group of esters (COO), whereas non-polar groups include a hydrocarbon chain. The vegetable oils are classified depending on the structure of the fatty acids. The chemical structure of the different fatty acids is depicted in Table 14.1. The most common fatty acids present in vegetable oils include palmitic, stearic, oleic, linoleic and linolenic fatty acids. Based on the predominant fatty acids, the vegetable oils are classified as lauric oils, oleic-linoleic oils, erucic acid oils and ricinoleic oils (Shafi et al., 2018b). The fatty acid composition of bio-oils is depicted in Table 14.2. The amount of free fatty acids (FFAs), the chain length of fatty acids and the degree of unsaturation collectively govern the thermo-physical and tribological properties of vegetable oil (Sharma and Sachan, 2019; Jayadas and Nair, 2006; Havet et al., 2001). The amount of FFAs is measured by acid value, whereas the saponification value indicates the length of the fatty acid chain (Sharma and Sachan, 2019; Sajeeb and Rajendrakumar, 2019a). The unsaturation of vegetable oils is measured by the iodine number. Saponification number is the amount of potassium hydroxide (KOH) in milligrams needed to neutralize fatty acids developed from complete hydrolysis of 1 g of fat (Salaji and Jayadas, 2020). A higher saponification number indicates shorter chain lengths. Short-chain-length fatty acids possess higher content of functional groups (carboxylic) per unit mass of fat, consequently requiring more potassium hydroxide for saponification of 1g of fat (Sajeeb and Rajendrakumar, 2019a). Iodine number is the amount of iodine (g) consumed by 100 g of oil (Salaji and Jayadas, 2020). The carbon double bonds react with the iodine and higher consumption of iodine indicates higher contents of carbon double bonds (Sajeeb and Rajendrakumar, 2019a; Salaji and Jayadas, 2020).

The different industrial and maintenance applications of bio-oils are depicted in Figure 14.1 (Madanhire and Mbohwa, 2016). Vegetable oils also have been added as an additive in various petroleum-based oils for lubrication purposes (Singh et al., 2016; Singh et al., 2018; Bahari et al., 2018b; Shahabuddin et al., 2013a; Shahabuddin et al., 2013b). Triacylglycerol molecules result in the formation of surface film by

TABLE 14.1

Conventional Fatty Acids Present in Bio-Oils (Gopinath et al., 2010; Sajjadi et al., 2016)

Fatty Acid	IUPAC Name	Molecular Formula	Number of Carbons/ Double Bonds	Structural Formula
Caprylic acid	Octanoic acid	$C_8H_{16}O_2$	C8:0	$CH_3(CH_2)_6COOH$
Capric acid	Decanoic acid	$C_{10}H_{20}O_2$	C10:0	$CH_3(CH_2)_8COOH$
Lauric acid	Dodecanoic acid	$C_{12}H_{24}O_2$	C12:0	$CH_3(CH_2)_{10}COOH$
Myristic acid	Tetradecanoic acid	$C_{14}H_{28}O_2$	C14:0	$CH_3(CH_2)_{12}COOH$
Palmitic acid	Hexadecanoic acid	$C_{16}H_{32}O_2$	C16:0	$CH_3(CH_2)_{14}COOH$
Palmitoleic Acid	(9Z)-Hexadec-9-enoic acid	$C_{16}H_{30}O_2$	C16:1	$CH_3(CH_2)_5CH=CH(CH_2)_7COOH$
Stearic acid	Octadecanoic acid	$C_{18}H_{36}O_2$	C18:0	$CH_3(CH_2)_{16}COOH$
Oleic acid	(9Z)-Octadec-9-enoic acid	$C_{18}H_{34}O_2$	C18:1	$CH_3(CH_2)_7CH=CH(CH_2)_7COOH$
Linoleic acid	(9Z,12Z)-octadeca-9,12-dienoic acid	$C_{18}H_{32}O_2$	C18:2	$CH_3(CH_2)_4CH=CHCH_2CH=CH(CH_2)_7COOH$
α - Linolenic acid	cis- 9,12,15-Octadeca-trienoic acid	$C_{18}H_{30}O_2$	C18:3	$CH_3CH_2CH=CHCH_2CH=CHCH_2CH=CH(CH_2)_7COOH$
Arachidic acid	Eicosanoic acid	$C_{20}H_{40}O_2$	C20:0	$CH_3(CH_2)_{18}COOH$
Behenic acid	Docosanoic acid	$C_{22}H_{44}O_2$	C22:0	$CH_3(CH_2)_{20}COOH$
Erucic acid	(Z)-Docos-13-enoic acid	$C_{22}H_{42}O_2$	C22:1	$CH_3(CH_2)_7CH=CH(CH_2)_{11}COOH$

attaching their polar molecular to the tribosurface, consequently decreasing friction and wear (Adhvaryu et al., 2005; Salih et al., 2011; Adhvaryu et al., 2004). However, oxidation stability and low-temperature properties restrict the application of vegetable oils (Shafi et al., 2018a). The higher content of polyunsaturated fatty acids present in vegetable oil improves the low-temperature properties and degrades the oxidation stability (Schneider, 2006). Consequently, some chemical modifications or the use of additives is necessary for their suitability as lubricants.

14.2 EFFECT OF FATTY ACIDS ON THE TRIBOLOGICAL PROPERTIES OF BIO-OILS

The chemical composition of vegetable oils plays an important role in defining their tribological properties as a lubricant. Polar ester groups of vegetable

TABLE 14.2
Commonly Used Bio-Oils for Lubrication Purposes

Bio-Oils	C8:0	C10:0	C12:0	C14:0	C16:0	C16:1	C18:0	C18:1	C18:2	C18:3 (n–3, n–6)	C20:0	C22:1
Olive oil	–	–	–	–	16.5	–	2.3	66.4	16.4	1.6	0.43	–
	–	–	–	–	12.1	0.8	2.6	72.5	9.4	0.6	0.4	–
Coconut oil	8	6	49	17	8	–	2.5	6.5	2	–	–	–
	9.5	4.5	51	18.5	7.5	–	3	5	1	–	–	–
Mustard oil	–	0.1	0.2	0.1	3	0.2	1	20.5	9.5	9	0.8	45
	–	0.1	0.2	0.1	2.6	0.2	1.1	12.9	13.8	8.8	0.7	48.7
Belize	0.6	–	–	–	3.3	–	3.5	72.1	1.3	10.3	–	3.5
Rapeseed oil	–	–	–	0.1	5.1	0.2	1.7	60.1	21.5	9.9	0.6	0.4
	–	–	–	–	6	–	2	61	19.5	8	–	–
SBO	–	–	–	0.1	10.7	–	4.4	25.9	50.2	7.2	–	–
	–	–	<0.1	<0.1	9.9	–	3	18.9	42.3	5.5	–	–
Hazelnut oil	–	–	–	–	5.11	0.15	2.27	77.25	13.59	0.19	0.14	–
	–	–	–	–	4.9	–	2.7	82.7	8.9	0.1	0.1	–
JO	–	–	0.4	–	14.6	1.2	6.6	40.6	36.2	0.3	0.22	–
	–	–	–	1.4	14.6	1.47	7.36	41.43	35.42	0.2	0.3	–
	–	–	–	–	15.3	1.1	5.6	37.1	40.4	0.3	–	–
Palm oil	0.08	0.06	0.36	1.13	42.31	0.17	4.27	40.9	10.07	0.28	0.31	–
	–	–	–	–	44	–	5	40	10	Traces	–	–
Safflower oil	–	–	–	0.12	7.41	0.04	2.36	14.37	75.17	0.08	0.08	–
	–	–	–	–	7.95	–	1.91	12.59	77.54	–	–	–
Sunflower oil	–	–	–	–	6.08	–	3.26	16.93	73.73	–	–	–
	–	–	–	0.04	6.35	0.07	3.92	20.91	67.58	0.17	0.22	–
Avocado oil	–	–	–	–	15.7	7.3	0.7	60.3	13.7	1.4	–	–
	–	–	–	0.06	12.87	3.86	1.45	60.87	18.70	0.92	0.31	0.16

Source: Orsavova et al. (2015), Dubois et al. (2007), Yalcin et al. (2012), Sajeeb and Rajendrakumar (2019a), Koshy et al. (2015), Schneider (2006), Singh et al. (2016), Sajjadi et al. (2016), Giakoumis (2018), Misra and Murthy (2010), Okullo et al. (2012), Goering et al. (1982), Bahari et al. (2018a), Choi et al. (1997)

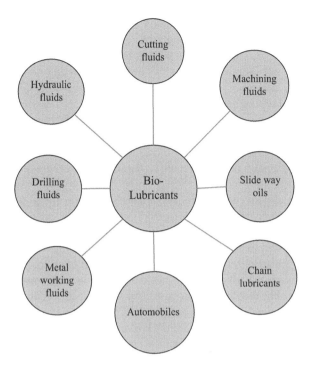

FIGURE 14.1   Applications of bio-lubricants.

oils bind themselves on the metal surfaces, thereby augmenting the strength of protective lubricant films. The non-polar fatty acid component results in the development of mono-multimolecular layers between the surfaces and prevents the metal–metal interaction. Adhvaryu et al. (2004) modified soybean oil (SBO) using heat treatments and chemical processes and compared their tribological properties with pure SBO. The coefficient of friction (COF) with chemically modified soybean oil (CMSBO) was nearly halved in comparison with the SBO and thermally modified SBO. The wear properties studied using SEM analysis indicated better wear characteristics with CMSBO than the pure oil and was ascribed to the higher percentage of polar groups obtained by the chemical modification. The polar groups bind to the metal surface, whereas the non-polar groups form a molecular layer and separate the surfaces in contact. The polar groups strongly interact with the metal surface resulting in a stable protective film. A similar investigation by Kashyap and Harsha (2016) demonstrated that the epoxidized rapeseed oil depicts better COF and wear properties than the pure rapeseed oil owing to more polar groups that resulted in strong chemical and physical adsorption on the tribosurfaces.

Sharma and Sachan (2019) investigated the lubricating properties of karanja oil (KO) on a four-ball tribometer using ASTM 4172 standards. The authors ascribed the effective tribological properties to the absorption of polar ester molecules on to the tribosurface. The ester groups increase the strength of protective film on the surface, thereby increasing the resistance to the shearing forces.

Noorawzi and Samion (2016) conducted an experimental study on the lubricating properties of palm oil and compared the results with hydraulic oil and SAE-40 engine oil. Palm oil showed lesser COF than hydraulic oil and nearly similar COF as that of engine oil. The maximum COF observed with engine oil, palm oil and hydraulic oil at a load and rpm of 100 N and 5 m/s is equal to 0.025, 0.024 and 0.1, respectively. Further, engine oil also depicted the minimum wear rate in comparison with palm oil and hydraulic oil. However, palm oil exhibited minimum surface roughness among all the lubricants and indicated its effectiveness in preserving the surface finish of tribopairs. The authors concluded that the improvement in tribological properties with palm oil is attributed to the monoprotective layer formed on the tribosurfaces.

However, the effectiveness of protective layers depends on the dominating fatty acids present in the vegetable oils. Saturated fatty acids with no double bonds align themselves in a linear chain and get closely packed to the tribosurfaces resulting in stronger protective films on the surface, consequently leading to a reduction in friction and wear (Siniawski et al., 2007). The monounsaturated fatty acids also form protective layers on the tribosurfaces, but the effectiveness is lesser than the saturated fatty acids. The strength of the protective layers keeps on decreasing with the increase in double bonds as it forces a bend in the fatty acid chain and results in the development of weaker protective films (Fox et al., 2004). The higher content of double bonds decreases the density of surface layers, thereby degrading the tribological properties of oil.

Fox et al. (2004) in their study reported the impact of FFAs on the wear properties of the oil. It was observed that the addition of monounsaturated fatty acids (oleic acid) and saturated fatty acid (stearic acid) in the sunflower oil improved its wear properties with maximum reduction in wear depicted by stearic acid. However, the addition of linoleic acid did not impact the wear properties of the bio-oil. The authors concluded that the increase in unsaturated fatty acids has a negative impact on wear.

Reeves et al. (2015) compared the tribological properties of eight bio-oils (avocado, olive, peanut, rapeseed, sesame, soybean, corn and safflower oil). The dominating fatty acids in all the bio-oils were oleic acid and linoleic acid, whereas the amount of saturated fatty acids was comparable to the different oils. It was observed that the bio-oils with high-oleic acid content depict less COF and wear rate except for sunflower oil. The different results of the study are depicted in Table 14.3. Avocado oil depicts minimum COF and wear rate equal to 0.02 and 0.1 mm, respectively, as contemplated from Table 14.3. The wear properties were found to be highly dependent on the unsaturation number, whereas saturated fatty acids also had a moderate impact on the wear properties. Higher oleic acid and lower linoleic fatty acids lead to an improvement in the frictional properties. The authors observed that the higher content of double bonds in linoleic acid and linolenic acid decreases the density of the protective monolayer formed on the surface, thereby degrading the frictional properties. However, olive oil and safflower oil are comprised of higher contents of oleic acid than the avocado oil and were still less effective than the avocado oil. It was concluded that the performance of oils with nearly similar oleic acid content also depends on the amount of other saturated, unsaturated and free fatty acids.

TABLE 14.3

Effect of Fatty Acids on Tribological Properties

Bio-Oils	C18:0	C16:0	C18:1	C18:2	COF	Wear (mm³)
Avocado oil	1	10.2	68	11	0.02	0.104
Olive oil	2	11	71	10	0.0778	0.175
Peanut oil	2	9	43	32	0.0975	0.168
Sesame oil	4	8	39	41	0.318	0.215
SBO	4	10	22	51	0.405	0.38
Corn oil	2	10	28	52	0.44	0.296
Safflower oil	2	5	75	12	0.49	0.195
Canola oil	2	5	61	29	0.0868	0.2

Koshy et al. (2015) compared the thermophysical and tribological properties of coconut oil and mustard oil. Lauric acid and myristic acid equal to 49% and 17%, respectively, are the dominating fatty acids of coconut oil, whereas the major fatty acids of mustard oil include oleic acid and erucic acid equal to 21% and 45%, respectively. The results showed that coconut oil depicts better frictional properties and wear properties than mustard oil. The COF observed with coconut oil and mustard oil is equal to 0.09 and 0.126, respectively. The specific wear rate observed with coconut oil and mustard oil is equal to 7.3×10^{-9} and 8.2×10^{-9} cm³/Nm, respectively.

In another investigation, Sajeeb and Rajendrakumar (2019a) reported that coconut oil depicts better frictional properties, whereas mustard oil depicts better wear properties. The observed WSD for coconut oil and mustard oil is equal to 597 and 478 μm, respectively. The effectiveness of mustard oil in improving the wear properties was attributed to the longer chain length of erucic fatty acids.

Cermak et al. (2013) determined the tribological and physiochemical properties of different crop oils (cuphea, pennycress, lesquerella and meadowfoam). It was observed that cuphea oil with dominating saturated fatty acid (68%) depicts minimum WSD in comparison with the other oils. The excellent wear properties of cuphea oil were attributed to the stronger protective layers formed by saturated fatty acids. Siniawski et al. (2007) investigated the impact of fatty acids on two vegetable oils (soybean and sunflower oil). It was noted that the SBO depicts a lower abrasion rate than the sunflower oil. The authors deduced that since both the oils have nearly similar saturated fatty acids, the lubricating properties must be governed by the unsaturated fatty acids.

Quinchia et al. (2014) conducted an experimental study on determining the tribological properties of SBO, high-oleic sunflower oil and castor oil. The authors reported that castor oil and high-oleic sunflower oil depict better tribological properties than the SBO whose main fatty acid is linoleic fatty acids. Castor oil and high-oleic sunflower oil formed thick boundary films, whereas no boundary film formation was observed with SBO. The authors also reported that castor oil depicts better tribological performance than the high-oleic sunflower oil owing to its hydroxyl functional group that increases the polarity and viscosity of the castor oil.

The other factors that govern the lubricating properties of vegetable oils are as follows:

1. **Chain length** – The oils consisting of fatty acids with longer chain lengths are associated with better wear properties.
2. **Acid value** – FFAs may improve the tribological properties by acting as a boundary lubricant. However, under extreme conditions, it may degrade the wear properties as it is more prone to oxidation.
3. **Iodine number** – The increase in the iodine number increases the formation of FFAs. The increase in FFAs is associated with high wear.

14.3 EFFECT OF FATTY ACIDS ON THE RHEOLOGICAL PROPERTIES OF BIO-OILS

The evaluation of the thermophysical characteristics of natural oils is important for their functioning as a lubricant. Different thermophysical properties governing the effectiveness of lubricants include flashpoint, viscosity, viscosity index and fire point. The viscosity of a lubricant affects the lubrication regimes at tribosurfaces. Low viscosity can cause the metal-to-metal interaction in boundary and lubrication regimes, thereby leading to the degradation of tribological properties. Further, the viscosity of a lubricant also governs the pumping efficiency where high viscosity can lead to higher power consumptions (Pantzali et al., 2009; Peng et al., 2009; Gherasim et al., 2011). Bio-oils depict good viscosity characteristics with high viscosity indices. High viscosity indices are one of the important properties of lubricating that make bio-oils desirable. This indicates a better resistance to the change in viscosity with an increase in temperature.

The viscosity of bio-oils is also governed by their fatty acid composition. The increase in double bonds in oils is associated with more flowability, therefore, less viscosity. Also, the increase in the saturated fatty acid concentration is associated with the increase in viscosity owing to their higher melting points. It has been delineated that the increase in monounsaturation increases the viscosity of oil, whereas the decrease in polyunsaturation decreases the viscosity. Further, the oils with nearly similar unsaturation are also dependent on the saturated fatty acids.

Yalcin et al. (2012) compared the impact of fatty acids on the viscosity of various bio-oils. The different oils involved in the study include hazelnut, soybean, cottonseed, sunflower, canola and olive oils. The rheological properties are measured at a temperature of 25°C and a shear rate of 0.1–100 s^{-1}. The fatty acid composition of different oils is depicted in Figure 14.2a, whereas the measured viscosity for different bio-oils is depicted in Figure 14.2b. The authors reported that the unsaturated fatty acids in the oils mainly govern the viscosity characteristic of oil. The decrease in monounsaturation and an increase in polyunsaturation resulted in a decrease in viscosity. However, saturated fatty acids also had an effect on the oils with similar monounsaturated fatty acid content. Hazelnut oil with a greater amount of monounsaturated fatty acids depicted less viscosity than the olive oil. The higher viscosity in the olive oil was accredited to the greater contents of saturated fatty acids,

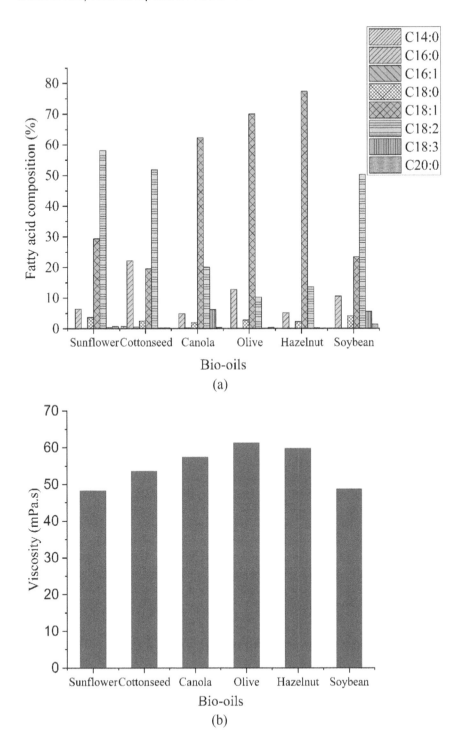

FIGURE 14.2   (a) Fatty acid composition of bio-oils and (b) viscosity of bio-oils.

which resulted in greater resistance to the flow, thereby increasing the viscosity. It is deduced that both mono- and saturated fatty acids govern the viscosity characteristics of vegetable oils. A similar observation is reported by other researchers.

Shafi and Charoo (2019a) investigated the rheological behaviour of three bio-oils, apricot oil, hazelnut oil and avocado oil, at a varying temperature of 40°C–100°C and shear rates of 1°C–4,000 s^{-1}. All the oils possess higher percentages of oleic fatty acids. The maximum viscosity among the three bio-oils was depicted by avocado oil that possesses a higher concentration of saturated fatty acids than the other two oils.

Fasina et al. (2006) studied the relationship between the viscosity and fatty acids of twelve vegetable oils. The authors found a positive correlation between mono-unsaturated fatty acids and viscosity, whereas a negative correlation was observed between polyunsaturated fatty acids and viscosity. This implies an increase in viscosity with an increase in monounsaturation and a decrease in viscosity with an increase in polyunsaturation. However, the correlation obtained between the fatty acids and viscosity was poor. The effect of saturated fatty acids was also noticeable in oils with similar unsaturation.

The evaluation of the flow behaviour of the oils is also necessary as it can affect the tribological and pumping properties. The bio-oils tend to show non-Newtonian behaviour at low shear rates 1–50 1/s and change their flow behaviour to Newtonian at higher shear rates. Sajeeb and Rajendrakumar (2019b) examined the flow behaviour of coconut oil at a temperature of 60°C and a shear rate of 0–500 s^{-1}. Further, the effect of temperature ranging from 30°C to 90°C on the viscosity of the oil is observed at a shear rate of 400 s^{-1}. It was observed that the bio-oil depicts shear thinning behaviour at very low shear rates of 5–20 s^{-1}. Afterward, the bio-oil behaves as a Newtonian fluid. Further, the viscosity of the oil decreases with an increase in temperature.

A similar investigation of the rheological properties of coconut oil is carried out by Cortes and Ortega (2019). The flow behaviour is determined for a varying shear rate of 1–100 s^{-1}. The authors noted that coconut oil depicts shear thinning for small shear rates up to 20 s^{-1} and the flow behaviour changes to Newtonian after 20 s^{-1}.

Shafi and Charoo (2019b) investigated the rheological properties of sesame at varying temperatures of 20°C, 50°C and 70°C. It is observed that the sesame oil depicts nearly Newtonian behaviour at all temperatures. The power-law index at 20°C, 50°C and 70°C is equal to 1.005, 1.01 and 101, respectively, implying the Newtonian behaviour of fluids. Abdelraziq and Nierat (2015) investigated the flow behaviour of castor oil at varying temperatures of 1.7°C to 62°C and varying shear rates of 6–56 s^{-1}. The authors report that the castor oil behaves as a Newtonian fluid with a power-law index close to 1.

14.4   EFFECT OF FATTY ACIDS ON THE THERMAL PROPERTIES OF BIO-OILS

14.4.1   HIGH-TEMPERATURE PROPERTIES

The flashpoint of a lubricant is the lowest temperature where its vapours start burning. However, the burning of vapours stops after the withdrawal of the ignition source. The fire point of a lubricant is the lowest temperature where its vapours continue

TABLE 14.4

Effect of Fatty Acids on Thermal Properties

Bio-Oils	C16:0	C18:0	C18:1	C18:1[a]	C18:2	C18:3	C20:0	Flashpoint (°C)	Pour Point (°C)	Cloud Point (°C)
Castor	1.09	3.1	4.85	89.6	1.27	–	–	260	–31.7	none
Corn	11.67	1.85	25.16	–	60.6	0.48	0.24	277	–4.0	–1.1
Cottonseed	28.33	0.89	13.27	–	57.51	–	–	234	–1.5	1.7
Crambe	2.07	0.7	18.86	–	9	6.85	2.09	274	–12.2	10
Linseed	4.92	2.41	19.7	–	18.03	54.94	–	241	–1.5	1.7
Peanut	11.38	2.39	48.28	–	31.95	0.93	1.32	271	–6.7	12.8
Rapeseed	3.49	0.85	64.4	–	22.3	8.23	–	246	–31.7	–3.9
Safflower	8.6	1.93	11.58	–	77.89	–	–	260	–6.7	18.3
H.O. Safflower	5.46	1.75	79.36	–	12.86	–	0.23	293	–20.6	–12.2
Sesame	13.1	3.92	52.84	–	30.14	–	–	260	–9.4	–3.9
Soybean	11.75	3.15	23.26	–	55.53	6.31	–	254	–12.2	–3.9
Sunflower	6.08	3.26	16.93	–	73.73	–	–	274	–1.5	7.2

[a] indicates ricinoleic acid.

to burn even after the removal of the ignition source. Goering et al. (1982) investigated the flashpoints of vegetable oils as per ASTM D93 standard. The oils with low acid values and low peroxide values tend to show higher flashpoints as depicted in Table 14.4. The maximum flashpoints are depicted by high-oleic safflower oil, corn oil and sunflower oil equal to 293, 277 and 274, respectively, which possess fairly low acid and peroxide values. However, peanut oil is an exception that shows a high flashpoint despite possessing high peroxide value.

Sneha et al. (2019) investigated the physicochemical and thermal properties of jatropha oil (JO), rice bran oil (RBO) and KO. The authors report that RBO depicts higher flashpoints equal to 312 in comparison with JO and KO. The acid values observed with RBO are lower than JO and KO and are attributed to the presence of lesser FFA in the RBO.

Mannekote et al. (2017) investigated the physicochemical properties of three different types of coconut oils. Refined coconut oil (RCO) had slightly greater flashpoints than the unrefined coconut oil (URCO) and virgin coconut oil (VCO). The flashpoint of RCO, URCO and VCO lies between 320°C–328°C, 313°C–320°C and 310°C–318°C, respectively. The lower flashpoints in URCO and VCO were attributed to the higher contents of FFAs and peroxides present in them. The FFAs and peroxides are more volatile compounds and result in lower flashpoints.

14.4.2 Low-Temperature Properties

Pour point and cloud point can define the progress of bio-oils as lubricants in cold conditions. The cloud point of a lubricant is the temperature at which some components of a lubricant begin separating from the oil. The separating components start crystallizing resulting in the formation of waxes. The pour point of a lubricant is the temperature below which the lubricant loses its ability to flow. The oils should possess high pour points to function as a lubricant at low temperatures. The low-temperature properties of a lubricant are governed by the amount of saturated fatty acids present in the oil. Saturated fatty acids like palmitic acid and stearic acid are solid at room temperatures, thereby affecting the flowability below room temperatures. Higher the saturated fatty acid composition, better the pour point (Zeman et al., 1995). The saturated bonds allow the packing between the fatty acid molecules close to each other leading to the decrease in the flowability of oils. Rodrigues Jr et al. (2006) reported that the location of double bonds also affects the low-temperature properties. The double bonds close to the end of the carbon chain depict higher cloud points than the double present near the middle of the chain.

Babu et al. (2018) investigated the low-temperature properties of pure coconut oil and chemically modified coconut oil. The chemical modification of coconut oil achieved by the alkali transesterification process depicted better pour points equal to −4.5°C in comparison to 26°C with pure coconut oil. The authors reported that the low-temperature properties depend on the motion of triglycerides which is further dependent on the intermolecular forces of attraction between the triglyceride molecules. The alkali transesterification process removes the glycerol molecules from the oil and enhances the free movement of fatty acids, consequently increasing the pour point of the oil.

Lanjekar and Deshmukh (2016) reported that rapeseed esters and soybean esters depict higher pour points and cloud points in comparison with palm methyl esters. The pour point with palm methyl esters, rapeseed esters and soybean esters are equal to 10°C, −4°C and −10°C, respectively, whereas the cloud point is equal to 14°C, 0°C and −5°C. The better low-temperature properties of the rapeseed and soybean esters were attributed to their higher saturated fatty acid concentration. The pour point and cloud point of various bio-oils are depicted in Table 14.4 (Goering et al., 1982).

14.5 EFFECT OF FATTY ACIDS ON THE OXIDATIVE PROPERTIES OF BIO-OILS

The oxidation stability is a prime factor that is preventing the use of vegetable oils as an industrial lubricant (Campanella et al., 2010). Unsaturated fatty acids govern the oxidation stability of bio-oils (Shafi and Charoo, 2020). Double bonds between the carbon atoms in a fatty acid chain are active spots for oxidation reactions. The oxidation of the vegetable oils is initiated by the generation of free radicals (Erhan et al., 2006). The free radicals react with the oxygen and form peroxy radicals. The peroxy radicals then attract other triglycerides to form hydroperoxides, simultaneously generating more free radicals. Hydroperoxides are the chief oxidation compounds of bio-oils. Hydroperoxides subsequently break down to generate more free radicals. The generation of free radicals increases the propagation of hydroperoxides in the oil. However, all the free radicals do not propagate the formation of hydroperoxides. Some free radicals react with one another and result in the formation of stable compounds. After sometime, the assemblage of hydroperoxides becomes unstable and it decomposes into

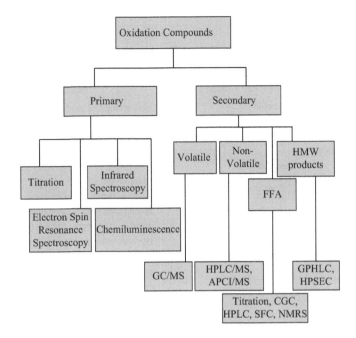

FIGURE 14.3 Measurement techniques for oxidation compounds (GC/MS, Gas chromatography/mass spectrometry; HPLC/MS, High-performance liquid chromatography; APCI/S, Atmospheric pressure chemical ionization; GPHLC, Gel permeation high-performance liquid chromatography, HPSEC, High-performance size-exclusion chromatography; CGC, Capillary gas chromatography; SFC, Supercritical fluid chromatography; NMRS, Nuclear magnetic resonance spectroscopy.)

volatile and non-volatile oxidation products (Fox and Stachowiak, 2007). Therefore, higher unsaturation in bio-oils decreases the oxidation stability of bio-oils. The different analytical techniques used for the measurement of primary and secondary oxidation compounds are depicted in Figure 14.3 (Fox and Stachowiak, 2007).

The chemical modifications of bio-oils are necessary for enhancing the high-temperature properties (Salih et al., 2011; Birova et al., 2002). The chemical modification of bio-oils can be achieved in two ways: (1) modifications at the double bonds of the fatty acid chain and (2) modifications at the carboxyl group of triglycerides (Cecilia et al., 2020). Various chemical modifications and the use of antioxidants are reported in the studies and are given in Table 14.5.

Jayadas and Nair (2006) made a comparative investigation of the thermal and oxidative properties of three vegetable oils (coconut oil, sesame oil and sunflower oil) and results are compared with commercial mineral oil (2T oil). It was observed that weight gain due to the oxidation products is lesser for coconut oil in comparison with the sesame oil and sunflower oil. Consequently, it was concluded that coconut oil depicts higher oxidation stability among the three bio-oils. Erhan and Asadauskas (2000) made a comparative investigation of the oxidation stability of vegetable oils, mineral oil and synthetic oil. The different oils involved in the study include SBO, high-oleic sunflower oil (HOSO), polyalphaolefin (PAO) oil and commercial mineral

TABLE 14.5
Methods Affecting the Oxidation Stability of Bio-Oils

Methods	Chemical Modification/Additives	Mechanism
Chemical modification	Epoxidation	• Acts on the C=C and converts them into oxiranes • Increases the oxidation stability and lubricity of oil
	Transesterification	• Conversion of one ester to a different by the addition of alcohol • Leads to better oxidation stability
	Selective hydrogenation	• Involves the modification of the C=C and decreases the degree of unsaturation • Hydrogenation of unsaturation of ester
	Estolide formation	• Involves the reaction between two fatty acids in the presence of acidic catalyst • Acidic functional group of one fatty acid reacts at the unsaturation sites of another fatty acid
Antioxidants	Chain-breaking radical scavengers	• Antioxidants react with free radicals and form stable compounds • Examples: butylated hydroxyanisole, propyl gallate, tocopherols, butylated hydroxytoluene
	Peroxide decomposers	• Decomposes hydroperoxides into stable compounds • Examples: catalase and glutathione

Source: Cecilia et al. (2020), Becker and Knorr (1996), Salimon et al. (2010), Schneider (2006), Aluyor and Ori-Jesu (2008), Adhvaryu and Erhan (2002), Kerni et al. (2019)

oil. It was observed that the vegetable oil depicts less oxidation stability in comparison with PAO and mineral oil. SBO depicts a faster rate of oxypolymerization in comparison with other oils due to higher polyunsaturated fatty acids. The authors concluded that the use of additive is a must for making the use of vegetable oils viable in industries.

Castro et al. (2006) investigated the oxidation and lubrication behaviour of SBO, high-oleic soybean oil (HOSO) and epoxidized soybean oil (ESBO). The tribological tests are performed in a four-ball tester, whereas oxidative products are evaluated on the gel permeation chromatography (GPC) and Fourier-transform infrared spectroscopy (FTIR). The GPC results indicated the rate of formation of high molecular weight (HMW) products is greater with SBO and HOSO. Lesser HMW products are formed with ESBO indicating better oxidation stability.

Hwang and Erhan (2001) modified the epoxidized soybean oil in two steps: (1) ring-opening reaction and (2) esterification of OH groups formed by ring-opened product. The ring-opening reaction was achieved using various alcoholic precursors such as methanol, 1-butanol, 2-butanol, cyclohexanol, 1-hexanol, 1-decanol and 2,2, dimethyl-1-propanol, whereas esterification of ring-opened products was carried out by acid anhydride, water and pyridine. However, the oxidation stability was determined only for hexyl-esterified products using micro-oxidation tests and was compared with soybean, epoxidized soybean and mineral oils. The oxidative evaporation of esterified products was higher than SBO, epoxidized soybean oil and mineral oil and was between 69% and 71% for the esterified products. The authors also reported that esterified SBO formed very less insoluble products after oxidation in comparison with the soybean and epoxidized soybean oil. The deposits were even lesser than most mineral oils except one.

El-Boulifi et al. (2015) chemically modified the jojoba oil using transesterification technique using various alcoholic precursors (methanol, 1-butanol, ethanol and 1-butanol) followed by the removal of jojobyl alcohols from fatty acid alkyl esters (FAAEs). The oxidation stability of FAAE was compared with transesterified jojoba oils using Rancimat instrument. The authors report that the oxidation stability of oils with similar unsaturation is governed by their molecular weights. It was observed that FAAEs depict better oxidation properties than their transesterified counterparts. The maximum induction time equals to 6.89 h was depicted by fatty acid butyl ester followed by its transesterified counterpart where induction time was equal to 6.2 h.

Adhvaryu and Erhan (2002) compared the oxidative stability of vegetable oils, i.e., SBO, epoxidized soybean oil (ESBO) and high-oleic soybean oil (HOSO). The thin microfilm test revealed that epoxidized soybean oil is most stable in comparison with the other oils. The induction time with epoxidized soybean oil for deposit formation is nearly double than that with HOSO at 175°C. Better results with ESBO were attributed to the removal of double bonds. Adhvaryu and Erhan (2002) in the same study added alkylated phenolic antioxidant in the oils. The pressure differential scanning calorimeter (PDSC) showed similar results where ESBO mixed with antioxidant depicts better results than addivated HOSO and addivated SBO. Further, it was observed that the increase in the antioxidant concentration increases the peak temperature of oils.

Wu et al. (2000) investigated the oxidation properties of rapeseed oil, epoxidized rapeseed oil and additive rapeseed oil. The antioxidant used in the study is a combination of different compounds like 2-2-methylene-bisphenol, methylene-thiocarbonate, alkylated diphenylamine and benzenetriazole. The study finds that the generation of hydroperoxides in the rapeseed oil increases very rapidly in comparison with the epoxidized rapeseed oil. The hydroperoxide value increases to 100 within 16 h, whereas for epoxidized rapeseed oil, it was equal to 17 after 20 h. The rotary bomb tests revealed that the addition of antioxidants in the epoxidized rapeseed oil and rapeseed oil further increases the oxidation stability of oils.

14.6 FUTURE SCOPE

The environmental and energy crisis is the driving force for the replacement of petroleum-based oils. The present study supports and encourages the use of renewable bio-oils for lubrication purposes. However, the performance of bio-oils as industrial lubricants depends on their fatty acid composition. The studies reported in the chapter have established the need for understanding the effects of fatty acid composition on the lubricating properties of bio-oils. The different shortcomings of the saturated and polyunsaturated fatty acids like low-temperature properties and oxidative properties, respectively, have been established.

1. The chapter can assist the researchers in the selection of vegetable oils as industrial lubricants. The study will help in selecting a specific bio-oil for a particular industrial application.
2. Different chemical modifications and the use of additives can be employed for improving the thermophysical and tribological properties of bio-oils.
3. The effect of different commercial additives on the biodegradability of the bio-oils can also be employed.

REFERENCES

Abdelraziq, I. & Nierat, T. 2015. Rheology properties of castor oil: Temperature and shear rate-dependence of castor oil shear stress. *Journal of Material Sciences and Engineering*, 5(220), pp 2169–0022.10002. Available at Doi: 10.4172/2169-0022.1000220.

Adhvaryu, A. & Erhan, S. 2002. Epoxidized soybean oil as a potential source of high-temperature lubricants. *Industrial Crops and Products*, 15(3), pp 247–254. Available at Doi: 10.1016/S0926-6690(01)00120-0.

Adhvaryu, A., Erhan, S. & Perez, J. 2004. Tribological studies of thermally and chemically modified vegetable oils for use as environmentally friendly lubricants. *Wear*, 257(3–4), pp 359–367. Available at: Doi: 10.1016/j.wear.2004.01.005.

Adhvaryu, A., Liu, Z. & Erhan, S. 2005. Synthesis of novel alkoxylated triacylglycerols and their lubricant base oil properties. *Industrial Crops and Products*, 21(1), pp 113–119. Available at: Doi: 10.1016/j.indcrop.2004.02.001

Agrawal, S. M., Lahane, S., Patil, N. & Brahmankar, P. 2014. Experimental investigations into wear characteristics of M2 steel using cotton seed oil. *Procedia Engineering*, 97, pp 4–14. Available at Doi: 10.1016/j.proeng.2014.12.218.

Aluyor, E. O. & Ori-Jesu, M. 2008. The use of antioxidants in vegetable oils–A review. *African Journal of Biotechnology*, 7(25), pp 4836–4842. Available at: http://www.academicjournals.org/AJB.

Baba, Z. U., Shafi, W. K., Haq, M. I. U. & Raina, A. 2019. Towards sustainable automobiles-advancements and challenges. *Progress in Industrial Ecology, an International Journal*, 13(4), pp 315–331. Available at Doi: 10.1504/PIE.2019.102840.

Babu, K. J., Kynadi, A. S., Joy, M. & Nair, K. P. 2018. Enhancement of cold flow property of coconut oil by alkali esterification process and development of a bio-lubricant oil. *Proceedings of the Institution of Mechanical Engineers, Part J: Journal of Engineering Tribology*, 232(3), pp 307–314. Available at Doi: 10.1177/1350650117713541.

Bahari, A., Lewis, R. & Slatter, T. 2018a. Friction and wear phenomena of vegetable oil–based lubricants with additives at severe sliding wear conditions. *Tribology Transactions*, 61 (2), pp 207–219. Available at Doi: 10.1080/10402004.2017.1290858.

Bahari, A., Lewis, R. & Slatter, T. 2018b. Friction and wear response of vegetable oils and their blends with mineral engine oil in a reciprocating sliding contact at severe contact conditions. *Proceedings of the Institution of Mechanical Engineers, Part J: Journal of Engineering Tribology*, 232(3), pp 244–258. Available at Doi: 10.1177/1350650117712344.

Becker, R. & Knorr, A. 1996. An evaluation of antioxidants for vegetable oils at elevated temperatures. *Lubrication Science*, 8(2), pp 95–117. Available at Doi: 10.1002/ls.3010080202.

Birova, A., Pavlovičová, A. & Cvenroš, J. 2002. Lubricating oils based on chemically modified vegetable oils. *Journal of Synthetic Lubrication*, 18(4), pp 291–299. Available at Doi: 10.1002/jsl.3000180405.
Campanella, A., Rustoy, E., Baldessari, A. & Baltanás, M. A. 2010. Lubricants from chemically modified vegetable oils. *Bioresource Technology*, 101(1), pp 245–254. Available at Doi: 10.1016/j.biortech.2009.08.035.
Castro, W., Perez, J. M., Erhan, S. Z. & Caputo, F. 2006. A study of the oxidation and wear properties of vegetable oils: Soybean oil without additives. *Journal of the American Oil Chemists' Society*, 83(1), pp 47–52. Available at Doi: 10.1007/s11746-006-1174-2.
Cecilia, J. A., Ballesteros Plata, D., Alves Saboya, R. M., Tavares de Luna, F. M., Cavalcante, C. L. & Rodríguez-Castellón, E. 2020. An overview of the biolubricant production process: Challenges and future perspectives. *Processes*, 8(3), p 257. Available at Doi: 10.3390/pr8030257.
Cermak, S. C., Biresaw, G., Isbell, T. A., Evangelista, R. L., Vaughn, S. F. & Murray, R. 2013. New crop oils—Properties as potential lubricants. *Industrial Crops and Products*, 44, pp 232–239. Available at Doi: 10.1016/j.indcrop.2012.10.035.
Choi, U., Ahn, B., Kwon, O. & Chun, Y. 1997. Tribological behavior of some antiwear additives in vegetable oils. *Tribology International*, 30(9), pp 677–683. Available at Doi: 10.1016/S0301-679X(97)00039-X.
Cortes, V. & Ortega, J. A. 2019. Evaluating the rheological and tribological behaviors of coconut oil modified with nanoparticles as lubricant additives. *Lubricants*, 7(9), p 76. Available at Doi: 10.3390/lubricants7090076.
Dubois, V., Breton, S., Linder, M., Fanni, J. & Parmentier, M. 2007. Fatty acid profiles of 80 vegetable oils with regard to their nutritional potential. *European Journal of Lipid Science and Technology*, 109(7), pp 710–732. Available at Doi: 10.1002/ejlt.200700040.
El-Boulifi, N., Sánchez, M., Martínez, M. & Aracil, J. 2015. Fatty acid alkyl esters and mono-unsaturated alcohols production from jojoba oil using short-chain alcohols for bio-refinery concepts. *Industrial Crops and Products*, 69, pp 244–250. Available at Doi: 10.1016/j.indcrop.2015.02.031.
Erhan, S. Z. & Asadauskas, S. 2000. Lubricant basestocks from vegetable oils. *Industrial Crops and Products*, 11(2–3), pp 277–282. Available at Doi: 10.1016/S0926-6690(99)00061-8.
Erhan, S. Z., Sharma, B. K. & Perez, J. M. 2006. Oxidation and low temperature stability of vegetable oil-based lubricants. *Industrial Crops and Products*, 24(3), pp 292–299. Available at Doi: 10.1016/j.indcrop.2006.06.008.
Fasina, O., Hallman, H., Craig-Schmidt, M. & Clements, C. 2006. Predicting temperature-dependence viscosity of vegetable oils from fatty acid composition. *Journal of the American Oil Chemists' Society*, 83(10), p 899. Available at Doi: 10.1007/s11746-006-5044-8.
Fox, N. & Stachowiak, G. 2007. Vegetable oil-based lubricants—a review of oxidation. *Tribology International*, 40(7), pp 1035–1046. Available at Doi: 10.1016/j.triboint.2006.10.001
Fox, N., Tyrer, B. & Stachowiak, G. 2004. Boundary lubrication performance of free fatty acids in sunflower oil. *Tribology Letters*, 16(4), pp 275–281. Available at Doi: 10.1023/B:TRIL.0000015203.08570.82.
Gherasim, I., Roy, G., Nguyen, C. T. & Vo-Ngoc, D. 2011. Heat transfer enhancement and pumping power in confined radial flows using nanoparticle suspensions (nanofluids). *International Journal of Thermal Sciences*, 50(3), pp 369–377. Available at Doi: 10.1016/j.ijthermalsci.2010.04.008.
Giakoumis, E. G. 2018. Analysis of 22 vegetable oils' physico-chemical properties and fatty acid composition on a statistical basis, and correlation with the degree of unsaturation. *Renewable Energy*, 126, pp 403–419. Available at Doi: 10.1016/j.renene.2018.03.057.

Goering, C., Schwab, A., Daugherty, M., Pryde, E. & Heakin, A. 1982. Fuel properties of eleven vegetable oils. *Transactions of the ASAE*, 25(6), pp 1472–1477. Available at: 10.13031/2013.33748.

Gopinath, A., Puhan, S. & Nagarajan, G. 2010. Effect of biodiesel structural configuration on its ignition quality. *Energy and Environment*, 1(2), pp 295–306.

Havet, L., Blouet, J., Valloire, F. R., Brasseur, E. & Slomka, D. 2001. Tribological characteristics of some environmentally friendly lubricants. *Wear*, 248(1–2), pp 140–146. Available at Doi: 10.1016/S0043-1648(00)00550-0.

Hwang, H.-S. & Erhan, S. Z. 2001. Modification of epoxidized soybean oil for lubricant formulations with improved oxidative stability and low pour point. *Journal of the American Oil Chemists' Society*, 78(12), pp 1179–1184. Available at Doi: 10.1007/s11745-001-0410-0.

Jayadas, N. & Nair, K. P. 2006. Coconut oil as base oil for industrial lubricants—evaluation and modification of thermal, oxidative and low temperature properties. *Tribology International*, 39(9), pp 873–878. Available at Doi: 10.1016/j.triboint.2005.06.006.

Kashyap, A. & Harsha, A. 2016. Tribological studies on chemically modified rapeseed oil with CuO and CeO$_2$ nanoparticles. *Proceedings of the Institution of Mechanical Engineers, Part J: Journal of Engineering Tribology*, 230(12), pp 1562–1571. Available at Doi: 10.1177/1350650116641328.

Kerni, L., Raina, A. & Haq, M. I. U. 2019. Friction and wear performance of olive oil containing nanoparticles in boundary and mixed lubrication regimes. *Wear*, 426, pp 819–827. Available at Doi: 10.1016/j.wear.2019.01.022.

Kodali, D. R. 2002. High performance ester lubricants from natural oils. *Industrial Lubrication and Tribology*. Available at Doi: 10.1108/00368790210431718.

Koshy, C. P., Rajendrakumar, P. K. & Thottackkad, M. V. 2015. Analysis of tribological and thermo-physical properties of surfactant-modified vegetable Oil-based CuO nano-lubricants at elevated temperatures-An experimental study. *Tribology Online*, 10(5), pp 344–353. Available at Doi: 10.2474/trol.10.344.

Lanjekar, R. & Deshmukh, D. 2016. A review of the effect of the composition of biodiesel on NO$_x$ emission, oxidative stability and cold flow properties. *Renewable and Sustainable Energy Reviews*, 54, pp 1401–1411. Available at Doi: 10.1016/j.rser.2015.10.034.

Madanhire, I. & Mbohwa, C. 2016. Development of biodegradable lubricants. *Mitigating Environmental Impact of Petroleum Lubricants*. Springer, 85–101.

Mannekote, J., Kailas, S., Venkatesh, K. & Kathyayini, N. 2017. A study on chemical and lubrication properties of unrefined, refined and virgin coconut oil samples. Available at Doi: 10.1007/978-3-319-31358-0.

Misra, R. & Murthy, M. 2010. Straight vegetable oils usage in a compression ignition engine—A review. *Renewable and Sustainable Energy Reviews*, 14(9), pp 3005–3013. Available at Doi: 10.1016/j.rser.2010.06.010.

Noorawzi, N. & Samion, S. 2016. Tribological effects of vegetable oil as alternative lubricant: A pin-on-disk tribometer and wear study. *Tribology Transactions*, 59(5), pp 831–837. Available at Doi: 10.1080/10402004.2015.1108477.

Nowak, P., Kucharska, K. & Kamiński, M. 2019. Ecological and health effects of lubricant oils emitted into the environment. *International Journal of Environmental Research and Public Health*, 16(16), p 3002. Available at Doi: 10.3390/ijerph16163002.

Okullo, A. A., Temu, A., Ogwok, P. & Ntalikwa, J. 2012. Physico-chemical properties of biodiesel from jatropha and castor oils. *International Journal of Renewable Energy Research (IJRER)*, 2(1), pp 47–52.

Orsavova, J., Misurcova, L., Ambrozova, J. V., Vicha, R. & Mlcek, J. 2015. Fatty acids composition of vegetable oils and its contribution to dietary energy intake and dependence of cardiovascular mortality on dietary intake of fatty acids. *International Journal of Molecular Sciences*, 16(6), pp 12871–12890. Available at Doi: 10.3390/ijms160612871.

Pantzali, M., Mouza, A. & Paras, S. 2009. Investigating the efficacy of nanofluids as coolants in plate heat exchangers (PHE). *Chemical Engineering Science*, 64(14), pp 3290–3300. Available at Doi: 10.1016/j.ces.2009.04.004.

Peng, H., Ding, G., Jiang, W., Hu, H. & Gao, Y. 2009. Measurement and correlation of frictional pressure drop of refrigerant-based nanofluid flow boiling inside a horizontal smooth tube. *International Journal of Refrigeration*, 32(7), pp 1756–1764. Available at Doi: 10.1016/j.ijrefrig.2009.06.005.

Quinchia, L., Delgado, M., Reddyhoff, T., Gallegos, C. & Spikes, H. 2014. Tribological studies of potential vegetable oil-based lubricants containing environmentally friendly viscosity modifiers. *Tribology International*, 69, pp 110–117. Available at Doi: 10.1016/j.triboint.2013.08.016.

Reeves, C. J., Menezes, P. L., Jen, T.-C. & Lovell, M. R. 2015. The influence of fatty acids on tribological and thermal properties of natural oils as sustainable biolubricants. *Tribology International*, 90, pp 123–134. Available at Doi: 10.1016/j.triboint.2015.04.021.

Rodrigues Jr, J. d. A., Cardoso, F. d. P., Lachter, E. R., Estevão, L. R., Lima, E. & Nascimento, R. S. 2006. Correlating chemical structure and physical properties of vegetable oil esters. *Journal of the American Oil Chemists' Society*, 83(4), pp 353–357. Available at Doi: 10.1007/s11746-006-1212-0.

Sajeeb, A. & Rajendrakumar, P. K. 2019a. Comparative evaluation of lubricant properties of biodegradable blend of coconut and mustard oil. *Journal of Cleaner Production*, 240, p. 118255. Available at Doi: 10.1016/j.jclepro.2019.118255.

Sajeeb, A. & Rajendrakumar, P. K. 2019b. Investigation on the rheological behavior of coconut oil based hybrid CeO$_2$/CuO nanolubricants. *Proceedings of the Institution of Mechanical Engineers, Part J: Journal of Engineering Tribology*, 233(1), pp 170–177. Available at Doi: 10.1177/1350650118772149.

Sajjadi, B., Raman, A. A. A. & Arandiyan, H. 2016. A comprehensive review on properties of edible and non-edible vegetable oil-based biodiesel: Composition, specifications and prediction models. *Renewable and Sustainable Energy Reviews*, 63, pp 62–92. Available at Doi: 10.1016/j.rser.2016.05.035.

Salaji, S. & Jayadas, N. 2020. Evaluation of physicochemical and tribological properties of chaulmoogra (Hydnocarpus wightianus) oil as green lubricant base stock. *Proceedings of the Institution of Mechanical Engineers, Part J: Journal of Engineering Tribology*, 1350650119899529. Available at Doi: 10.1177/1350650119899529.

Salih, N., Salimon, J. & Yousif, E. 2011. The physicochemical and tribological properties of oleic acid based triester biolubricants. *Industrial Crops and Products*, 34(1), pp 1089–1096. Available at Doi: 10.1016/j.indcrop.2011.03.025.

Salimon, J., Salih, N. & Yousif, E. 2010. Biolubricants: Raw materials, chemical modifications and environmental benefits. *European Journal of Lipid Science and Technology*, 112(5), pp 519–530. Available at Doi: 10.1002/ejlt.200900205.

Schneider, M. P. 2006. Plant-oil-based lubricants and hydraulic fluids. *Journal of the Science of Food and Agriculture*, 86(12), pp 1769–1780. Available at Doi: 10.1002/jsfa.2559.

Shafi, W. K. & Charoo, M. 2019a. Experimental study on rheological properties of vegetable oils mixed with titanium dioxide nanoparticles. *Journal of the Brazilian Society of Mechanical Sciences and Engineering*, 41(10), p 431. Available at Doi: 10.1007/s40430-019-1905-6.

Shafi, W. K. & Charoo, M. 2019b. Rheological properties of sesame oil mixed with H-Bn nanoparticles as industrial lubricant. *Materials Today: Proceedings*, 18, pp 4963–4967. Available at Doi: 10.1016/j.matpr.2019.07.488.

Shafi, W. K., & Charoo, M. S. 2020. An overall review on the tribological, thermal and rheological properties of nanolubricants. *Tribology-Materials, Surfaces & Interfaces*, pp 1–35. Available at Doi: 10.1080/17515831.2020.1785233.

Shafi, W. K., Raina, A. & Haq, M. I. U. 2018a. Tribological performance of avocado oil containing copper nanoparticles in mixed and boundary lubrication regime. *Industrial Lubrication and Tribology*. Available at Doi: 10.1108/ILT-06-2017-0166.

Shafi, W. K., Raina, A. & Haq, M. I. U. 2018b. Friction and wear characteristics of vegetable oils using nanoparticles for sustainable lubrication. *Tribology-Materials, Surfaces & Interfaces*, 12(1), pp 27–43. Available at Doi: 10.1080/17515831.2018.1435343.

Shafi, W. K., Raina, A. & Haq, M. I. U. 2019. Performance evaluation of hazelnut oil with copper nanoparticles-a new entrant for sustainable lubrication. *Industrial Lubrication and Tribology*. Available at Doi: 10.1108/ILT-07-2018-0257.

Shahabuddin, M., Masjuki, H. & Kalam, M. 2013a. Development of eco-friendly biodegradable biolubricant based on jatropha oil. *Tribology in Engineering*, pp 135–146. Available at Doi: 10.5772/51376.

Shahabuddin, M., Masjuki, H., Kalam, M., Bhuiya, M. & Mehat, H. 2013b. Comparative tribological investigation of bio-lubricant formulated from a non-edible oil source (Jatropha oil). *Industrial Crops and Products*, 47, pp 323–330. Available at Doi: 10.1016/j.indcrop.2013.03.026.

Sharma, U. C. & Sachan, S. 2019. Friction and wear behavior of karanja oil derived biolubricant base oil. *SN Applied Sciences*, 1(7), p 668. Available at Doi: 10.1007/s42452-019-0706-y.

Singh, Y., Garg, R. & Kumar, S. 2016. Comparative tribological investigation on EN31 with pongamia and jatropha as lubricant additives. *Energy Sources, Part A: Recovery, Utilization, and Environmental Effects*, 38(18), pp 2756–2762. Available at Doi: 10.1080/15567036.2015.1105326.

Singh, Y., Singla, A., Singh, A. K. & Upadhyay, A. K. 2018. Tribological characterization of Pongamia pinnata oil blended bio-lubricant. *Biofuels*, 9(4), pp 523–530. Available at Doi: 10.1080/17597269.2017.1292017.

Siniawski, M. T., Saniei, N., Adhikari, B. & Doezema, L. A. 2007. Influence of fatty acid composition on the tribological performance of two vegetable-based lubricants. *Journal of Synthetic Lubrication*, 24(2), pp 101–110. Available at Doi: 10.1002/jsl.32.

Sneha, E., Rani, S. & Arif, M. 2019. Evaluation of lubricant properties of vegetable oils as base oil for industrial lubricant. *IOP Conference Series: Materials Science and Engineering*, IOP Publishing, 012022. Available at Doi: 10.1088/1757-899X/624/1/012022.

Woma, T. Y., Lawal, S. A., Abdulrahman, A. S., MA, O. & MM, O. 2019. Vegetable oil based lubricants: Challenges and prospects. *Tribology Online*, 14(2), pp 60–70. Available at Doi: 10.2474/trol.14.60.

Wu, X., Zhang, X., Yang, S., Chen, H. & Wang, D. 2000. The study of epoxidized rapeseed oil used as a potential biodegradable lubricant. *Journal of the American Oil Chemists' Society*, 77(5), pp 561–563. Available at Doi: 10.1007/s11746-000-0089-2.

Yalcin, H., Toker, O. S. & Dogan, M. 2012. Effect of oil type and fatty acid composition on dynamic and steady shear rheology of vegetable oils. *Journal of Oleo Science*, 61(4), pp 181–187. Available at Doi: 10.5650/jos.61.181.

Zeman, A., Sprengel, A., Niedermeier, D. & Späth, M. 1995. Biodegradable lubricants—studies on thermo-oxidation of metal-working and hydraulic fluids by differential scanning calorimetry (DSC). *Thermochimica Acta*, 268, pp 9–15. Available at Doi: 10.1016/0040-6031(95)02512-X.

15 Multi-Lobe Journal Bearings Analysis with Limited Texture

T.V.V.L.N. Rao
SRM Institute of Science and Technology

Ahmad Majdi Abdul Rani, Norani Muti Mohamed, Hamdan Haji Ya, Mokhtar Awang, and Fakhruldin Mohd Hashim
Universiti Teknologi PETRONAS

CONTENTS

15.1 INTRODUCTION

The sustainability of rotating machinery running at higher speeds is attributed to the design of multi-lobe journal bearings. Multi-lobe journal bearings provide enhanced load capacity and stability of rotor-bearing systems. The multi-lobe journal bearings have shown the capability to influence the performance (static and dynamic) characteristics under the influence of the number of lobes, preload and offset factors as the geometric design parameters. The dynamic characteristics have a prominent effect on the instability of bearings.

Recent literature is focused on the properties of texture/slip on the static and dynamic operation of fluid film bearings. Limited (partial) texture/slip surface patterns resulted in significant enhancements of steady-state-bearing operating features under low-load conditions.

The performance (static and dynamic) characteristics of multi-lobe (journal) bearings are significantly influenced by limited (partial) texture/slip patterns on the lobed configurations. Therefore, the analysis of lobed configuration with limited (partial) texture for improvement of the bearing (multi-lobe) performance is presented. The analysis of texturing to improve the static, dynamic and stability characteristics of multi-lobe journal bearings encompasses tribology for sustainability.

15.1.1 Multi-Lobe Journal Bearings

Lund and Thomsen (1978) derived data for bearing (multi-lobe) configurations from solution of the Reynolds equation based on infinitesimal perturbation method. Li et al. (1980) and Allaire et al. (1980) analysed characteristics (stability and transient) of bearings (multi-lobe). Rao et al. (2001) evaluated a semianalytical approach for the determination of stiffness and damping constants and transient variation of bearings (multi-lobe). Mehrjardi et al. (2016) investigated the effects of preload on the instability operation of noncircular journal bearings.

Sinhasan and Chandrawat (1988) presented performance (static and dynamic) characteristics of two-axial-groove bearings based on elastohydrodynamic analysis. Kostrzewsky et al. (1996) presented a comparison of experimental and calculated functioning of two-axial-groove bearing. The experimental and predicted data of coefficients (dynamic) with Sommerfeld number are in good concurrence for two-axial-groove (journal) bearing.

Kumar et al. (1980) and Malik (1983) presented design data for two-lobe journal bearings. Crosby and Chetti (2009) presented performance characteristics of couple-stress fluid-lubricated two-lobe journal bearing with improved stability. Nair et al. (1987) investigated the elastohydrodynamic effects on the operating characteristics in a two-lobe journal bearing. The stability limit increases using a deforming liner material in an elliptical journal bearing.

Malik et al. (1981) and Sinhasan et al. (1981) provided data for three-lobe journal bearings. Flack and Lanes (1982) investigated stability characteristics of three-lobe bearings. Goyal and Sinhasan (1991) studied elastohydrodynamic bearings (three-lobe) with non-Newtonian (shear stress with cubic law) lubricants. Kostrzewsky et al. (1998) presented characteristics (dynamic) of a bearing (three-lobe). Nair et al. (1987) presented the EHD effects in a bearing (three-lobe). The liner flexibility effects on the features of three-lobe journal bearing are presented for a wide range of various eccentricities. Chasalevris (2015) presented an analytical method of determination of the characteristics of finite-length bearings (three-lobe). The closed-form analytical solutions are provided for the evaluation of dynamic characteristics.

15.1.2 Limited (Partial) Texture Bearings

Lin et al. (2015) investigated the consequence of texture/slip location on improvement in bearing performance. Gui and Meng (2019) compared the effects of dimple (spherical) and bump on the bearing performance. The location of dimples at pressure rise in journal bearing results in enhanced load capacity and reduced friction force. Matele and Pandey (2018) analysed the consequence of texturing using dimples on the dynamic features of bearing. The improvement in stability characteristics is influenced by location and geometry of dimples. Yamada et al. (2018) presented the dynamic features of bearings with dimple (square) parameters using numerical and experimental investigations. The effect of dimples is to reduce the cross-coupled stiffness coefficients due to lower tangential force. The reduction in tangential force results in an improvement in threshold speed of stability.

15.1.3 Limited (Partial) Texture Multi-Lobe Journal Bearings

Khatri and Sharma (2018) presented the joint effect of electrorheological lubricant and texturing on the operation of hybrid bearing (two-lobe/circular nonrecessed) systems. Rao et al. (2020) presented the steady-state and unsteady-state pressure gradients for each lobe (limited (partial) slip texture) derived based on a large number of grooves theory (Vohr & Chow, 1965).

This paper presents an overview of limited (partial) texture influence on the performance of multi-lobe journal bearings. The results analysed in this work are based on Rao et al. (2020) considering the limited (partial) texture influence with a low magnitude of slip parameter. Performance characteristics are analysed based on the coordinate (angular) from the line of lobe and journal centre for lobe texture θ_{lt}, land-to-unit (cell) ratio (γ), nondimensional depth of recess (ζ).

15.2 METHODOLOGY

A method based on narrow groove (a large number of grooves) theory (NGT) to determine the load and dynamic analysis of multi-lobe bearings with limited (partial) texture is presented (Rao et al., 2020). Figure 15.1 shows multi-lobe bearings. The configuration of limited (partial) texture is located on the bearing lobe sharing the major load and extends from inlet of a groove in the fluid flow path.

15.2.1 Dynamic Reynolds (Modified) Model

Based on NGT for large (infinite) cells in limited (partial) texture regions, the local distribution of pressure over land (p_{tl}) and recess (p_{tr}) is expressed as

$$q_{tl} = \frac{u_j h_l}{2} - \frac{h_l^3}{12\mu} \frac{dp_{tl}}{Rd\theta} \tag{15.1}$$

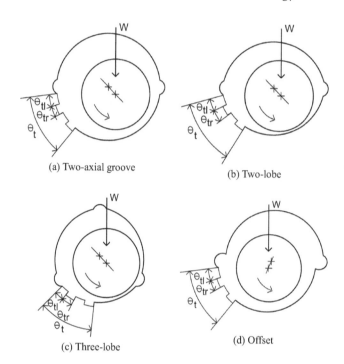

(a) Two-axial groove

(b) Two-lobe

(c) Three-lobe

(d) Offset

FIGURE 15.1   Multi-lobe journal bearings with limited (partial) texture.

$$q_{tr} = \frac{u_j\left(h_l + h_{tr}\right)}{2} - \frac{\left(h_l + h_{tr}\right)^3}{12\mu} \frac{dp_{tr}}{Rd\theta} \tag{15.2}$$

Using the pressure (local) gradients, the pressure (overall) gradient is obtained as

$$\frac{dp_l}{d\theta} = \gamma \frac{dp_{tl}}{d\theta} + \left(1 - \gamma\right)\frac{dp_{tr}}{d\theta} \tag{15.3}$$

The flow (nondimensional) along circumferential direction is expressed in terms of the overall nondimensional pressure gradient as

$$Q_l = \frac{H_l}{2} \left(\frac{\gamma + \left(1 - \gamma\right)\dfrac{\Delta_{1t}}{\Delta_{2t}}}{\gamma + \left(1 - \gamma\right)\dfrac{1}{\Delta_{2t}}}\right) - \frac{H_l^3}{12} \frac{1}{\left(\gamma + \left(1 - \gamma\right)\dfrac{1}{\Delta_{2t}}\right)} \frac{dP_l}{d\theta} \tag{15.4}$$

where

$$\Delta_{1t} = 1 + \zeta, \Delta_{2t} = \left(1 + \zeta\right)^3 \tag{15.5}$$

The dynamic modified Reynolds model (nondimensional) for bearing (multi-lobe) with limited (partial) texture orientation is

$$\Omega \frac{dH_l}{dT} + \frac{d}{d\theta}\left(\frac{A_{sl}H_l}{2} - \frac{A_{pl}H_l^3}{12}\frac{dP_l}{d\theta} \right) = 0 \tag{15.6}$$

where the coefficients are

$$A_{sl} = \frac{\left(\gamma + (1-\gamma)\frac{\Delta_{1t}}{\Delta_{2t}}\right)}{\left(\gamma + (1-\gamma)\frac{1}{\Delta_{2t}}\right)}, A_{pl} = \frac{1}{\left(\gamma + (1-\gamma)\frac{1}{\Delta_{2t}}\right)} \tag{15.7}$$

15.2.2 STEADY-STATE ANALYSIS

The pressure distribution (nondimensional) for lobe with limited (partial) texture region is expressed as

$$P_{0l}(\theta_{1l} \le \theta \le \theta_l) = \left[\frac{A_{sl}}{A_{pl}} \int_{\theta_{1l}}^{\theta} \frac{6}{H_{0l}^2} d\theta - \frac{1}{A_{pl}} \int_{\theta_{1l}}^{\theta} \frac{12c_{011l}}{H_{0l}^3} d\theta \right] \tag{15.8}$$

$$P_{0l}(\theta_l \le \theta \le \theta_{2l}) = P_{0l}|_{\theta=\theta_l} + \left[\int_{\theta_l}^{\theta} \frac{6}{H_{0l}^2} d\theta - \int_{\theta_l}^{\theta} \frac{12c_{011l}}{H_{0l}^3} d\theta \right] \tag{15.9}$$

where the Reynolds boundary conditions result in

$$c_{011l} = 0.5 H_{0l}|_{\theta=\theta_{2l}}, c_{011l} = \frac{0.5\frac{A_{sl}}{A_{pl}} \int_{\theta_{1l}}^{\theta_l} \frac{1}{H_{0l}^2} d\theta + 0.5 \int_{\theta_l}^{\theta_{2l}} \frac{1}{H_{0l}^2} d\theta}{\frac{1}{A_{pl}} \int_{\theta_{1l}}^{\theta_l} \frac{1}{H_{0l}^3} d\theta + \int_{\theta_l}^{\theta_{2l}} \frac{1}{H_{0l}^3} d\theta} \tag{15.10}$$

The pressure (nondimensional) for each lobe (P_{ol}) is integrated to determine the load capacity (nondimensional) components (W_{el} and $W_{\phi l}$) for each lobe. The load capacity (nondimensional) (W) for multi-lobe journal bearing is evaluated from the nondimensional load components (W_{el} and $W_{\phi l}$) for all the lobes. The load capacity (nondimensional) (W) of bearing (multi-lobe) is attained by attitude angle iteration until resultant load is in the vertical line.

15.2.3 DYNAMIC ANALYSIS

The pressure (nondimensional) gradients for lobe with limited (partial) texture orientation for the centre of the journal perturbation displacements are

$$P_{il}(\Theta_{1l} \le \Theta \le \Theta_l) = \int_{\Theta_{1l}}^{\Theta} \left(-\frac{A_{sl}}{A_{pl}}\frac{12H_i}{H_{0l}^3} + \frac{36H_ic_{011l}}{A_{pl}H_{0l}^4} - \frac{12c_{ki11l}}{A_{pl}H_{0l}^3} \right) d\Theta \tag{15.11}$$

$$P_{il}\left(\Theta_l \le \Theta \le \Theta_{2l}\right) = P_{il}\big|_{\Theta=\Theta_l} + \int_{\Theta_l}^{\Theta}\left(-\frac{12H_i}{H_{0l}^3} + \frac{36H_i c_{011}}{H_{0l}^4} - \frac{12c_{ki11}}{H_{0l}^3}\right)d\Theta \quad (15.12)$$

where the Reynolds pressure-gradient boundary conditions result in

$$c_{ki11l} = \frac{-\dfrac{\Delta_{sl}}{\Delta_{pl}}\displaystyle\int_{\Theta_{1l}}^{\Theta_l}\frac{H_i}{H_{0l}^3}d\Theta + \frac{3c_{011l}}{\Delta_{pl}}\displaystyle\int_{\Theta_{1l}}^{\Theta_l}\frac{H_i}{H_{0l}^4}d\Theta - \displaystyle\int_{\Theta_l}^{\Theta_{2l}}\frac{H_i}{H_{0l}^3}d\Theta + 3c_{011l}\displaystyle\int_{\Theta_l}^{\Theta_{2l}}\frac{H_i}{H_{0l}^4}d\Theta}{\dfrac{1}{\Delta_{pl}}\displaystyle\int_{\Theta_{1l}}^{\Theta_l}\frac{1}{H_{0l}^3}d\Theta + \displaystyle\int_{\Theta_l}^{\Theta_{2l}}\frac{1}{H_{0l}^3}d\Theta}$$

$$(15.13)$$

for $i = x, y$, $H_x = \cos\Theta$ and $H_y = \sin\Theta$

The pressure (nondimensional) gradients for lobe with limited (partial) texture orientation for the centre of journal perturbation velocities are

$$P_{il}\left(\Theta_{1l} \le \Theta \le \Theta_l\right) = \left[\int_{\Theta_{1l}}^{\Theta}\left(\frac{12\Omega}{\Delta_{pl}H_{0l}^3}\int H_i d\Theta - \frac{12c_{bi11l}}{\Delta_{pl}H_{0l}^3}\right)d\Theta\right] \quad (15.14)$$

$$P_{il}\left(\Theta_l \le \Theta \le \Theta_{2l}\right) = P_{il}\big|_{\Theta=\Theta_l} + \left[\int_{\Theta_l}^{\Theta}\left(\frac{12\Omega}{H_{0l}^3}\int H_i d\Theta - \frac{12c_{bi11l}}{H_{0l}^3}\right)d\Theta\right] \quad (15.15)$$

where the Reynolds pressure-gradient boundary conditions result in

$$c_{bi11l} = \frac{\displaystyle\int_{\Theta_{1l}}^{\Theta_l}\left(\frac{\Omega}{\Delta_{pl}H_{0l}^3}\int H_i d\Theta\right)d\Theta + \displaystyle\int_{\Theta_l}^{\Theta_{2l}}\left(\frac{\Omega}{H_{0l}^3}\int H_i d\Theta\right)d\Theta}{\displaystyle\int_{\Theta_{1l}}^{\Theta_l}\left(\frac{1}{\Delta_{pl}H_{0l}^3}\right)d\Theta + \displaystyle\int_{\Theta_l}^{\Theta_{2l}}\left(\frac{1}{H_{0l}^3}\right)d\Theta} \quad (15.16)$$

for $i = \dot{x}, \dot{y}$, $H_{\dot{x}} = \cos\Theta$ and $H_{\dot{y}} = \sin\Theta$

The pressure (nondimensional) and pressure gradients (nondimensional) P_0, P_x, P_y, $P_{\dot{x}}$ and $P_{\dot{y}}$ for each lobe without limited (partial) texture (for plain orientation) are obtained by substituting $\Delta_{sl} = \Delta_{pl} = 1$ in Eqs. (15.8 and 15.9), Eqs. (15.11 and 15.12) and Eqs. (15.14 and 15.15).

The pressure gradients (nondimensional) for each lobe (P_{xl}, P_{yl}, $P_{\dot{x}l}$ and $P_{\dot{y}l}$) are integrated along and perpendicular to the load line to determine the dynamic coefficients (nondimensional) ($\bar{K}_{xxl}$, $\bar{K}_{yxl}$, $\bar{K}_{xyl}$, $\bar{K}_{yyl}$, $\bar{B}_{xxl}$, $\bar{B}_{yxl}$, $\bar{B}_{xyl}$ and $\bar{B}_{yyl}$) for each lobe. The dynamic coefficients (nondimensional) ($\bar{K}_{xx}$, $\bar{K}_{yx}$, $\bar{K}_{xy}$, $\bar{K}_{yy}$, $\bar{B}_{xx}$, $\bar{B}_{yx}$, $\bar{B}_{xy}$ and $\bar{B}_{yy}$) for bearing (multi-lobe) are calculated from the coefficients (nondimensional) ($\bar{K}_{xxl}$, $\bar{K}_{yxl}$, $\bar{K}_{xyl}$, $\bar{K}_{yyl}$, $\bar{B}_{xxl}$, $\bar{B}_{yxl}$, $\bar{B}_{xyl}$ and $\bar{B}_{yyl}$) for all the lobes.

The whirl ratio (critical) and the instability speed threshold (nondimensional) derived based on Lund's methodology (Lund, 1987) result in

$$\Omega_s = \sqrt{\frac{(K_{xx} - \kappa_o)(K_{yy} - \kappa_o) - K_{xy}K_{yx}}{B_{xx}B_{yy} - B_{xy}B_{yx}}} \qquad (15.17)$$

$$\omega_s = \sqrt{M} = \frac{\sqrt{\kappa_o}}{\Omega_s} \qquad (15.18)$$

where $K_{ij} = \bar{K}_{ij} / W$ and $B_{ij} = \bar{B}_{ij} / W$ for $i = x, y$,

$$\kappa_o = \frac{B_{xx}K_{yy} + B_{yy}K_{xx} - B_{xy}K_{yx} - B_{yx}K_{xy}}{B_{xx} + B_{yy}}$$

15.3 RESULTS AND DISCUSSION

The performance characteristics considering the effect of limited (partial) texture are analysed based on the investigations of Rao et al. (2020) on multi-lobe journal bearings. Results of W, ω_s and Ω_s are analysed based on Sommerfeld/long bearing (one-dimensional) analysis (Rao et al., 2020). The bearing (multi-lobe) parameters are shown in Table 15.1.

15.3.1 TWO-AXIAL-GROOVE BEARING

Figures 15.2–15.4 show how the load (nondimensional) capacity (W), speed (nondimensional) of instability threshold (ω_s) and whirl (critical) ratio (Ω_s), respectively, of two-axial-groove bearing change corresponding to limited (partial) extent of texture (θ_t), land-to-unit (cell) ratio (γ) and depth (nondimensional) of recess (ζ).

Figure 15.2 shows W variation with limited (partial) extent of texture ($\theta_t = 0°$–$90°$) and land-to-unit cell ratio ($\gamma = 0.2$–0.8). W is invariant with increasing limited (partial) extent of texture ($\theta_t = 0°$–$90°$) for land-to-unit (cell) ratio (γ) of 0.8 at $\varepsilon = 0.2$–0.3 (Figure 15.2a). As depicted in Figure 15.2a, W is lower for land-to-unit cell ratio (γ) of 0.2 compared to 0.8 with limited (partial) extent of texture ($\theta_t = 0°$–$90°$) at depth (nondimensional) of recess (ζ) of 1. W is lower for depth (nondimensional) of recess (ζ) of 1 compared to 0.1 with land-to-unit (cell) ratio ($\gamma = 0.2$–0.8) at limited (partial) extent of texture (θ_t) of 90° (Figure 15.2b).

Figure 15.3 shows the deviation in ω_s with limited (partial) extent of texture ($\theta_{t=} 0°$–$90°$) and land-to-unit (cell) ratio ($\gamma = 0.2$–0.8) for with depth (nondimensional)

TABLE 15.1
Parameters of Bearings (Multi-Lobe)

Parameter	Two-Axial Groove	Two-Lobe	Three-Lobe	Offset
Preload parameter (δ)	–	0.5	0.5	0.5
Offset parameter (ψ)	0.5	0.5	0.5	1.0
Limited (partial) extent of texture (θ_t)	0°–90°	0°–90°	0°–60°	0°–90°
Eccentricity ratio (ε)	0.2–0.4	0.1–0.3	0.1–0.3	0.3–0.5

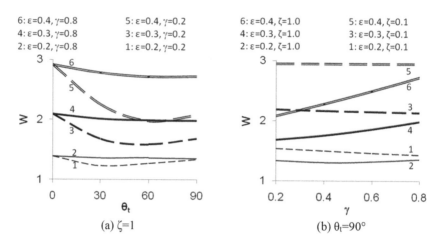

FIGURE 15.2 Load capacity (nondimensional) (W) of two-axial-groove bearing with limited (partial) texture.

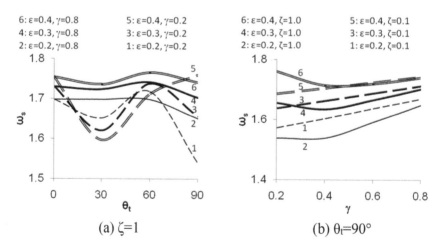

FIGURE 15.3 Instability threshold speed (nondimensional) (ω_s) of two-axial-groove bearing with partial texture.

of recess (ζ) of 1. Greater ω_s is attained for land-to-unit (cell) ratio (γ) of 0.2 correlated to 0.8 at partial extent of texture (θ_t) from 60° to 90° (Figure 15.3a). Superior ω_s is obtained for depth (nondimensional) of recess (ζ) of 1 related to 0.1 at land-to-unit (cell) ratio (γ) of 0.2 and at limited (partial) extent of texture (θ_t) of 90° (Figure 15.3b). Superior linear stability with higher ω_s is attained for two-axial-groove bearing with limited (partial) texture ($\zeta = 1$) at a lesser value of land-to-unit (cell) ratio ($\gamma = 0.2$).

Figure 15.4 shows the variation of Ω_s of two-axial-groove bearing with limited (partial) texture. Ω_s is lower for land-to-unit (cell) ratio (γ) of 0.2 correlated to 0.8 at

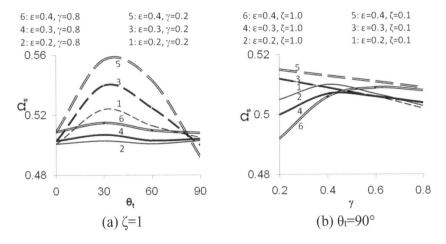

FIGURE 15.4   Whirl (critical) ratio (Ω_s) of two-axial-groove bearing with limited (partial) texture

---

TABLE 15.2

Limited (Partial) Texture Two-Axial-Groove Bearing Performance Characteristics

Parameter	Limited (Partial) Texture Two-Axial-Groove Journal Bearing Characteristics
W	W is lower γ of 0.2 compared to $\gamma=0.8$ for $\zeta=1$ with $\theta_t=0°–90°$ for $\varepsilon=0.2–0.4$
ω_s	Higher ω_s is obtained at $\gamma=0.2$ (80% recess extent) for $\zeta=1$, $\theta_t=90°$ for $\varepsilon=0.4$
Ω_s	Ω_s is lower at $\gamma=0.2$ (80% recess extent) for $\zeta=1$, $\theta_t=90°$

---

limited (partial) extent of texture (θ_t) of 90° and depth (nondimensional) of recess (ζ) of 1 (Figure 15.4a). Ω_s is lower for depth (nondimensional) of recess (ζ) of 1 compared to $\zeta=0.1$ at land-to-unit (cell) ratio (γ) of 0.2 and at limited (partial) extent of texture (θ_t) of 90° (Figure 15.4b).

Table 15.2 shows the performance of two-axial-groove bearing with limited (partial) texture. W obtained with larger recess extent ($\gamma=0.2$) is lower compared to larger land extent ($\gamma=0.8$) for ε of 0.2–0.4. Higher stability is obtained for the larger extent of recess region ($\gamma=0.2$) for ε of 0.4.

15.3.2   TWO-LOBE JOURNAL BEARING

Figures 15.5–15.7 show how the load (nondimensional) capacity (W), speed (nondimensional) of instability threshold (ω_s) and whirl (critical) ratio Ω_s, respectively, of two-lobe bearing with limited (partial) texture change corresponding to limited (partial) extent of texture (θ_t), land-to-unit (cell) ratio (γ) and depth (nondimensional) of recess (ζ).

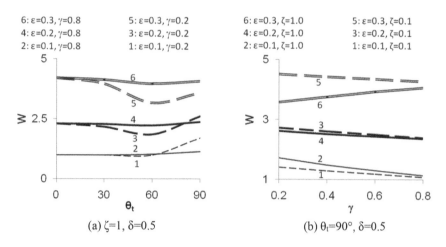

(a) $\zeta=1$, $\delta=0.5$

(b) $\theta_t=90°$, $\delta=0.5$

FIGURE 15.5 Load capacity (nondimensional) (W) of two-lobe bearing with limited (partial) texture.

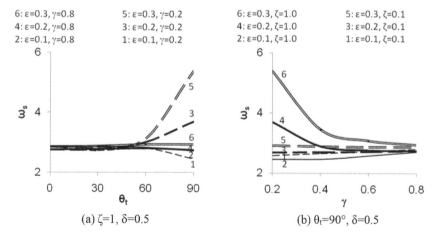

(a) $\zeta=1$, $\delta=0.5$

(b) $\theta_t=90°$, $\delta=0.5$

FIGURE 15.6 Instability threshold speed (nondimensional) (ω_s) of two-lobe bearing with limited (partial) texture.

Figure 15.5 shows W variation with limited (partial) extent of texture ($\theta_t=0°$–$90°$) and land-to-unit (cell) ratio ($\gamma=0.2$–0.8) with depth (nondimensional) of recess (ζ) of 1. W is invariant with increasing limited (partial) extent of texture ($\theta_t=0°$–$90°$) for ratio of land to the unit (cell) region (γ) of 0.8 at ε of 0.1–0.3 (Figure 15.5a). W increases with limited (partial) extent of texture from $\theta_t=60°$–$90°$ at depth (nondimensional) of recess (ζ) of 1 for land-to-unit (cell) ratio (γ) of 0.2. W is higher for depth (nondimensional) of recess (ζ) of 1 at land-to-unit (cell) ratio $\gamma=0.2$ correlated to $\gamma=0.8$ at limited (partial) extent of texture (θ_t) of $90°$ (Figure 15.5b). Enhancements in load are obtained for greater limit (extent) of limited (partial)

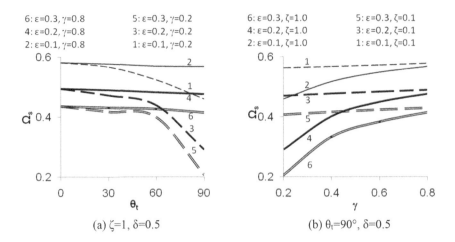

FIGURE 15.7   Whirl (critical) ratio (Ω_s) of two-lobe bearing with limited (partial) texture.

texture ($\theta_t = 60°–90°$). Load capacity of two-lobe (journal) bearing is enriched under limited (partial) texture influences at lesser eccentricity ratio ($\varepsilon = 0.1$).

Figure 15.6 shows the deviation in ω_s with limited (partial) extent of texture ($\theta_{t=}0°–90°$) and land-to-unit (cell) ratio ($\gamma = 0.2–0.8$) for two-lobe bearing with depth (dimensionless) of recess (ζ) of 1. The magnitude of ω_s increases significantly at limited (partial) extent of texture (θ_t) from 60° to 90° and land-to-unit (cell) ratio (γ) of 0.2 for eccentricity ratios (ε) of 0.2–0.3 (Figure 15.6a). Higher ω_s is obtained for depth (nondimensional) of recess (ζ) of 1 for the land-to-unit (cell) ratio (γ) of 0.2 and at limited (partial) extent of texture (θ_t) of 90° (Figure. 15.6b). Higher ω_s of limited (partial)-texture two-lobe journal bearing with a larger extent of recess ($\gamma = 0.2$) enriched stability compared to a greater extent of land ($\gamma = 0.8$). Limited (partial) texture effects enrich the stability for eccentricity ratios (ε) of 0.2–0.3.

Figure 15.7 shows the deviation of Ω_s of a two-lobe bearing with limited (partial) texture. Ω_s significantly decreases with limited (partial) extent of texture (θ_t) from 60° to 90° for land-to-unit (cell) ratio (γ) of 0.2 and depth (dimensionless) of recess (ζ) of 1 for ε of 0.2–0.3 (Figure 15.7a). Ω_s is lower for depth (dimensionless) of recess (ζ) of 1 compared to $\zeta = 0.1$ at land-to-unit (cell) ratio (γ) of 0.2 and at limited (partial) extent of texture (θ_t) of 90° (Figure 15.7b).

Table 15.3 shows the performance of limited (partial) texture two-lobe bearing. The load capacity increases with a larger extent of recess ($\gamma = 0.2$) for ε of 0.1. Higher stability is obtained for the larger extent of recess ($\gamma = 0.2$) for ε of 0.2–0.3.

15.3.3   THREE-LOBE JOURNAL BEARING

For the three-lobe bearing with limited (partial) texture change corresponding to limited (partial) extent of texture (θ_t), land-to-unit (cell) ratio (γ) and depth (nondimensional) of recess (ζ), Figures 15.8–15.10 reveal the variation of load (nondimensional) capacity (W), speed (nondimensional) of instability threshold (ω_s) and whirl (critical) ratio (Ω_s), respectively.

TABLE 15.3
Limited (Partial) Texture Two-Lobe Bearing Performance Characteristics

Parameter	Limited (Partial) Texture Two-Lobe Bearing Characteristics
W	W is higher for $\zeta=1$, at lower $\gamma=0.2$ compared to $\gamma=0.8$ at $\varepsilon=0.1$, $\theta_t=90°$, $\delta=0.5$
ω_s	Higher ω_s is obtained at γ of 0.2 (80% recess extent) for $\zeta=1$, $\theta_t=90°$, $\varepsilon=0.2$–0.3, $\delta=0.5$
Ω_s	Ω_s is lower at γ of 0.2 (80% recess extent) for $\zeta=1$, $\theta_t=90°$, $\varepsilon=0.2$–0.3, $\delta=0.5$

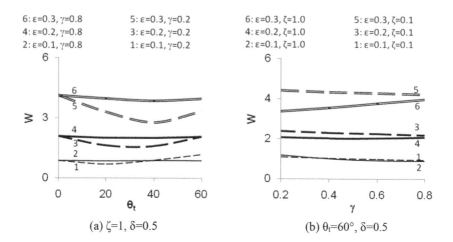

6: ε=0.3, γ=0.8   5: ε=0.3, γ=0.2   6: ε=0.3, ζ=1.0   5: ε=0.3, ζ=0.1
4: ε=0.2, γ=0.8   3: ε=0.2, γ=0.2   4: ε=0.2, ζ=1.0   3: ε=0.2, ζ=0.1
2: ε=0.1, γ=0.8   1: ε=0.1, γ=0.2   2: ε=0.1, ζ=1.0   1: ε=0.1, ζ=0.1

(a) ζ=1, δ=0.5   (b) θ_t=60°, δ=0.5

FIGURE 15.8 Load capacity (nondimensional) (W) of three-lobe bearing with limited (partial) texture.

The variation of W with limited (partial) extent of texture ($\theta_{t=}0°$–$60°$) and land-to-unit (cell) ratio ($\gamma=0.2$–0.8) with depth (nondimensional) of recess (ζ) of 1 is shown in Figure 15.8. Similar to two-axial-groove and two-lobe bearing, W is invariant with increasing limited (partial) extent of texture ($\theta_t=0°$–$90°$) for land-to-unit (cell) ratio (γ) of 0.8 at ε of 0.1–0.3 (Figure 15.8a). W increases with limited (partial) extent of texture from $\theta_{t=}40°$–$60°$ at depth (nondimensional) of recess (ζ) of 1 for land-to-unit (cell) ratio (γ) of 0.2 and for ε of 0.1 (Figure 15.8a). W for depth (nondimensional) of recess (ζ) of 1 is lower compared to $\zeta=0.1$ at limited (partial) extent of texture (θ_t) of 60° (Figure 15.8b).

The deviation in ω_s with limited (partial) extent of texture ($\theta_t=0°$–$60°$) and land-to-unit (cell) ratio ($\gamma=0.2$–0.8) for three-lobe bearing with depth (nondimensional) of recess (ζ) of 1 is shown in Figure 15.9. The magnitude of ω_s increases significantly at limited (partial) extent of texture (θ_t) from 40° to 60° and land-to-unit (cell) ratio (γ) of 0.2 for ε of 0.3 (Figure 15.9a). Higher ω_s is obtained for depth (nondimensional) of recess (ζ) of 1 for the land-to-unit (cell) ratio (γ) of 0.2 and at limited (partial) extent of texture (θ_t) of 90° for ε of 0.3 (Figure 15.9b). The linear stability is enriched with higher ω_s of three-lobe bearing with limited (partial) texture for a larger extent of recess ($\gamma=0.2$).

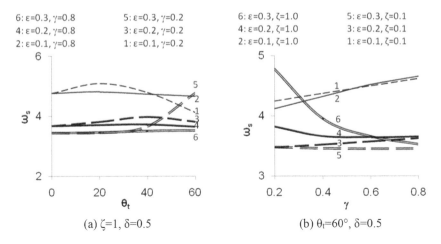

FIGURE 15.9   Instability threshold speed (nondimensional) (ω_s) of three-lobe bearing with limited (partial) texture.

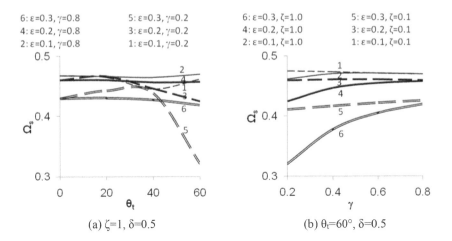

FIGURE 15.10   Whirl (critical) ratio (Ω_s) of three-lobe bearing with limited (partial) texture.

The variation of Ω_s of a three-lobe bearing with limited (partial) texture is shown in Figure 15.10. Ω_s significantly decreases with limited (partial) extent of texture (θ_t) from 40° to 60° for land-to-unit (cell) ratio (γ) of 0.2 and depth (nondimensional) of recess (ζ) of 1 for ε of 0.3 (Figure 15.10a). Ω_s is lower for depth (nondimensional) of recess (ζ) of 1 at land-to-unit (cell) ratio (γ) of 0.2 and at limited (partial) extent of texture (θ_t) of 60° (Figure 15.10b).

Table 15.4 shows the performance of three-lobe bearing with limited (partial) texture. The load of three-lobe bearing is lower for limited (partial) texture orientation. ω_s of three-lobe bearing is higher for the larger extent of recess ($\gamma=0.2$) at ε of 0.3.

TABLE 15.4
Limited (Partial) Texture Three-Lobe Bearing Characteristics

Parameter	Limited (Partial) Texture Three-Lobe Bearing Characteristics
W	W for ζ of 1 is lower compared to $\zeta = 0.1$ at $\gamma = 0.2$–0.8 for $\zeta = 1$, $\theta_t = 60°$, $\delta = 0.5$
ω_s	Higher ω_s is obtained for $\varepsilon = 0.3$ at γ of 0.2 for $\zeta = 1$, $\theta_t = 60°$, $\delta = 0.5$
Ω_s	Ω_s is lower for $\varepsilon = 0.3$ at γ of 0.2 for $\zeta = 1$, $\theta_t = 60°$, $\delta = 0.5$

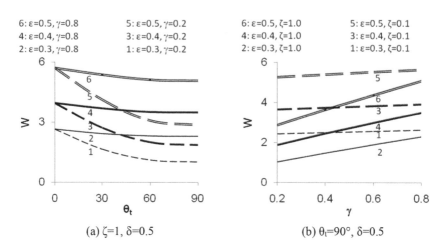

6: ε=0.5, γ=0.8   5: ε=0.5, γ=0.2   6: ε=0.5, ζ=1.0   5: ε=0.5, ζ=0.1
4: ε=0.4, γ=0.8   3: ε=0.4, γ=0.2   4: ε=0.4, ζ=1.0   3: ε=0.4, ζ=0.1
2: ε=0.3, γ=0.8   1: ε=0.3, γ=0.2   2: ε=0.3, ζ=1.0   1: ε=0.3, ζ=0.1

(a) $\zeta = 1$, $\delta = 0.5$

(b) $\theta_t = 90°$, $\delta = 0.5$

FIGURE 15.11 Load capacity (nondimensional) (W) of offset bearing with limited (partial) texture.

15.3.4 OFFSET JOURNAL BEARING

For the offset bearing with limited (partial) texture change corresponding to limited (partial) extent of texture (θ_t), land-to-unit (cell) ratio (γ) and depth (nondimensional) of recess (ζ), Figures 15.11–15.13 reveal the variation of load (nondimensional) capacity (W), speed (nondimensional) of instability threshold (ω_s) and whirl (critical) ratio (Ω_s), respectively.

The variation of W for offset bearing with limited (partial) extent of texture ($\theta_{t=}0°$–$90°$) and land-to-unit cell ratio ($\gamma = 0.2$–0.8) with depth (nondimensional) of recess (ζ) of 1 is shown in Figure 15.11. W decreases with land-to-unit (cell) ratio (γ) for 0.2 compared to $\gamma = 0.8$ at limited (partial) extent of texture (θ_t) in the range of $0°$–$90°$ for ε of 0.3–0.5 with depth (nondimensional) of recess (ζ) of 1 (Figure 15.11a). W for depth (nondimensional) of recess (ζ) of 1 is lower compared to $\zeta = 0.1$ at limited (partial) extent of texture (θ_t) of $90°$ for offset bearing (Figure 15.11b).

The deviation in ω_s with limited (partial) extent of texture ($\theta_{t=}0°$–$90°$) for offset journal bearing with depth (nondimensional) of recess (ζ) of 1 is shown in Figure 15.12. The magnitude of ω_s increases significantly for ε of 0.3 (Figure 15.12a). Higher ω_s of limited (partial) texture offset bearing is attained with a larger extent

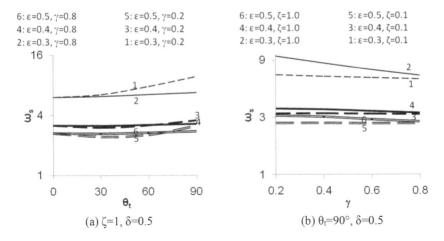

FIGURE 15.12   Instability threshold speed (nondimensional) (ω_s) of offset bearing with limited (partial) texture.

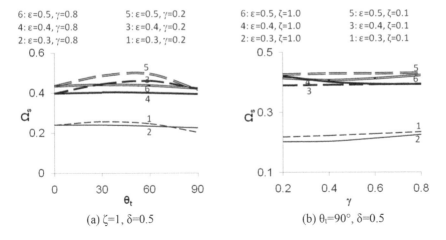

FIGURE 15.13   Whirl (critical) ratio (Ω_s) of offset bearing with limited (partial) texture.

of recess ($\gamma=0.2$) and at limited (partial) extent of texture (θ_t) of 90° for ε of 0.3 (Figure 15.12b).

The variation of Ω_s of an offset journal bearing with limited (partial) texture is shown in Figure 15.13. Ω_s significantly decreases depth (nondimensional) of recess (ζ) of 1 for ε of 0.3 (Figure 15.13a). Ω_s is lower for ε of 0.3 compared to $\varepsilon=0.4$–0.5 for depth (nondimensional) of recess (ζ) of 1 and at limited (partial) extent of texture (θ_t) of 90° (Figure 15.13b).

Table 15.5 shows the performance of offset bearing with limited (partial) texture. W improves significantly for limited (partial) texture at medium eccentricity ratios. ω_s of offset journal bearing increases with decreasing land-to-unit (cell) ratio ($\gamma=0.8$–0.2) for $\varepsilon=0.2$–0.3.

TABLE 15.5

Limited (Partial) Texture Offset Bearing Performance Characteristics

Parameter	Limited (Partial) Texture Offset Bearing Characteristics
W	W decreases for ζ of 1 compared to $\zeta = 0.1$ for $\varepsilon = 0.3–0.5$ with $\gamma = 0.2–0.8$ for $\theta_t = 90°$, $\delta = 0.5$
ω_s	Higher ω_s is obtained for $\varepsilon = 0.3$ at γ of 0.2 for $\zeta = 1$, $\theta_t = 90°$, $\delta = 0.5$
Ω_s	Ω_s is lower for $\varepsilon = 0.3$ but does not vary substantially with variation of $\gamma = 0.2–0.8$ for $\zeta = 1$, $\theta_t = 90°$, $\delta = 0.5$

15.5 CONCLUSIONS

The present study examines the pronounced consequence of limited (partial) texture on multi-lobe bearing configurations. The effects of limited (partial) extent of texture (θ_t), land-to-unit (cell) ratio (γ) and depth (nondimensional) of recess (ζ) on load and linear stability of multi-lobe bearings performance are discussed. The modified Reynolds (nondimensional) dynamic equation for multi-lobe bearings with limited (partial) texture is examined based on Sommerfeld (long) bearing approximation. The expressions for pressure and pressure gradients for multi-lobe bearings have been derived. The analysis of load (nondimensional), speed of threshold and whirl frequency (critical) ratio with limited (partial) texture is presented. The multi-lobe bearing performance is appreciably influenced by limited (partial) texture region.

The two-lobe bearing with limited (partial) texture ($\zeta = 1$, $\theta_t = 90°$) with low land-to-unit (cell) ratio ($\gamma = 0.2$) has a remarkable effect on load and linear stability for low-to-medium eccentricity ratios ($\varepsilon = 0.2–0.3$).

REFERENCES

Allaire, P.E., Li, D.F. & Choy, K.C. 1980. Transient unbalance response of four multilobe journal bearings. *J LubrTechnol*, 102, pp. 300–307.

Chasalevris, A. 2015. Analytical evaluation of the static and dynamic characteristics of three-lobe journal bearings with finite length. *J Tribol*, 137(4), p. 041701.

Crosby, W.A. & Chetti B. 2009. The static and dynamic characteristics of a two-lobe journal bearing lubricated with couple-stress fluid. *Tribol Trans*, 52(2), pp. 262–268.

Flack, R. D. & Lanes, R. F. 1982. Effects of three-lobe bearing geometries on rigid-rotor stability. *ASLETrans*, 25(2), pp. 221–228.

Goyal, K.C. & Sinhasan, R. 1991. Elastohydrodynamic studies of three-lobe journal bearings with non-Newtonian lubricants. *ProcIMechE, Part C: J Mech Eng Sci*, 205, pp.379–388.

Gui, C. & Meng, F. 2019. Comparative study of spherical dimple and bump effects on the tribological performances of journal bearing. *Proc IMechE, Part J: J EngTribol*, 233 (1), pp.139–157.

Khatri, C.B. & Sharma, S.C. 2018. Analysis of textured multi-lobe non-recessed hybrid journal bearings with various restrictors. *Int J Mech Sci*, 145, pp. 258–286.

Kostrzewsky, G.J., Flack, R.D. & Barrett, L.E. 1996. Comparison between measured and predicted performance of two-axial groove journal bearing. *Tribol Trans*, 39(3), pp. 571–578.

Kostrzewsky, G.J., Taylor, D.V., Flack, R.D. & Barrett, L.E. 1998. Theoretical and experimental dynamic characteristics of a highly preloaded three-lobe journal bearing. *Tribol Trans*, 41(3), pp. 392–398.

Kumar, A., Sinhasan, R. & Singh, D.V. 1980. Performance characteristics of two-lobe hydrodynamic journal bearings. *J LubrTechnol*, 102, pp. 425–429.

Lin, Q., Wei, Z., Wang, N. & Chen, W. 2015. Effect of large-area texture/slip surface on journal bearing considering cavitation. *Ind Lubr Tribol*, 67(3), pp. 216–226.

Li, D.F., Choy, K.C. & Allaire P.E. 1980. Stability and transient characteristics of four multi lobe bearing configurations. *J LubrTechnol*, 102, pp. 291–299.

Lund, J.W. 1987. Review of the concept of dynamic coefficients for fluid film journal bearings. *J Tribol*, 109, pp. 37–41.

Lund, J.W. & Thomsen, K.K., 1978. A calculation method and data for the dynamic coefficients of oil lubricated journal bearings. In *Topics in Fluid Film Bearing and Rotor Bearing System*, ASME New York, pp. 1–28.

Malik, M. 1983. A comparative study of some two-lobed journal bearing configurations. *ASLE Trans*, 26(1), pp. 118–124.

Malik, M., Sinhasan, R. & Chandra, M. 1981. Design data for three-lobe bearings. *ASLE Trans*, 24(3), pp.345–353.

Matele, S. & Pandey, K.N. 2018. Effect of surface texturing on the dynamic characteristics of hydrodynamic journal bearing comprising concepts of green tribology. *ProcIMechE, Part J: J EngTribol*, 232(11), pp. 1365–1376.

Mehrjardi, M.Z., Rahmatabadi, A.D. & Meybodi, R.R. 2016. A comparative study of the preload effects on the stability performance of noncircular journal bearings using linear and nonlinear dynamic approaches. *Proc IMechE, Part J: J EngTribol*, 230(7), pp. 797–816.

Nair, K.P., Sinhasan, R. & Singh, D.V. 1987. Elastohydrodynamic effects in elliptical bearings. *Wear*, 118(2), pp. 129–145.

Nair, K.P., Sinhasan, R. & Singh, D.V. 1987. A study of elastohydrodynamic effects in a three-lobe journal bearing. *TribolInt*, 20(3), pp. 125–132.

Rao, T.V.V.L.N., Biswas, S. & Athre, K. 2001. A methodology for dynamic coefficients and nonlinear response of multi-lobe journal bearings. *Tribol Trans*, 44(1), pp. 111–117.

Rao, T.V.V.L.N., Rani, A.M.A., Mohamed, N.M., Ya, H.H., Awang, M. & Hashim, F.M. 2020. Static and stability analysis of partial slip texture multi-lobe journal bearings. *Proc IMechE, Part J: J EngTribol*, 234(4), pp. 567–587.

Sinhasan, R. & Chandrawat, H.N. 1988. An elastohydrodynamic study on two-axial-groove journal bearings. *TribolInt*, 21(6), pp. 341–351.

Sinhasan, R., Malik, M. & Chandra, M. 1981. A comparative study of some three-lobe bearing configurations. *Wear*, 72, pp. 277–286.

Vohr, J.H & Chow C.Y. 1965. Characteristics of herringbone-grooved, gas-lubricated journal bearings. *J Basic Eng*, 9, pp. 568–578.

Yamada, H., Taura, H. & Kaneko S. 2018. Numerical and experimental analyses of the dynamic characteristics of journal bearings with square dimples. *J Tribol*, 140, pp. 011703–1-13.

16 Minimum Quantity Lubrication for Sustainable Manufacturing

Rahul Anand, Ankush Raina, and Mir Irfan Ul Haq
Shri Mata Vaishno Devi University

Mohd Fadzli Bin Abdollah
Universiti Teknikal Malaysia Melaka

CONTENTS

16.1 INTRODUCTION

The growing need for greener technologies and fast product development involved in almost all engineering applications has become very difficult to manage the resources with the rising demand of the products all across the globe (Baba et al. 2019). Along with the improvement in the quality of goods, emphasis on the reduction of cost and increase in productivity is being laid. Machining plays a vital role in ramping up the production rate in the industries as it is the most versatile manufacturing process being applied in almost all the manufacturing industries for material removal (Figure 16.1). Material is removed by providing direct contact and relative motion between tool and workpiece. This relative motion along with the force exerted by the machine tool causes plastic deformation of workpiece. The energy transferred back to the cutting tool raises the temperature of tool which may lead to the wear and failure of cutting tool. Using a blunt tool for further machining may lead to excessive power consumption and poor surface texture, and it also limits the production rate. Thus, there arises a need to constantly minimize the tool temperature by extracting the heat generated due to plastic deformation. Cutting fluids along with different techniques are thus developed in recent past to deal with the extraction of heat from cutting interface. These cutting fluids provide many advantages in terms of reduction in friction between tool workpiece interface, reducing chipping of chips over the flank face of the tool, and maintaining the dimensional accuracy but along with it they are costly to purchase and have disposability issues. These negative effects of cutting fluid paved a way for researches to investigate more on vegetable oils and mineral oils along with different techniques for their applications which would minimize the use of cutting fluids without compromising with the quality of finished product. Different biodegradable vegetable oils and mineral oils have potentially proved their significance in improving the tool life and surface texture in different machining operations (Anand, Raina, et al. 2020; Shafi et al. 2018), and in addition to the conventional cooling methods, other methods like cryogenic cooling, mist lubrication, minimum quantity lubrication, solid lubrication and high-pressure lubrication have been developed so as to make the cutting process more eco-friendly,

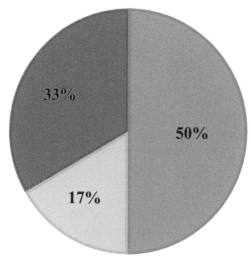

FIGURE 16.1   Pie chart representations of proportionate machining costs (Tai et al. 2017).

thereby taking a step forward towards sustainable manufacturing. Many researchers have reported the studies on different cooling techniques along with the usage of different biodegradable oils in different conventional machining techniques. The aim of this chapter is to present the investigations done using MQL along with different lubricants in conventional machining processes.

16.2 BASICS OF TOOL-CHIP TRIBOLOGY

The term 'Tribology' means the science of interacting surfaces which involves physics of contact and mechanics of moving interfaces. In metal cutting, the word tribology has a broader meaning of assessment of tool wear and its by-product (Astakhov 2006). The sole purpose of the various techniques adopted in cutting tribology is the reduction of energy utilized in the cutting operation along with the optimization of the cutting parameters (King and Hahn 2012). Reduction in tool wear and failure leads to enhanced tool life, improved surface quality and cost-efficient product (Anand et al. 2016). This is possible by improving the tribological conditions at the chip-tool and tool-workpiece interface (Figure 16.2). In metal cutting, 30%–40% of the energy supplied is consumed for the separation of the material from the workpiece and remaining 20%–60% is wasted which is due to non-optimization of the parameters related to tribology at the chip-tool and tool-workpiece interfaces (Astakhov and Xiao 2008). These non-optimized conditions thus lead to the increase in power consumption, tool interface temperature and cutting forces. These consequences thus require the improvement in lubrication technology.

The basic parameters of the chip-tool interface are relative velocity or sliding velocity, tool-chip contact length, mean shear stress, mean contact temperature, normal force at the tool-chip interface and shear stress over the contact length. At the time of contact, the tool deforms elastically and plastically with the formation of deformation zone on the cutting tool (Anand, Haq, et al. 2020). As the full contact is developed between the tool and the chip, a combined bending and compressive state of stress gets developed. When this stress reaches to the maximum level, the chip starts to flow over the rake surface of the tool with some velocity (Khajuria and Wani 2017). As the tool progresses further, the resistance offered to the tool by the deformed material decreases. There is a formation of an additional deformation zone over the tool-chip interface where the chip continues to flow till the cutting forces get reduced (Shivpuri et al. 2002). The cutting force reduces as the fresh part of the chip is exposed and attracts the cutting forces. Thus, the formation of continuous chips takes place. Cutting speed has major influence on the type and shape of chip formation that takes place during cutting operation. There are two methods defined for improving the tribological aspects related to the cutting tools viz. *Component Means* which involve selection of application-based material for cutting tool, coatings of cutting tools, and design and quality considerations that are application specific (Kopac et al. 2001), and *Systemic Means* that involve the improvement during the cutting operations such as flow of cutting fluid at the cutting interface, improvement in tool holder design (Astakhov 2010), and introduction of ultrasonic vibrations that reduce the cutting forces.

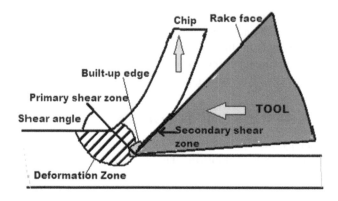

FIGURE 16.2   Schematic showing shear flow of chip over rake surface of the tool.

16.3   METHODS OF CUTTING FLUID

16.3.1   Wet Cooling Technique

The most commonly used application of cutting fluids during the machining operations is wet cooling or flood lubrication technique (Figures 16.3 and 16.4). In this technique, a continuous stream of lubricant is impinged over the chip-tool interface with flow measuring around 10 l/min for single point cutting tool and 225 l/min for multipoint cutting tools (Schmid et al. 2014). The authors Jayal et al. (2007) and Imran et al. (2014) have concluded that wet cooling technique has significant

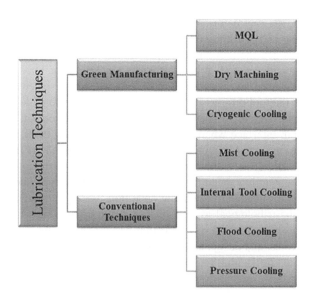

FIGURE 16.3   Techniques of cutting fluid application in machining operations (Debnath et al. 2014).

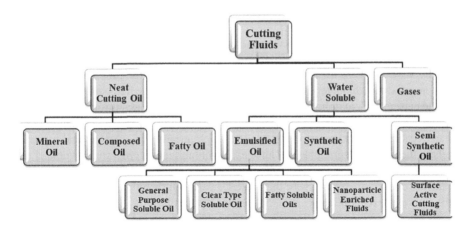

FIGURE 16.4 Cutting fluid classification (El Baradie et al. 1996).

advantages in terms of cutting forces' reduction, dimensional accuracy and temperature rise in comparison with dry machining. Flooded lubrication also minimizes the catastrophic failure of tool while machining (Rajaguru and Arunachalam 2020).

16.3.2 HIGH PRESSURE COOLING

The method of high pressure cooling technique is used to minimize the interference temperature of cutting zone. It finds the application in machining operations where the chip-tool interface temperature is high and rate of heat removal becomes significant. In this technique, the lubricant jet with specially designed nozzle is imparted at the tool clearance face with 5.5–35 MPa pressure and impinging speed of 350–500 km/h (Sharma et al. 2009). This technique efficiently increases the productivity by allowing better penetration of high-pressure lubricant in the cutting zone and by reducing the contact area and length of the chip over the tool surface. The authors Naves et al. (2013), Fang and Obikawa (2020), and Lu et al. (2020) have reported that the use of high pressure cooling technique with longer surface distance significantly improves tool life and improves surface roughness along with reduced cutting forces.

16.3.3 CRYOGENIC MACHINING

This is a recently developed application of fluids using liquid nitrogen and helium as a lubricant. In this technique, liquid nitrogen is sprayed in the cutting zone with specially designed nozzles at a temperature measuring around −200°C. The nitrogen in liquid form absorbs the interface heat and transforms into a gas which act as a cushion between the tool and workpiece interface, at lower cutting speed (Gajrani and Sankar 2020). At higher cutting speeds, the penetration of liquid nitrogen is hindered due to fully plastic deformation at the shear zone. The authors Dhananchezian and Kumar (2011), and Yıldırım et al. (2020) have opined that the cryogenic lubrication technique lowers the cutting forces and improves the surface finish as compared to conventional machining.

16.3.4 Minimum Quantity Lubrication (MQL)

Minimum quantity lubrication technique is introduced as an economically and environmentally friendly technique in context to the sustainable development (Dixit et al. 2012). MQL uses fine air-oil aerosol mixture sprayed at a high pressure of 600 kPa. The nozzles with diameter of around 1mm are placed over the flank and the rake face of the cutting tool, and the pressurized mist of lubricant is flown through the nozzles in the cutting zone at flow rates of 50 ml/h–2 l/h (Schmid et al. 2014). The lubricants like vegetable oil and ester oils are generally preferred over synthetic and mineral oils as they are biodegradable and have excellent lubrication properties (Woma et al. 2019). MQL is also known as near dry machining because the aerosol mixture evaporates on coming in contact with the heated shear zone thereby cooling the tool and the workpiece (Gajrani et al. 2017; Simunovic et al. 2015).

16.3.4.1 Types of MQL Systems

MQL delivery setup mainly consists of atomizer, oil sump, nozzles, delivery timed units, pressure valves, and compressors. When the high-pressurized air of almost 5–6 bar pressure is passed through the mixing chamber, it divides the lubricant droplets into very fine aerosols which are ejected out at the cutting zones with the help of atomizer (Simunovic et al. 2015). This mist of coolant and air provides cooing and lubricating effect at the cutting zone.

MQL delivery system is basically divided into two systems (1) External feed and (2) Internal feed systems based on the mixing of air-oil and creation of mist (Attanasio et al. 2006). In external feed system, both air and oil are mixed externally and are then fed through the spindle from where it is sprayed towards the cutting zone. The advantage of external feed system is easy to maintain as it does not have any critical part inside the spindle. In external feed system, however there is a disadvantage of droplet separation and dispersion in actual travel from nozzle towards the cutting zone which is overcome by internal feed system. In internal feed system, there is two-way delivery to bring the air and oil which are routed inside the spindle to the mixing chamber inside the spindle. Internal feed system thus delivers the mist with larger droplet size and without dispersion of droplets while spraying with less lag (Varadarajan et al. 2000). Internal delivery system is however difficult to maintain as it contains critical parts inside the spindle.

16.4 MQL USING DIFFERENT CUTTING FLUIDS IN CONVENTIONAL MACHINING

16.4.1 MQL in Drilling

Owing to the continuous sticking in conventional drilling operations which becomes a cause of increased frictional coefficient and reduced heat transfer rates, it becomes extremely important to enhance the performance of drilling operations. Many researchers have investigated the influence of the drilling parameters with MQL lubrication technique and have reported significant and improved results in terms of tool life, surface finish of drilled holes and metal adhesion on flutes.

Rahim and Sasahara (2011) examined drilling of Inconel 718 with MQL using synthetic ester and palm oil. It was observed that microhardness increases with the feed rate (0.1 mm/rev) and increase in cutting speed. Due to high viscosity and cooling effect of palm oil, it shows improved cutting speed, feed rate, surface finish and microhardness than synthetic ester. Continuous MQL supply with better penetrating power of low viscosity fluid benefits in improving tool life of twist drills due to improved cooling and heat transfer rates (Heinemann et al. 2006). Tasdelen et al. (2008) examined axial forces after drilling 105 and 315 holes wherein the maximum axial force was observed for air cooling followed by MQL (5 ml/h). Meena and Mansori (2011) observed an increase in surface roughness in dry conditions as compared to wet and flooded lubrication. MQL significantly reduces drilling torque and axial thrusts as compared to dry and flooded lubrication. Flank wear values recorded for wet, MQL and dry machining conditions were 0.12, 0.25 and 0.41 mm, respectively. Flank wear values were observed almost similar for MQL and flooded lubrication with soluble oil while drilling with uncoated K10 drill (Braga et al. 2002). Biermann et al. (2012) analysed feed rate as a determining factor for both mechanical and thermal loading of workpiece. Bhowmick et al. (2010) on comparing MQL drilling conditions over dry lubrication reported an increase in tool life, reduction in thrust forces and average torque in case of MQL. The results also state MQL drilling as effective as flooded drilling because the temperature softening strain effect was minimized by strain hardening. The application of MQL reduced the temperature at chip and tool interface resulting in the decrease of magnesium adhesion on the drill and reduced the Built-up edge (BUE) formation. MQL thus lowers the requirement of cutting forces and torque (Bhowmick and Alpas 2011). MQL technique reduces adhesion of chip on tool interface, and tribological properties obtained with MQL lubrication were comparable with flood-H_2O lubrication with reduced thrust and cutting forces (Bhowmick and Alpas 2008). Thus, it is clear that compared to the other techniques, MQL proves to be an effective technique for sustainable lubrication in drilling.

16.4.2 MQL IN TURNING

Turning is a very common machining operation. In recent past, significant studies have been reported by the researchers to present the comparison between dry, flood and MQL lubrication techniques in turning operation. MQL lubrication technique in particular has been suggested by most of the researchers due to minimal use of cutting fluid. Comprehensive research by varying various turning parameters like feed rate, depth of cut, cutting speed and nozzle orientation to examine their influence on chip-tool interface temperature, chip colour, chip shape, surface roughness and tool wear with respect to dry, flood and MQL techniques has been undertaken. In this regard, Hwang and Lee (2010) using design of experiments investigated the parameters that influence the performance of MQL over dry turning. It was observed that cutting speed and depth of cut inversely affect cutting forces and surface roughness with MQL providing better cutting performance as compared to wet turning. Out of the three turning types (dry, wet and MQL), the MQL with ester oil for 4140 alloy steel provides better machining conditions with reduced cutting temperature and cutting forces (Hadad and Sadeghi 2013).

TABLE 16.1

Summary of Research Pertaining to Drilling Assisted with MQL Technique

Ref.	Cutting Fluid	Workpiece Material	Machining Parameters		Mode of Lubrication	Process Findings
			Cutting Speed (m/min)	Feed Rate (mm/rev)		
Heinemann et al. (2006)	Synthetic ester + 20% alcohol + additives, Oil free synthetic lubricant + 40% water	Plain carbon steel (carbon + 0.45%)	26	0.26	MQL (14 bar 18 ml/h)	• Continuous MQL supply is effective than obstructed one in terms of tool life. • Higher water content and low viscosity of oil enhance the drilling depth.
Braga et al. (2002)	Air + Oil (MQL) Soluble oil (flood)	Aluminium–silicon alloy (7% silicon) SAE323	300	0.1	MQL, Flood (4.5 bar, 10 ml/h)	• Value of flank wear was almost similar in both types of cooling. • Better quality holes were obtained by MQL
Bhowmick and Alpas (2008)	Distilled water spray	Sand cast (319 Al-Si alloy)	50	0.25	MQL, Dry (30 ml/h)	• Less torque and cutting forces were observed in MQL as compared to dry machining. • BUE due to adhesion of aluminium on tool bit was reduced in H_2O-MQL as observed by electron and optical microscopy.
Tasdelen et al. (2008)	Emulsion oil	Steel (hardened)	155	0.11	MQL (5, 15, 23 ml/h) emulsion, air	• Chip contact length on rake face reduced with increase of oil in lubrication. • Torque requirement was lower than dry drilling but higher than wet drilling.

(Continued)

TABLE 16.1 (*Continued*)
Summary of Research Pertaining to Drilling Assisted with MQL Technique

Ref.	Cutting Fluid	Workpiece Material	Cutting Speed (m/min)	Feed Rate (mm/rev)	Mode of Lubrication	Process Findings
			Machining Parameters			
Rahim and Sasahara (2011)	Palm oil + synthetic ester	Inconel 718	30, 40, 50	0.5, 0.1	MQL (10.3 ml/h)	• Palm oil exhibited better characteristics in comparison with synthetic ester in terms of surface hardness, surface roughness and surface defects. • Palm oil provides better cooling effects due to high viscosity.
Biermann et al. (2012)	Water (distilled)	Aluminium cast alloy EN AC – 46,000 (AlSi$_9$Cu$_3$)	140, 200	0.1–0.3	MQL (14 bar)	• Pressure of 14 bar provided the required cooling effect. • In case of deep hole drilling, inversion of heat from heated MQL into workpiece borehole occurred.
Bhowmick and Alpas (2011)	Water (distilled)	AZ91 (Mg-Al-Zn) alloy	Cutting speed = 1,000, 1,500, 2,000, 2,500	Feed = 0.1, 0.15, 0.20, 0.25 mm/rev	MQL (30 ml/h), Dry, Flood	• MQL restricted the adhesion of Mg on tool face due to temperature rise, thereby increasing the tool life and torque reduction.
Bhowmick et al. (2010)	Water (distilled), Fatty acid spray	Magnesium-6%Aluminium alloy (AM60)	Cutting speed = 50 m/min	Feed = 0.25 mm/rev	MQL (10 ml/h)	• BUE formation, temperature generation and cutting forces along with torque were reduced to large extent than in dry milling.
Meena and El Mansori (2011)	Swiss cut frisco 6122 I	ADI (Austempered Ductile Iron)	Cutting speed = 60 m/min	Feed = 0.15 mm/rev	MQL (10 ml/h), Flooded	• MQL outperformed flooded drilling in cutting performance, tool life and surface finish • Crater wear was dominant in flood and MQL while flank wear reduced in flooded drilling.

A study by Khan et al. (2009) on turning of alloy steel (AISI 9310) with MQL technique using vegetable recorded a decrease in chip-tool interface temperature by 10% along with the improved surface finish than wet turning. Dhar et al. (2006) analysed turning performance of ductile AISI-1040 steel using MQL. It was observed that effective penetration of cutting fluid using MQL jet reduces the chip-tool interface temperature which reduces tool wear and improves chip-tool interaction. Cutting forces were reduced by 5%–15% as compared to dry turning with reduced friction between chip and tool interface which may be ascribed to decrease in cutting temperature using MQL (Dhar et al. 2007). Bruni et al. (2006) investigated that surface roughness depends on both flank wear and shearing mechanisms during machining. Surface roughness and flank wear were reduced by using wiper inserts and MQL technique of lubrication. The surface roughness obtained at 200 ml/h MQL or flood lubrication and power consumed has no significant variation (Davim et al. 2007). Physical vapor deposition (PVD) coated inserts enhanced cutting performance and reduced tip tool temperature using MQL technique. Cutting forces were reduced by 17.07% as compared to chemical vapor deposition (CVD) coated insert 13.25% under identical cutting conditions (Saini et al. 2014).

In a similar investigation, Sarıkaya and Gullu (2015) using grey rational analysis reported the flow rate of 180 ml/h and cutting speed of 30 m/min as optimized machining parameters which resulted in the minimum tool wear and surface roughness using vegetable oil. MQL using rapeseed oil has small lubricating effect and is used under low loading conditions due to weak boundary film formation whereas MQL with water mixed synthetic ester showed better turning performance (Itoigawa et al. 2006). Supercritical CO_2 lubrication system enhanced tool life and rate of material removal by around 40% with improved lubricity when compared to aqueous flood cooling (Stephenson et al. 2014). Masoudi et al. (2018) experimentally analysed parameters such as workpiece hardness and nozzle angle on turning of AISI 1045 using MQL technique. It was observed that MQL technique decreases surface roughness, cutting forces and chip contact length significantly than dry and wet lubrication. Furthermore, low tool wear rates and improved surface finish were observed with tungsten carbide tool which attributes to low coefficient of friction and better wettability of carbide tool than HSS tool. Jamaludin et al. (2018) investigated the influence of MQL using cold air on turning AISI 316 stainless steel. It was observed that there is reduction in thrust force and feed force by 60 and 30 N, respectively, with cold air (−13.6°C) for hydroscopic oil as compared to dry machining. In conclusion, it can be inferred from the above discussion that MQL unlike dry machining or flooded lubrication is an efficient way to improve the tool life and other machining parameters.

16.4.3   MQL in Grinding

With enhancement in technology, engineers and researchers are in pursuit of an effective lubrication setup that minimizes the heat generation at the interface of workpiece and grinding wheel because it directly influences the working life of grinding wheel, abrasive loss and wear. To enhance the cooling capacity and lubrication effects, researchers have replaced flood cooling with dry and MQL lubrication technique in grinding operations. Tawakoli et al. (2009) in their investigations on grinding observed that oil-assisted MQL lubrication technique produced better surface finish than water-assisted MQL grinding. Tawakoli et al. (2010) reported that nozzle angle between 10° and 20° along with critical distance of 80 mm between nozzle and workpiece yielded optimum results.

Sadeghi et al. (2009) found MQL with vegetable oil improves surface texture and reduces cutting forces. Reduction in grinding energy and cutting forces retains the sharpness of the grinding wheel, thereby increasing the life of the grinding wheel (Sadeghi et al. 2010). Temperature at the chip-tool interface was recorded similar to as that of flood lubrication as compared to MQL mode of lubrication for EN 31 steel and softer material (Morgan et al. 2012). Batako and Tsiakoumis (2015) observed 15% reduction in normal and tangential cutting forces in their investigations on steel and nickel alloys.

Emami et al. (2013) recorded the nozzle angle of 15° and minimum distance of 30 mm from workpiece as optimum nozzle parameters along with flow rate of 150 ml/h. Rabiei et al. (2015) concluded that with effective penetration of spray between chip-tool interface, friction coefficient reduced along with 0.42 μm reduction in surface roughness. Thus, it can be stated that the use of MQL in grinding operations helps to improve various properties such as surface roughness and COF at tool-chip interface, apart from the saving in energy.

16.4.4 MQL in Milling

Research studies carried out to reduce the milling forces, power consumed, surface roughness and tool wear by the use of MQL-assisted milling have shown improved results as compared to dry, flood and cryogenic cooling techniques. Liao and Lin (2007) observed the formation of protective layer at the interface which improves the tool life significantly. In another investigation by Liao et al. (2007), MQL lubrication improved the tool life and surface roughness significantly as compared to flood and dry lubrication. Adhesion of workpiece material over the tool surface was significantly reduced by the mist lubrication (MQL). Tosun and Huseyinoglu (2010) noticed that the reduction of flank wear by 60% and surface roughness reduction of 0.2 μm compared to dry milling under similar conditions of feed and cutting speed in milling AA7075-T6 alloy using TiCN cutting tool.

Kang et al. (2008) recorded minimum flank wear in MQL-assisted end milling as compared to flood and dry lubrication techniques. Zhang et al. (2012) recorded 1.78 times increase in tool life on MQL-assisted milling using vegetable oil for Inconel alloy as compared to dry milling. Fratila and Caizar (2011) reported that MQL-assisted milling can replace flood and wet lubrication without affecting the results of machining in terms of surface roughness and power consumption. In view of above discussion, it can be concluded that the use of MQL in milling can help to reduce wear of the tool considerably thereby improving the tool life and improving other characteristics such as surface roughness of the workpiece.

TABLE 16.2

Summary of Research Pertaining to Turning Assisted with MQL Technique

Ref.	Cutting Fluid	Workpiece Material	Cutting Speed (m/min)	Feed Rate (mm/rev)	Mode of Lubrication	Process Findings
			Machining Parameters			
Dhar et al. (2006)	Emulsion oil	AISI-1040 Steel	60, 80, 110, 130	0.1, 0.13, 0.16, 0.2	MQL (60 ml/h), Dry, Wet	• Less Rise in temperature with favourable chip colour and cutting edges of the tool were maintained sharp in MQL.
Davim et al. (2007)	Emulsion oil (Microtrend 231L)	Brass (CuZn39Pb3)	100, 200, 400	0.05, 0.1, 0.15, 0.2	MQL (50, 100, 200 ml/h), Wet	• Surface roughness results were better at higher cutting speeds in MQL.
Khan et al. (2009)	Vegetable oil	AISI 9310 Steel	223, 246, 348, 483	0.10, 0.13, 0.16, 0.18	MQL (100 ml/h), Dry, Wet	• Surface finish was improved due to less tool wear in MQL. • Ribbon type chips at lower feed rates and tubular type at higher feed rates with reduced tool-chip interface temperature up to 10%.
Hwang and Lee (2010)	Supercritical CO₂	Carbon steel AISI 1045	100, 300	0.1, 0.3	MQL, Wet	• MQL was more advantageous than wet lubrication in terms of cutting speed and surface roughness
Stephenson et al. (2014)	Supercritical CO₂	Inconel 750	45.7, 50.3	0.25, 0.30	MQL, Wet	• Tool life was improved, and material removal rate was increased to 40 % in MQL lubrication.
Hadad and Sadeghi (2013)	Ester oil (Rs 1642)	42CrMo4 soft steel AISI 4140	50.2, 100.4, 141.4	0.09, 0.22	MQL, Dry, Wet	• MQL had improved surface finish due to decrease in cutting forces and less worn out tool edges.
Saini et al. (2014)	Mineral oil	AISI 4340 steel	39.75, 55.91	0.066, 0.08, 0.1, 0.133, 0.6	MQL (300 ml/h), Dry	• MQL maintained the sharpness of insert by reducing the cutting forces and temperature considerably. • Reduction in fumes results in better machining conditions and ecology.

(Continued)

TABLE 16.2 (Continued)
Summary of Research Pertaining to Turning Assisted with MQL Technique

Ref.	Cutting Fluid	Workpiece Material	Machining Parameters		Mode of Lubrication	Process Findings
			Cutting Speed (m/min)	Feed Rate (mm/rev)		
Leppert (2011)	Accu-lube LB8000 (Biodegradable vegetable oil)	Carbon structural steel (AISI 1045)	76, 190, 237	0.08, 0.27, 0.47	MQL (50 ml/h), Dry	• Surface topography, cutting forces at higher feed rates and surface roughness were satisfactory in MQL and dry lubrication.
Khan and Maity (2018)	Vegetable oil	Pure Titanium (CP-Ti) Grade2	51, 67, 87	0.12	MQL, Dry, Flood	• Tool wear and flank wear are reduced by 57% and 34% in MQL. • No surface hardening and better surface finish were achieved.
Senevirathne and Punchihewa (2018)	Emulsion oil	AISI P20 And D2	150	0.5	MQL, Flooded	• Improved machining characteristics were achieved with MQL as compared to conventional cooling. • Best surface finish was obtained at 5°C temperature for both workpieces.
Dhar et al. (2007)	Mobil Cut 102	AISI 1040	72, 94, 139, 164	1.5	MQL, Dry	• Cutting forces reduced by 5%–15 % due to reduced chip-tool interface friction.
Bruni et al. (2006)	Vegetable oil	AISI 420B steel	235	0.08, 0.16 rev/min	MQL, Dry, Wet	• MQL proves less efficient in reducing flank wear and improving the surface quality, when related to dry machining.
Masoudi et al. (2018)	Ester oil	AISI 1045 steel	110 m/min	0.008 mm/rev	MQL, Dry, Wet	• Tungsten carbide tool shows better performance than HSS in terms of cutting forces and surface roughness.
Jamaludin et al. (2018)	Vegetable oil	AISI 316 stainless steel	50–200 m/min	0.3 mm/rev	MQL, Dry	• Thrust force and feed force were reduced by 60 and 30N, respectively.

TABLE 16.3

Summary of Research Pertaining to Grinding Assisted with MQL Technique

Ref.	Cutting Fluid	Workpiece Material	Cutting Parameters			Mode of Lubrication	Process Findings
			Depth of Cut (μm)	Wheel Speed (m/s)			
Tawakoli et al. (2009)	LB8000-MQL oil	Hardened steel (100Cr6), 42CrMo4 soft steel	5, 10, 15, 25	20, 25, 30		MQL (60 ml/h), Dry, Flooded	• MQL reduces friction coefficient and specific energy considerably for both hard and soft steel. • Higher Material removal rate (MRR) is achieved with less forces and improved surface finish in MQL than flood and dry grinding.
Sadeghi et al. (2009)	Synthetic oil, Vegetable oil, Behran cutting oil 53, Behran cutting oil 34	Ti-6Al-4V	0.002, 0.005, 0.007 mm	15		MQL (15, 20, 30, 40, 50, 60, 100, 33 ml/h) Wet	• MQL with vegetable oil produced improved cooling effects but best surface morphology, surface roughness and reduced cutting forces were achieved with synthetic oil.
Tawakoli et al. (2010)	HAKUFORM 20-34 Oil	Steel (Hardened) (100Cr6)	20	30, 45		MQL (20, 50, 100 ml/h), Dry, Flood	• 10°–20° of nozzle angle to workpiece and a critical distance of 80mm form workpiece results in reduction of cutting forces.
Barczak et al. (2010)	5% Castrol Hysol XF	Mild steel (EN8), tool steel (M8), bearing steel (EN31)	5, 15	25, 45		MQL (33 ml/h) Dry, Wet	• MQL grinding countered comparatively less frictional force than dry and wet modes. • Reasonable MRR at less temperature and cutting speed was achieved.
Li and Lin (2012)	Bluebe lubricant LB-1 (vegetable oil)	SK3 with HRC18	50	30,000, 39,000. 48,000 RPM		MQL, Dry	• Improved surface morphology after chip removal without any burned surface marks was obtained. • Tool life in MQL increased by 7% than dry and 5% by air cooling.

(Continued)

TABLE 16.3 (*Continued*)
Summary of Research Pertaining to Grinding Assisted with MQL Technique

Ref.	Cutting Fluid	Workpiece Material	Cutting Parameters Depth of Cut (μm)	Wheel Speed (m/s)	Mode of Lubrication	Process Findings
Morgan et al. (2012)	Castor carecut ES1 oil	M2, EN31 and EN8 steel	5, 15	25, 45	MQL, Dry, Wet	• MQL and wet grinding produce almost similar thermal characteristics under given conditions of thermal model.
Balan et al. (2013)	Cimtech D14 MQL oil	Inconel-751 superalloy	30	47.6	MQL (60, 80, 1,000)	• MQL grinding at high air pressure and minimum oil flow requires less grinding force.
						• MQL reduces surface roughness and chip-tool interface temperature considerably.
Emami et al. (2013)	Mineral oil	Al₂O₃ Engineering ceramic	18	30	MQL (60, 80, 100 ml)	• Nozzle angle 15°, Nozzle distance 30 m, gas flow rate 150 ml/h and 30 l/min were achieved as optimum grinding conditions for minimizing the cutting forces and surface roughness.
Sadeghi et al. (2010)	Behran cutting oil 53, synthetic oil, Behran cutting oil 34 and Vegetable oil,	AISI 4140 low alloy steel	0.005, 0.010, 0.015 mm	30 m/s	MQL, Dry, Wet	• Behran 53 and vegetable oil show improved cooling at cutting surfaces. Grinding energy and cutting forces reduced with MQL which attributes to retained grit sharpness of the grinding wheel.
Emami et al. (2014)	Vegetable oil, synthetic oil, mineral oil and hydrocracked oil	High purity alumina (Al₂O₃) ceramic	8, 12 18, 27	30	MQL (150 ml/h),	• Synthetic oil and hydrocracked oil showed high thermal and oxidative stability at high temperatures.
						• Addition of sulphur and phosphorous improved surface roughness and reduced cutting forces.

(*Continued*)

TABLE 16.3 (Continued)

Summary of Research Pertaining to Grinding Assisted with MQL Technique

Ref.	Cutting Fluid	Workpiece Material	Cutting Parameters		Mode of Lubrication	Process Findings
			Depth of Cut (μm)	Wheel Speed (m/s)		
Batako and Tsiakoumis (2015)	Castrol care cut E1 oil	BS534A99 Steel and Nickel alloy 718	10–30	30	MQL (30 ml/h), Dry	• Reduction of 15% in tangential and normal forces was obtained on introducing workpiece to oscillations. • Reduction of 31% was achieved at a particular speed of workpiece and wheel.
Rabiei et al. (2015)	RS 1642 Behran oil	Tool steel (HSS), Mild carbon steel (CK45), Bearing steel (100cr6), Raw HSS (S305)	5, 20, 35, 50	30	MQL (120 ml/h), Dry, Fluid	• Effective dispersion of oil in the contact zone and cutting forces along with frictional coefficient in MQL decrease. • A minimum surface roughness of 0.42 μm was achieved through GA optimization.
Wang et al. (2016)	Soybean oil, paraffin oil, peanut oil maize oil, rapeseed oil, palm oil, castor oil and sunflower oil	Nickel GH4169 alloy	10	30	MQL (50 ml/h)	• Castor oil and palm oil have low friction coefficient (0.30 and 0.33) and specific energy (73.47 and 78.8 J/mm³). • Lubrication follows: maize oil < rapeseed oil < soybean oil < sunflower oil < peanut oil < palm oil < castor oil. • Castor oil gave minimum roughness value (Ra = 0.366)
Li et al. (2016)	Castor oil + (soybean oil, rapeseed oil, palm oil, sunflower, maize oil and peanut oil)	Nickel GH4169	10	30	MQL (50 ml/h)	• Mixed oil MQL improves cutting response parameters instead of machining with castor oil alone. • Best lubricity was provided by soybean with castor oil among all combinations. • Normal force (F_n) and tangential forces (F_t) in palm / castor oil were reduced by 0.91% and 2.454%.

TABLE 16.4
Summary of Research Pertaining to Milling Assisted with MQL Technique

Ref.	Cutting Fluid	Workpiece	Cutting Parameters			Mode of Lubrication	Process Findings
			Feed Rate (mm)/tooth	Cutting Speed (m/min)			
Liao and Lin (2007)	Synthetic ester, castor carecut ES3	Hardened steel	0.10, 0.15, 0.20	300, 400, 500		MQL, Dry	• A protective oxide layer formation takes place at chip-tool interface in MQL which enhances tool life at optimum cutting speed. • Extreme high speed MQL milling is not suggested as it may give abrupt temperature rise which may result in thermal cracks.
Liao et al. (2007)	Synthetic ester, castor carecut ES3	NAK80 Hardened die steel	0.10, 0.15 0.20	150, 200, 250		MQL (10 ml/h), Dry Wet	• Tool-chip contact time is reduced which improves tool life. • Due to MQL, high speed milling at high feed rate was performed.
Kang et al. (2008)		AISI D2 cold worked Die steel				MQL (6 ml/h), Dry, Wet	• MQL end milling enhanced the cutting performance. • Least flank wear was recorded in MQL among all three lubricating techniques.
Li and Chou (2010)	Blube lubricant LB-1	SKD 61 steels	1, 1.5, 2 µm/rev	20,000, 30,000, 40,000 rpm		MQL, (25, 40 l/min), Dry	• A reduction of 60% in flank wear along with reduced burr formation in MQL milling. • Surface roughness less than $R_a = 0.2$ µm was achieved.
Liew (2010)	Liquid paraffin (93%) and cyclomethicone (7%)	STAVAX (modified AISI 420 Steel)	0.04, 0.2, 0.4	88		MQL (0.2 l/h), flood, mist	• Good surface finish with less tool wear (flank wear) was achieved in MQL milling.

(Continued)

TABLE 16.4 (Continued)

Summary of Research Pertaining to Milling Assisted with MQL Technique

Ref.	Cutting Fluid	Workpiece	Cutting Parameters		Mode of Lubrication	Process Findings
			Feed Rate (mm)/tooth	Cutting Speed (m/min)		
Tosun and Huseyinoglu (2010)	Boron oil-water solution	AA 7075-T6 aluminium alloy	20, 40, 80 mm/min	260, 780, 1330 rpm	MQL (5 ml/min), conventional cooling (cc)	• MQL 1:10 ratio of cutting fluid showed good cooling properties and surface roughness over 9:10 ratio. • Conventional cooling produces poor surface roughness at higher spindle speed.
Fratila and Caizar (2011)	LUBRIMAX oil, SAROL 474 EP emulsion oil	AlMg$_3$ (EN AW 5754)			MQL (0.030 ml/h), Dry, Flood	• Nano-lubrication successfully does face milling at higher feed rates and at a much higher cutting velocity. • MQL can be successfully applied in order to achieve less surface roughness and cutting power.
Zhang et al. (2012)	Bescut 173 synthetic vegetable oil	Inconel 718	0.1	55	MQL, Dry, Flood	• Machinability of Inconel alloy improved with reduced cutting forces. • Tool life doubled as compared to dry milling.
Kishawy et al. (2005)	BM2000 (phosphate ester)	A356 aluminium alloy	–	1,500, 2,000, 5,000, 5,225 m/min	MQL (30 ml/h), flood, dry	• Improved machining properties were obtained with optimum cutting speed of 5,200 m/min. • Lowest cutting forces were observed in flood cooling.

16.5 CHALLENGES IN USING MQL FOR CONVENTIONAL MACHINING

Apart from the benefits mentioned in this chapter, the widespread use of MQL in machining also offers some challenges such as the inefficient removal of chips from the interface when compared to the flooded lubrication. Also, this technique is not efficient in case of specific applications such as drilling of deep holes, small holes and honing operations of hard materials like titanium and nickel. Further, the application of this technique needs alterations and changes in the machine tool setup and may require additional cost input and a relatively skilled manpower.

16.6 CONCLUSION AND FUTURE SCOPE

MQL technique has been investigated by various authors in recent years in terms of robustness and capability. In this chapter, from the different studies it was observed that a very few authors have contributed towards the optimization of the physical parameters which includes nozzle orientation angle, standoff distance of nozzle, flow rate variation and the pressure applied. More material surface texture issues need to be addressed like in case of aluminium and low carbon alloy steels which are more prone to build up edge formation during machining. The mist formation and optimization of droplet size along with optimum flow rate are a new field which should be explored. MQL technique paves a way for more economical, cost-effective and green machining as concluded by the authors but these comments hold good only if MQL is applied for mass production at industrial level. Furthermore, application of MQL technique can be investigated along with the use of various vegetable oils, biodegradable nano-fluids and hybrid nano-fluids for future applications. MQL along with these lubricants can produce effective results with varying parameters of speed, feed, depth of cut, nozzle angle and varying concentration of nano-fluids. The concept of hybrid nano-fluids, wherein two different types of nanoparticles are used, can also be explored for a wider range of base fluids. The ideas related to MQL summarized in this chapter shall help the manufacturers and researchers to explore the potential of MQL technique for sustainable machining.

REFERENCES

Anand, R., Haq, M.I.U., and Raina, A., 2020. Bio-based nano-lubricants for sustainable manufacturing. In: *Nanomaterials and Environmental Biotechnology*. Springer, Cham, 333–380.

Anand, R., Raina, A., Ul Haq, M.I., Mir, M.J., Gulzar, O., and Wani, M.F., 2020. Synergism of TiO_2 and graphene as nano-additives in bio-based cutting fluid-an experimental investigation. *Tribology Transactions*, 64 (2), 1–21.

Anand, A., Vohra, K., Ul Haq, M.I., Raina, A., and Wani, M.F., 2016. Tribological considerations of cutting fluids in machining environment: A review. *Tribology in Industry*, 38 (4), 463–474.

Astakhov, V.P., 2006. *Tribology of Metal Cutting*. Elsevier, London.

Astakhov, V.P., 2010. *Geometry of Single-Point Turning Tools and Drills: Fundamentals and Practical Applications*. Springer Science & Business Media, New York.

Astakhov, V.P. and Xiao, X., 2008. A methodology for practical cutting force evaluation based on the energy spent in the cutting system. *Machining Science and Technology*, 12 (3), 325–347.

Attanasio, A., Gelfi, M., Giardini, C., and Remino, C., 2006. Minimal quantity lubrication in turning: Effect on tool wear. *Wear*, 260 (3), 333–338.

Baba, Z.U., Shafi, W.K., Haq, M.I.U., and Raina, A., 2019. Towards sustainable automobiles-advancements and challenges. *Progress in Industrial Ecology, An International Journal*, 13 (4), 315–331.

Balan, A.S.S., Vijayaraghavan, L., and Krishnamurthy, R., 2013. Minimum quantity lubricated grinding of Inconel 751 alloy. *Materials and Manufacturing Processes*, 28 (4), 430–435.

Barczak, L.M., Batako, A.D.L., and Morgan, M.N., 2010. A study of plane surface grinding under minimum quantity lubrication (MQL) conditions. *International Journal of Machine Tools and Manufacture*, 50 (11), 977–985.

Batako, A.D.L. and Tsiakoumis, V., 2015. An experimental investigation into resonance dry grinding of hardened steel and nickel alloys with element of MQL. *The International Journal of Advanced Manufacturing Technology*, 77 (1–4), 27–41.

Bhowmick, S. and Alpas, A.T., 2008. Minimum quantity lubrication drilling of aluminium--silicon alloys in water using diamond-like carbon coated drills. *International Journal of Machine Tools and Manufacture*, 48 (12–13), 1429–1443.

Bhowmick, S. and Alpas, A.T., 2011. The role of diamond-like carbon coated drills on minimum quantity lubrication drilling of magnesium alloys. *Surface and Coatings Technology*, 205 (23–24), 5302–5311.

Bhowmick, S., Lukitsch, M.J., and Alpas, A.T., 2010. Dry and minimum quantity lubrication drilling of cast magnesium alloy (AM60). *International Journal of Machine Tools and Manufacture*, 50 (5), 444–457.

Biermann, D., Iovkov, I., Blum, H., Rademacher, A., Taebi, K., Suttmeier, F.T., and Klein, N., 2012. Thermal aspects in deep hole drilling of aluminium cast alloy using twist drills and MQL. *Procedia CIRP*, 3, 245–250.

Braga, D.U., Diniz, A.E., Miranda, G.W.A., and Coppini, N.L., 2002. Using a minimum quantity of lubricant (MQL) and a diamond coated tool in the drilling of aluminum--silicon alloys. *Journal of Materials Processing Technology*, 122 (1), 127–138.

Bruni, C., Forcellese, A., Gabrielli, F., and Simoncini, M., 2006. Effect of the lubrication-cooling technique, insert technology and machine bed material on the workpart surface finish and tool wear in finish turning of AISI 420B. *International Journal of Machine Tools and Manufacture*, 46 (12–13), 1547–1554.

Davim, J.P., Sreejith, P.S., and Silva, J., 2007. Turning of brasses using minimum quantity of lubricant (MQL) and flooded lubricant conditions. *Materials and Manufacturing Processes*, 22 (1), 45–50.

Debnath, S., Reddy, M.M., and Yi, Q.S., 2014. Environmental friendly cutting fluids and cooling techniques in machining: A review. *Journal of Cleaner Production*, 83, 33–47.

Dhananchezian, M. and Kumar, M.P., 2011. Cryogenic turning of the Ti--6Al--4V alloy with modified cutting tool inserts. *Cryogenics*, 51 (1), 34–40.

Dhar, N.R., Ahmed, M.T., and Islam, S., 2007. An experimental investigation on effect of minimum quantity lubrication in machining AISI 1040 steel. *International Journal of Machine Tools and Manufacture*, 47 (5), 748–753.

Dhar, N.R., Islam, M.W., Islam, S., and Mithu, M.A.H., 2006. The influence of minimum quantity of lubrication (MQL) on cutting temperature, chip and dimensional accuracy in turning AISI-1040 steel. *Journal of Materials Processing Technology*, 171 (1), 93–99.

Dixit, U.S., Sarma, D.K., and Davim, J.P., 2012. *Environmentally Friendly Machining*. Springer Science & Business Media, London.

El Baradie, M.A., 1996. Cutting fluids: Part I. characterisation. *Journal of Materials Processing Technology*, 56 (1–4), 786–797.

Emami, M., Sadeghi, M.H., and Sarhan, A.A.D., 2013. Investigating the effects of liquid atomization and delivery parameters of minimum quantity lubrication on the grinding process of Al2O3 engineering ceramics. *Journal of Manufacturing Processes*, 15 (3), 374–388.

Emami, M., Sadeghi, M.H., Sarhan, A.A.D., and Hasani, F., 2014. Investigating the Minimum Quantity Lubrication in grinding of Al_2O_3 engineering ceramic. *Journal of Cleaner Production*, 66, 632–643.

Fang, Z. and Obikawa, T., 2020. Influence of cutting fluid flow on tool wear in high-pressure coolant turning using a novel internally cooled insert. *Journal of Manufacturing Processes*, 56, 1114–1125.

Fratila, D. and Caizar, C., 2011. Application of Taguchi method to selection of optimal lubrication and cutting conditions in face milling of $AlMg_3$. *Journal of Cleaner Production*, 19 (6–7), 640–645.

Gajrani, K.K., Ram, D., and Sankar, M.R., 2017. Biodegradation and hard machining performance comparison of eco-friendly cutting fluid and mineral oil using flood cooling and minimum quantity cutting fluid techniques. *Journal of Cleaner Production*, 165, 1420–1435.

Gajrani, K.K. and Sankar, M.R., 2020. Role of eco-friendly cutting fluids and cooling techniques in machining. In: *Materials Forming, Machining and Post Processing*. Springer, London, 159–181.

Hadad, M. and Sadeghi, B., 2013. Minimum quantity lubrication-MQL turning of AISI 4140 steel alloy. *Journal of Cleaner Production*, 54, 332–343.

Heinemann, R., Hinduja, S., Barrow, G., and Petuelli, G., 2006. Effect of MQL on the tool life of small twist drills in deep-hole drilling. *International Journal of Machine Tools and Manufacture*, 46 (1), 1–6.

Hwang, Y.K. and Lee, C.M., 2010. Surface roughness and cutting force prediction in MQL and wet turning process of AISI 1045 using design of experiments. *Journal of Mechanical Science and Technology*, 24 (8), 1669–1677.

Imran, M., Mativenga, P.T., Gholinia, A., and Withers, P.J., 2014. Comparison of tool wear mechanisms and surface integrity for dry and wet micro-drilling of nickel-base superalloys. *International Journal of Machine Tools and Manufacture*, 76, 49–60.

Itoigawa, F., Childs, T.H.C., Nakamura, T., and Belluco, W., 2006. Effects and mechanisms in minimal quantity lubrication machining of an aluminum alloy. *Wear*, 260 (3), 339–344.

Jamaludin, A.S., Hosokawa, A., Furumoto, T., Koyano, T., and Hashimoto, Y., 2018. Study on the effectiveness of Extreme Cold Mist MQL system on turning process of stainless steel AISI 316. In: *IOP Conference Series: Materials Science and Engineering*, London, 12054.

Jayal, A.D., Balaji, A.K., Sesek, R., Gaul, A., and Lillquist, D.R., 2007. Machining performance and health effects of cutting fluid application in drilling of A390. 0 cast aluminum alloy. *Journal of Manufacturing Processes*, 9 (2), 137–146.

Kang, M.C., Kim, K.H., Shin, S.H., Jang, S.H., Park, J.H., and Kim, C., 2008. Effect of the minimum quantity lubrication in high-speed end-milling of AISI D2 cold-worked die steel (62 HRC) by coated carbide tools. *Surface and Coatings Technology*, 202 (22–23), 5621–5624.

Khajuria, G. and Wani, M.F., 2017. High-temperature friction and wear studies of Nimonic 80A and Nimonic 90 against Nimonic 75 under dry sliding conditions. *Tribology Letters*, 65 (3), 100.

Khan, A. and Maity, K., 2018. Influence of cutting speed and cooling method on the machinability of commercially pure titanium (CP-Ti) grade II. *Journal of Manufacturing Processes*, 31, 650–661.

Khan, M.M.A., Mithu, M.A.H., and Dhar, N.R., 2009. Effects of minimum quantity lubrication on turning AISI 9310 alloy steel using vegetable oil-based cutting fluid. *Journal of materials processing Technology*, 209 (15–16), 5573–5583.

King, R.I. and Hahn, R.S., 2012. *Handbook of Modern Grinding Technology*. Springer Science & Business Media, London.

Kishawy, H.A., Dumitrescu, M., Ng, E.-G., and Elbestawi, M.A., 2005. Effect of coolant strategy on tool performance, chip morphology and surface quality during high-speed machining of A356 aluminum alloy. *International Journal of Machine Tools and Manufacture*, 45 (2), 219–227.

Kopac, J., Sokovic, M., and Dolinsek, S., 2001. Tribology of coated tools in conventional and HSC machining. *Journal of Materials Processing Technology*, 118 (1–3), 377–384.

Leppert, T., 2011. Effect of cooling and lubrication conditions on surface topography and turning process of C45 steel. *International Journal of Machine Tools and Manufacture*, 51 (2), 120–126.

Li, K.-M. and Chou, S.-Y., 2010. Experimental evaluation of minimum quantity lubrication in near micro-milling. *Journal of Materials Processing Technology*, 210 (15), 2163–2170.

Li, K.-M. and Lin, C.-P., 2012. Study on minimum quantity lubrication in micro-grinding. *The International Journal of Advanced Manufacturing Technology*, 62 (1–4), 99–105.

Li, B., Li, C., Zhang, Y., Wang, Y., Jia, D., and Yang, M., 2016. Grinding temperature and energy ratio coefficient in MQL grinding of high-temperature nickel-base alloy by using different vegetable oils as base oil. *Chinese Journal of Aeronautics*, 29 (4), 1084–1095.

Liao, Y.S. and Lin, H.M., 2007. Mechanism of minimum quantity lubrication in high-speed milling of hardened steel. *International Journal of Machine Tools and Manufacture*, 47 (11), 1660–1666.

Liao, Y.S., Lin, H.M., and Chen, Y.C., 2007. Feasibility study of the minimum quantity lubrication in high-speed end milling of NAK80 hardened steel by coated carbide tool. *International Journal of Machine Tools and Manufacture*, 47 (11), 1667–1676.

Liew, W.Y.H., 2010. Low-speed milling of stainless steel with TiAlN single-layer and TiAlN/ AlCrN nano-multilayer coated carbide tools under different lubrication conditions. *Wear*, 269 (7–8), 617–631.

Lu, Z., Zhang, D., Zhang, X., and Peng, Z., 2020. Effects of high-pressure coolant on cutting performance of high-speed ultrasonic vibration cutting titanium alloy. *Journal of Materials Processing Technology*, 279, 116584.

Masoudi, S., Vafadar, A., Hadad, M., and Jafarian, F., 2018. Experimental investigation into the effects of nozzle position, workpiece hardness, and tool type in MQL turning of AISI 1045 steel. *Materials and Manufacturing Processes*, 33 (9), 1011–1019.

Meena, A. and El Mansori, M., 2011. Study of dry and minimum quantity lubrication drilling of novel austempered ductile iron (ADI) for automotive applications. *Wear*, 271 (9–10), 2412–2416.

Morgan, M.N., Barczak, L., and Batako, A., 2012. Temperatures in fine grinding with minimum quantity lubrication (MQL). *The International Journal of Advanced Manufacturing Technology*, 60 (9–12), 951–958.

Naves, V.T.G., Da Silva, M.B., and Da Silva, F.J., 2013. Evaluation of the effect of application of cutting fluid at high pressure on tool wear during turning operation of AISI 316 austenitic stainless steel. *Wear*, 302 (1–2), 1201–1208.

Rabiei, F., Rahimi, A.R., Hadad, M.J., and Ashrafijou, M., 2015. Performance improvement of minimum quantity lubrication (MQL) technique in surface grinding by modeling and optimization. *Journal of Cleaner Production*, 86, 447–460.

Rahim, E.A. and Sasahara, H., 2011. An analysis of surface integrity when drilling inconel 718 using palm oil and synthetic ester under MQL condition. *Machining Science and Technology*, 15 (1), 76–90.

Rajaguru, J. and Arunachalam, N., 2020. A comprehensive investigation on the effect of flood and MQL coolant on the machinability and stress corrosion cracking of super duplex stainless steel. *Journal of Materials Processing Technology*, 276, 116417.

Sadeghi, M.H., Haddad, M.J., Tawakoli, T., and Emami, M., 2009. Minimal quantity lubrication-MQL in grinding of Ti--6Al--4V titanium alloy. *The International Journal of Advanced Manufacturing Technology*, 44 (5--6), 487–500.

Sadeghi, M.H., Hadad, M.J., Tawakoli, T., Vesali, A., and Emami, M., 2010. An investigation on surface grinding of AISI 4140 hardened steel using minimum quantity lubrication-MQL technique. *International Journal of Material Forming*, 3 (4), 241–251.

Saini, A., Dhiman, S., Sharma, R., and Setia, S., 2014. Experimental estimation and optimization of process parameters under minimum quantity lubrication and dry turning of AISI-4340 with different carbide inserts. *Journal of Mechanical Science and Technology*, 28 (6), 2307–2318.

Sarıkaya, M. and Güllü, A., 2015. Multi-response optimization of minimum quantity lubrication parameters using Taguchi-based grey relational analysis in turning of difficult-to-cut alloy Haynes 25. *Journal of Cleaner Production*, 91, 347–357.

Schmid, S.R., Hamrock, B.J., and Jacobson, B.O., 2014. *Fundamentals of Machine Elements: SI Version*. CRC Press, New York.

Senevirathne, S. and Punchihewa, H.K.G., 2018. Reducing surface roughness by varying aerosol temperature with minimum quantity lubrication in machining AISI P20 and D2 steels. *The International Journal of Advanced Manufacturing Technology*, 94 (1--4), 1009–1019.

Shafi, W.K., Raina, A., and Ul Haq, M.I., 2018. Friction and wear characteristics of vegetable oils using nanoparticles for sustainable lubrication. *Tribology-Materials, Surfaces & Interfaces*, 12 (1), 27–43.

Sharma, V.S., Dogra, M., and Suri, N.M., 2009. Cooling techniques for improved productivity in turning. *International Journal of Machine Tools and Manufacture*, 49 (6), 435–453.

Shivpuri, R., Hua, J., Mittal, P., Srivastava, A.K., and Lahoti, G.D., 2002. Microstructure-mechanics interactions in modeling chip segmentation during titanium machining. *CIRP Annals*, 51 (1), 71–74.

Simunovic, K., Simunovic, G., and Saric, T., 2015. Single and multiple goal optimization of structural steel face milling process considering different methods of cooling/lubricating. *Journal of Cleaner Production*, 94, 321–329.

Stephenson, D.A., Skerlos, S.J., King, A.S., and Supekar, S.D., 2014. Rough turning Inconel 750 with supercritical CO2-based minimum quantity lubrication. *Journal of Materials Processing Technology*, 214 (3), 673–680.

Tai, B., Stephenson, D., Furness, R., and Shih, A., 2017. Minimum quantity lubrication for sustainable machining. 477–485.

Tasdelen, B., Wikblom, T., and Ekered, S., 2008. Studies on minimum quantity lubrication (MQL) and air cooling at drilling. *Journal of Materials Processing Technology*, 200 (1--3), 339–346.

Tawakoli, T., Hadad, M.J., and Sadeghi, M.H., 2010. Influence of oil mist parameters on minimum quantity lubrication--MQL grinding process. *International Journal of Machine Tools and Manufacture*, 50 (6), 521–531.

Tawakoli, T., Hadad, M.J., Sadeghi, M.H., Daneshi, A., Stöckert, S., and Rasifard, A., 2009. An experimental investigation of the effects of workpiece and grinding parameters on minimum quantity lubrication—MQL grinding. *International Journal of Machine Tools and Manufacture*, 49 (12--13), 924–932.

Tosun, N. and Huseyinoglu, M., 2010. Effect of MQL on surface roughness in milling of AA7075-T6. *Materials and Manufacturing Processes*, 25 (8), 793–798.

Varadarajan, A.S., Philip, P.K., and Ramamoorthy, B., 2000. Investigations on hard turning with minimal pulsed jet of cutting fluid. In: *Proceedings of the International Seminar on Manufacturing Technology Beyond*, London, 173–179.

Wang, Y., Li, C., Zhang, Y., Yang, M., Li, B., Jia, D., Hou, Y., and Mao, C., 2016. Experimental evaluation of the lubrication properties of the wheel/workpiece interface in minimum quantity lubrication (MQL) grinding using different types of vegetable oils. *Journal of Cleaner Production*, 127, 487–499.

Woma, T.Y., Lawal, S.A., Abdulrahman, A.S., MA, O., and MM, O., 2019. Vegetable oil based lubricants: Challenges and prospects. *Tribology Online*, 14 (2), 60–70.

Yıldırım, Ç.V., Kıvak, T., Sarıkaya, M., and Şirin, S., 2020. Evaluation of tool wear, surface roughness/topography and chip morphology when machining of Ni-based alloy 625 under MQL, cryogenic cooling and CryoMQL. *Journal of Materials Research and Technology*, 9, 2079–2092.

Zhang, S., Li, J.F., and Wang, Y.W., 2012. Tool life and cutting forces in end milling Inconel 718 under dry and minimum quantity cooling lubrication cutting conditions. *Journal of Cleaner Production*, 32, 81–87.

Section III

Biotribology

17 Biomedical Tribology

Ali Sabae Hammood
University of Kufa

Rob Brittain
University of Leeds

CONTENTS

17.1   INTRODUCTION

Tribology is defined as the science and technology of interacting surfaces in relative motion. It includes the study and application of the principles of friction, lubrication and wear. The science of tribology is not limited to mechanical machinery but it also finds application in the medical field. The human body possesses a wide variety of sliding and frictional interfaces both those that naturally occur in the tissues or organs and those that arise after implantation of an artificial device, including artificial joints, artificial teeth, dental implants, blood vessels, heart, tendons, ligaments, cardiovascular valves, orthodontic appliances, contact lenses, artificial limbs, surgical instruments and skin systems. Moreover, the ability to measure properties such as friction, levels of tension and liquid cohesion has enabled the comfortable wearability of the common contact lens.

Biomedical tribology is the application of tribology in biological systems, is a rapidly growing field and extends well beyond the conventional boundaries.

The current chapter focuses on the tribology of medical devices; then, the important tribological issues are discussed and future developments of these medical devices are also highlighted from a tribological point of view, together with the underlying mechanisms.

Materials used for biomedical practices shield a wide range and ought to have specific properties. The most basic property of materials that is utilized for manufacturing implants is biocompatibility trailed by corrosion resistance [1]. Biocompatibility can be defined as 'the capability of a material on the way to perform with a suitable host reaction in a particular use'. Since the implantations and tissue interfacing devices can corrode in an in vivo condition, the biocompatibility of materials must be given considerable attention. The corrosion of the implant can prompt loss of load-bearing quality and resulting degradation into toxic products inside the tissue. Applications of biomaterials involve span from prostheses (e.g. artificial heart valves and hip implants), medical devices, tissue regeneration to drug delivery [2]. Different metallic materials are used in joint replacements, with the most commonly known being hip replacements. More recently, spinal fixation devices, trauma, replacement spinal discs and cardiovascular stents have begun to make use of metallic materials. The metallic materials' grade involves stainless steel (SS), Co-Cr-Mo alloys, titanium alloys and other more particular alloys, for example Au-Pd [3]. Economically, titanium (pure titanium) and its alloys, for example Ti-6Al-4V, have been generally utilized [4]. Superb corrosion protection and superior biocompatibility of titanium and its alloys increased its use in biomedical application as biomaterials because of the good mechanical properties and thin surface oxide layer, as a low density and specific flexible modulus that make these metals show mechanical properties similar to bones' properties [5].

Corrosion is identified frequently as rust and undesirable phenomena lead to destruction of the shine and attractiveness of substances and reduce their life. Numerous definitions of corrosion have been known which are as follows: corrosion is the damaging attack of metal by chemical or electrochemical response with the surroundings; corrosion is the weakening of materials because of response with its environment; and near the 1960s, the corrosion phenomenon did not combine ceramics, polymers, composites and semiconductors but was limited to individual metals and their alloys. The corrosion phenomenon now is not confined to metals and alloys alone but covers completely the types of natural and synthetic materials containing nanomaterials and biomaterials [6,7].

Wear including dynamic loss of material, due to movement between the contact surfaces. This leads to destruction of the solid surface [8]. There are no less than four principle mechanisms by which wear and surface harm can happen between solids in relative movement: adhesive wear, abrasive wear, fatigue wear and corrosive wear for systems containing communal materials, for example metals, polymers and ceramics. Fretting wear and fretting corrosion combine elements of more than one mechanism [9].

Tribocorrosion has been characterized as 'an irreversible transformation of a material because of concurrent mechanical and physicochemical interactions that happen in a tribological interaction' [10]. Tribocorrosion involves common effect between wear and corrosion, and this effect leads to release of wear debris and corrosion into the body which cause implant failure and opposing biological reactions [11]. The inactive film (oxide film) on material surfaces are subject to wear and the changes of these surfaces by friction or some other type of mechanical loading consider the source of tribocorrosion. Generally, the inactive film is thought to be grabbed between the contact surfaces. 'Debris' referred to oxide particles that are released from the contacting materials [12].

17.2   BIOMEDICAL MATERIALS

17.2.1   APPLICATION AND USES OF BIOMEDICAL MATERIALS

A starting definition of biomedical materials is a 'non-viable material used in a medical device, intended to interact with biological systems'[13]. These materials include polymers, ceramics, metals and composites, and all will be in close contact with human tissue. Biocompatibility with this material is one of the most fundamental concerns. When any material is implanted in the body, the surrounding tissue will induce a response, and this response has been graded by Hench [14]:

  I. toxic
 II. non-toxic and biologically inactive (bio-inert)
III. non-toxic and biologically active (bioactive)
IV. non-toxic and dissolves

A non-toxic biologically active biomaterial will allow bonds to form between the implant and the surrounding tissue increasing the anchorage of implants, is especially important in load-bearing devices such as a femoral stem.

Biomedical materials vary in form and sizes ranging from particles (tens of nanometres), all the way up, to hip prosthesis (tens of centimetres). The best commonly known examples of biomedical materials is for hip replacements, but there are other examples including sutures, implants, scaffolds and prosthetic heart valves (Figure 17.1).

17.2.2   MATERIAL PROPERTIES OF BIOMEDICAL MATERIALS

Biomaterials function in environments that have high pH levels, high loads, cyclic loading or simply that many are enclosed in a liquid environment. To be considered successful biomedical materials, it must be able to satisfy a range of material properties without catastrophic failure or adverse reactions to the body and these properties are displayed in Table 17.2.

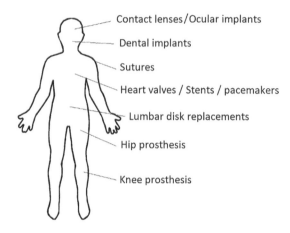

FIGURE 17.1   Common examples of biomedical devices used in the human body [15].

TABLE 17.1

Examples of Some Applications and Materials Used in Biomedical Devices

Application	Common Materials Used
Contact lenses / Ocular implants	Silicone hydrogel, poly(2-hydroxyethyl methacrylate) [16]
Dental implants	Titanium, Titanium alloys: Ti-Zr, Ti-Nb-Zr [17]
Sutures	Polyglycolic acid, Polydioxanone (PDS), Polytrimethylene carbonate, Surgical silk, Polypropylene [18,19]
Heart valves / Stents / Pacemakers	Polymers: Estane, Tecothane [20]
	Metals: Titanium, Titanium alloys (TiN, NiTi), Co-Cr [21,22]
Lumbar disc replacements	CoCrMo, polyethylene [23,24]
Hip and knee prosthesis	Metals: Titanium, Titanium alloys (Ti-Nb-Ta-Zr, Ti-Mo-Nb-Al), 316L-Stainless steel, [25]
	Polymers: PMMA, PEEK, PTFE, UHWMPE [25]
	Ceramics: Zirconia, Alumina, Bio-glass, Hydroxyapatite [26]

17.2.3   MATERIALS USED IN BIOMEDICAL DEVICES

Biomaterials are required to work in liquid corrosive environments with high loads and cyclic loading, so it is a challenge to find suitable materials that can be used long term without revision. There are four main categories of materials that are used:

- **Polymeric**
    Polymeric materials are widely used biomedical materials and can be found in applications such as sutures, prosthetics, dental implants, bone cement and encapsulants. They are generally cheaper and easier to manufacture compared to metals or ceramics [32]. Recent advances in hydrogel-based

TABLE 17.2
Desired Key Properties of Biomedical Materials

Desired Property	Example
Biocompatibility	The ability for the biomaterial to be compatible with the living biological system without causing adverse reactions, such as immune responses or toxicity [27].
Mechanical properties	Many failures of hip implants is due to mismatching of the modulus and bone, or lacking sufficient strength to match the load [22].
Low friction and wear	Natural joint friction values are around 0.01 or less, and wear that is often manageable to last lifetime [28]. Artificial joints need low friction to avoid torque loosening.
Corrosion resistance	Implanted materials are subjected to harsh environments of 7.4 pH saline solution, with ions of chlorine [29]. Metal ions released from low corrosion resistant materials can lead to toxic reactions [30].
Osseointegration	Successful implants will form a structural or functional connection with the load-bearing surface of the bone. Surface texture, coatings and chemistry can all promote successful osseointegration [31].
Manufacturability	There are many materials that are biocompatible, but production is hindered by the ability to manufacturing processes.

polymeric biomaterials for use in skin scaffolds can mimic the mechanical response of real skin while promoting skin tissue regeneration [33].

- **Metals**

  Considerable care must be taken when choosing metallic objects for in vivo environments due to biocompatibility issues arising from corrosion. Corrosion can lead to failure by disintegration and release harmful by-products into nearby tissue. The combination of biocompatibility, fatigue resistance, high corrosion resistance and specific strength have made titanium and titanium alloys the most common type of material used in biomedical applications [34]. Metallic biomaterials are one of the easiest materials to shape due to their ability to be machined or cast from a molten state.

- **Ceramics**

  The hardness of ceramic biomedical materials has led them to be used to replace damaged skeletal function (bones, joints, teeth). As with any foreign material that is placed into the body, a biological response is generated which can be separated into three different classes:

  1. **Bio-inert** ceramic materials form a minimal / negligible response to surrounding tissue. Examples such as alumina and zirconia are often used in orthopaedic application for articulating surfaces due to their low friction and wear. Coating surfaces with bio-inert materials such as zirconia by plasma spraying improves corrosion and abrasion resistance [35].

  2. **Bioactive** ceramic materials form chemical bonds with soft and hard tissues in the body. Bioactive ceramics such as hydroxyapatite $(Ca10(PO_4)_6(OH)_2)$ will bond directly to bone and surrounding tissue, promoting new bone growth through an osteoconductive process [36].

Hydroxyapatite is similar chemically to mineral bone, but is unsuitable for load-bearing applications due to its low strength and toughness. A practical way to harness hydroxyapatites is to use it as a bioactive coating on metallic implants. This combines the strength and toughness of the metal with the bioactive nature of the coating, enhancing bonding between the bone and the implant ensuring 'early prosthesis stabilization and superior fixation of the prosthesis to the surrounding tissues' [37].

3. **Bioresorbable** ceramic materials after placement into the body will start to dissolve and then be replaced by advancing tissue. Calcium phosphates, carbonates and sulphates are common bioresorbable ceramic materials used [38].

Ceramic materials are harder to machine than their metallic counterparts, due to their ionic, covalent or mixed bonding, in which the lattice structure resists dislocations. High-temperature sintering is the only way to produce ceramic biomaterials, and powders are heated to two-thirds of their melting point and then with a driving force (diffusion) they consolidate [39].

• **Composites**

A composite biomaterial is the combination of two or more constituents into one physical mixture, with resultant properties distinct from each of the constituents separately. There is no chemical reaction or alloying but there are some interfacial reactions present for good bonding between each constituent. A composite will have two phases: a matrix phase and a reinforcing phase. The reinforcement phase will usually be the strongest of the two, consisting of fibres or particulates. Biomedical composites allow tailoring of the material properties according to the application. It is common to classify composites by either the reinforcement or matrix phase. The reinforcement phase is classified into:

• **Continuous fibres** have an aspect ratio greater than 10^5
• **Short (chopped) fibres** have an aspect ratio 5–200
• **Particulates (powders)** have an aspect ratio which is between 1 and 2

Classification using the matrix material is dependent on the matrix constituent and classified as a metal, polymer or ceramic matrix composite. Polymer matrix composites are typically used in biomedical applications due to the easy fabricating processes, where polymers can be thermoset or thermoplastic. Polymer matrix ceramic composites are common in dental applications such as fillings and dental prosthesis (crowns and bridges). In the mouth, biomaterials need high fatigue resistance due to cyclic and thermal loading, and composite materials can provide the mechanical properties to be able to cope with these conditions [40].

17.2.4 Biomineralization of Metallic Biomaterials

Metallic implants for prosthesis have gained attention in recent years due to their high mechanical strength and toughness compared to that of ceramics which are inherently brittle and weak. Young's modulus of an implant should closely match natural bone to reduce the stress shielding effect present with harder materials [41]. Metallic implants can provide a source of harmful ions that can be released by corrosion processes (pitting and crevice) or from wear debris. Stainless steel (SS)-based biomaterials have a distinct disadvantage to other materials used in implants as it does not easily form bonds with surrounding tissue [42]. The surface-tissue interface is arguably the most important relationships to consider for an implanted biomaterial. If the surface is adequately biocompatible and bioactive, it will allow adequate bonding, ensuring it is anchored reducing the chance of loosening. The biocompatibility between an implanted material and the surrounding tissue is dependent on a number of factors [43]:

- Wettability
- Surface charge
- Smoothness
- Porosity
- Hydrophobicity
- Chemical composition [44]

One method to increase the biocompatibility of a metallic implant is to coat the surface with a bioactive film [45]. The best bioactive films contain pores that allow the surrounding tissue to grow into. The size of the pores is important due to the size of microphages (50 μm); therefore, pore sizes greater than 60 μm reduce the chance of infection [46].

Hydroxyapatite (HAp) as a coating on metallic implants has shown to be successful at improving corrosion resistance and osseointegration [47]. HAp has a similar chemical, crystal and structural composition to human bone, which promotes the formation of new bone growth and also provides a barrier against corrosion [48]. Biomineralization of metallic implants has been achieved by a variety of methods:

- Sol-gel
- Electrochemical deposition
- Electrophoretic deposition
- Plasma spraying process
- Biomimetic deposition

Currently, the only method that is approved commercially by the American Food and Drug Administration (FDA) is plasma spraying, but this method can produce coatings that have poor adhesion to the substrate, poor uniformity thickness and low crystallinity [49].

Recently, Hammood [50] has demonstrated electrophoretic deposition (EPD) as an attractive method for providing a quick, low-cost, thickness-controlled HAp coating onto the duplex stainless steel (DSS) (Figure 17.2). The failure of over 90% of SS implants is due to pitting and crevice corrosion [37,49–51,53]. This HAp coating was achieved using a two-step electrophoresis and deposition method, with a 2 min deposition voltage of 30 V. The corrosion behaviour of EPD deposited HAp displays a lower corrosion than pure DSS. This EPD method provided a homogenous smooth

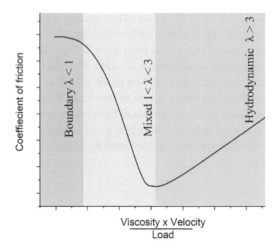

FIGURE 17.2   The Stribeck curve divided into three distinct lubrication regimes. The viscosity, velocity and load are the three main mechanisms which determine the thickness of the film between the two surfaces [15].

surface with low porosity, providing excellent bioactivity for bone growth, with Ca/P ratio similar to that of natural bone, which improves osseointegration [37].

17.3   TRIBOLOGY OF BIOMEDICAL MATERIALS

17.3.1   HISTORY OF TRIBOLOGY IN BIOMEDICAL MATERIALS

The human body is the accumulation of many years of evolution, turning itself into a tribological masterpiece which is hard to replicate. Throughout the human body, there are many tribological systems, ranging from the contact between inside of the eyelids to the eyeball, to the joints that allow us to move, and many systems in between.

Injury or illness has led to the development of biomedical implants such as those used in dentistry, arthroplasty (hip and knee) and artificial heart valves that have improved the lives of millions. These biomedical implants all encounter friction and wear, and thus, the study of these interfaces has developed into what is now known as 'biotribology'. The aim of biotribology is to evaluate how natural biological systems work with such efficiency and then to develop artificial tribological systems that can function to improve the quality of life of these that will require these implants. One of the most understood tribological systems is those used in the hip.

Total hip arthroplasty (THA) has undergone significant developments since the first recorded attempt in 1891 using an ivory femoral head and socket, but long-term use of this material was not sustainable [52]. A glass femoral head was developed in 1925 with a smooth biocompatible surface but the forces that are encountered led to catastrophic failure [53]. The consideration of wear was recognized by the Judet brothers [54], where an acrylic was used on the articulating surfaces, and this leads to failure of the devices. The first metal-on-metal (MoM) THA was developed in 1953 at Norwich Hospital by George McKee, using a cobalt-chrome socket [55]. These MoM implants provided an improved survival rate over earlier devices but the metal wear particles that were generated were a concern [56]. The current standard for THA is based upon the design from Sir John Charnley, and he wanted to reduce friction that was produced from the McKee MoM implant, and displaced a device consisting of a metal head on a polymer polytetrafluorethylene (PTFE) acetabular cup. High friction torque is often the cause of acetabular cup loosening, and by reducing the size of the femoral head and increasing the size of the PTFE cup, a low friction of 0.04 was achieved [57]. PTFE unfortunately proved not to be a suitable material due to excessive wear, even in some cases wearing all the way through the cup wall. This excessive wear led to the switch from PTFE to ultra-high molecular weight polyethylene (UHMWPE) which presented lower wear but higher friction. These metals on UHMWPE devices dominated the THA for the 20th century.

17.3.2 TRIBOLOGICAL CONSIDERATIONS OF TOTAL HIP ARTHROPLASTY

The hip joint is a complex system filled with synovial fluid which undergoes various loads and speeds allowing it to move between the full range of lubrication regimes. The lubrication regime for an artificial joint is important and will determine the friction and wear encountered. Understanding how to reduce the friction and wear to manageable levels will both increase the time before failure occurs and reduce wear particles produced which can cause an adverse response from surrounding tissue. The lubrication regime can be determined by different methods:

1. The calculation of the Lambda ratio (λ). The Lambda ratio is calculated as follows:

$$\text{Lambda ration } (\lambda) = \frac{\text{minimum film thickness } (h_{\min})}{\left(\text{Surface roughness of head}^2 + \text{Surface roughness of cup}^2\right)^{1/2}} \quad (17.1)$$

   If $\lambda < 1$, it can be assumed to be in the boundary lubrication regime, where asperity contacts will lead to high friction and wear; values between 1 and 3 will indicate a mixed lubrication regime; and values greater than 3 will indicate hydrodynamic lubrication where shearing of the lubricant will dominate the friction.

2. **Direct measurement** – This method can prove difficult in practice and can be achieved using electrical, optical, mechanical and ultrasound techniques.

The boundary lubrication regime is the most severe operating condition encountered in a THA device, and this can be achieved from either a high load or a low speed. In this regime, asperity contact will be encountered, with friction determined by the properties at the surface such as the material properties (including any surface films formed). The mixed lubrication regime will support the load through asperity contact and the action of the fluid film. The lubrication regime can change rapidly with only small variations of speed. Dowson and Jin [58] took into account entrainment speed, and shear-dependent viscosity along with the load, and cyclic speed to evaluate the 'effective film thickness' using an elastohydrodynamic model, and this recently has been improved upon taking into account the non-Newtonian lubricant rheology of synovial fluid which has improved calculations at low shear rates [59].

17.3.3 TRIBOLOGY OF MATERIALS IN TOTAL HIP ARTHROPLASTY

As life expectancy has increased in recent years, the number of people who require THA has also increased. There is currently various combination of materials used all with different benefits. The general components for artificial hip replacements are shown in Figure 17.3.

17.3.3.1 Metal on Metal

The tribological performance of a MoM implant is affected by a number of variables some of that are fixed, loading cycles, motion, viscosity (synovial fluid) and others that can be controlled radius, clearance and surface finish. MoM implants display both mixed and boundary lubrication in joint simulator contacts in voltage drop experiments [60]. The high friction and wear, that is inherent with the boundary lubrication regime, leads to the objective to separate the surfaces to reduce the

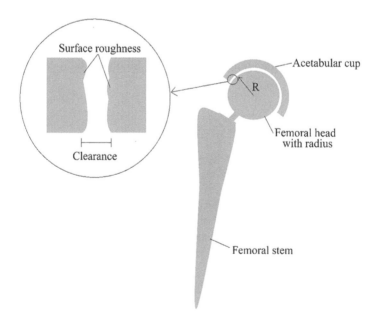

FIGURE 17.3 The components of an artificial hip joint. The main components are labels with the clearance between the femoral head, surface roughness and femoral head radius shown [15].

MoM contact. One of the ways to increase the separation between MoM contacts is increasing the film thickness. Increasing the radius of the femoral head has been found to move from a boundary to mixed lubrication through isoviscous elastohydro-dynamic lubrication (EHL) [58]. The high pressure generated causes elastic defor-mation, increasing the contact area, and a lubricant film is formed due to increased difficulty of the film escaping. The hydrodynamic action reduces the wear rate for larger radius heads, due to the increased load support from the trapped fluid. EHL in MoM implants has been shown to be present during the walking phase with film thicknesses of 10 and 15 nm, and although this thickness is noted to be smaller than the combined surface roughness, wear through asperity contact will occur [61]. Corrosion is one of the biggest factors of material loss in MoM implants, with metal ions and debris released leading to adverse effects in local tissues such as peripros-thetic osteolysis [62].

17.3.3.2   Ceramic on Ceramic

The excellent biocompatibility, scratch resistance and low friction and wear dis-played by ceramic-on-ceramic (CoC) implants make them an excellent candidate for THA. The incidences of 'squeaking' and failure from catastrophic shattering can be a concern but is attributed to poor surgical technique or in the case of squeaking, staved or poor lubrication [63]. Poor surgical technique has also been blamed for the edge loading resulting in a 'stripe' seen in CoC implants [64].

Ceramic hip joints display many desirable properties, such as high hardness, and the ability to be polished to a low surface roughness. The low surface roughness allows achieve an adequate film thickness to be maintained in the hydrodynamic lubrication regime in activities like walking where typically boundary lubrication will present [65].

The wear rate of CoC hip joints is typically lower than that of MoM or metal on polymer (MoP), with the debris being regarded as lower risk of osteolysis. The CoC although considered by many to be inert ceramic debris should still be treated with caution due to Mochida et al [66] noting macrophage response.

17.3.3.3   Metal on Polymer

The MoP hip joint was considered the gold standard for many years due to the low friction and wear compared to alternatives at the time, but lately has competi-tion from CoC and MoM pairs. The main materials considered for are UHMWPE and crosslinked UHMWPE. The crosslinked UHMWPE was introduced as it was believed to produce less wear, and with almost all reported results displaying this to be true, there is great disparity in the reported values [67].

The lubricant film thickness for crosslinked UHMWPE is smaller than the sur-face roughness leading to a greater degree of asperity contact, and thus, more wear and wear particles will be formed [68]. These wear particles are of particular concern due to biological responses implicated in aseptic loosening of THA, although even nanometre-sized particles which are believed to be more biologically active failed to contribute to osteolysis when measured experimentally [69].

Polyether ether ketone (PEEK) and related composites have recently been under scrutiny due to the exceptional thermomechanical properties displayed in tribological applications [70]. Some early experimental results for carbon fibre-reinforced PEEK (CFR-PEEK) displayed increased wear for CoCr femoral heads, but more recent work by Scholes et al. [71] using medical grade PAN and Pitch-based CFR-PEEK Optima has shown lower wear [72]. Although the wear displayed by Scholes et al. [71] made it a prime candidate for THR, Kandemir et al. [73] was able to reproduce the early high wear results for the counter-body which would indicate it would not be recommended against orthopaedic metals.

17.4   TRIBOCORROSION AND BIOTRIBOCORROSION

17.4.1   Tribocorrosion

Tribocorrosion is generally acknowledged as an interdisciplinary region of research, and such investigations on various substances bring increasingly studied by researchers because of its economical and practical effect in broad kinds of usages [74]. Fretting corrosion is an important subdivision of tribocorrosion. This subdivision refers to the degradation of a substance that is subjected to mechanical damage and chemical dissolution due to slight amplitude movement. The fretting corrosion type of medical devices is not completely understood [75]. The presence of corrosive human solutions leads to fretting loss at the implant bone contact. A fretting corrosion is a situation where material degradation causes by immediate presence of micro-movements and of an aggressive surroundings [76]. The mechanisms and possible locations for tribocorrosion to occur in total hip replacement are shown in Figure 17.4 [77].

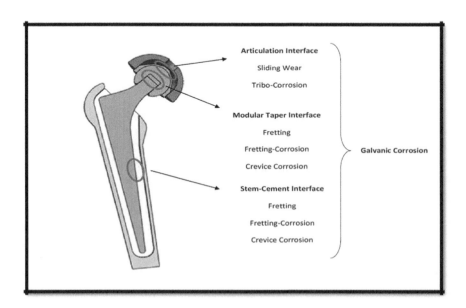

FIGURE 17.4   Tribocorrosion mechanisms in the total hip replacement (THR) [77].

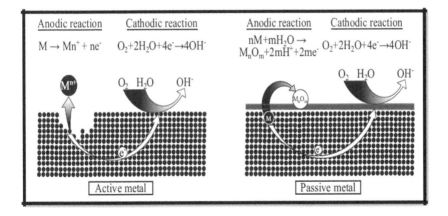

FIGURE 17.5 Generic representation of the electrochemical reactions involved in the corrosion of an active metal (left) and a passive metal (right) [79].

17.4.2 CORROSION

Corrosion refers to materials' reaction with atmosphere and leads to destructive attack [78,26]. The destructive attack of the material can be produced by physical, chemical or electrochemical reactions, or by a mixture of them. Electrochemical cell of corrosion includes a cathodic electron depletion reaction, an anodic oxidation reaction, and the electrons moved between the two reactions as shown in Figure 17.5 (Figure 17.6).

17.4.3 WEAR

Wear is characterized as deterioration to a solid materials surface, commonly its include loss of substance, because of particular relative movement among the contact surfaces material or materials [82]. Wear particles are formed as shown in Figure 17.7 [83].

17.4.3.1 Classification of Wear Mechanisms

17.4.3.1.1 Abrasive Wear

Abrasive wear occurs in the softer substance that is removed from the path followed by the asperity through the movement of the harder surface. It is the result of surfaces with dissimilar relative hardness [84].

17.4.3.1.2 Adhesive Wear

Adhesive wear is imagined to happen when the pressure applied between the sliding or contact surfaces is sufficiently enough to lead to local plastic deformation and welding between the contact severities [85].

17.4.3.1.3 Oxidative (Corrosive) Wear

Corrosive wear is the combined effects of corrosion and wear that lead to substance losses that are greater than the another effects that are the result of each process alone, which demonstrates a synergism between corrosion and wear processes [86].

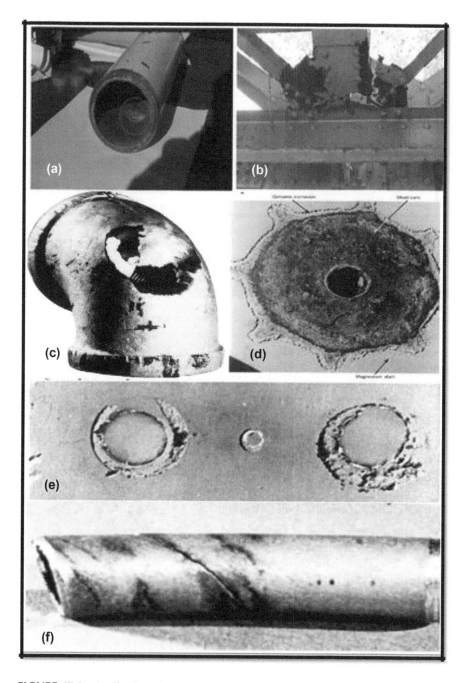

FIGURE 17.6 Application of corrosion: (a) An internal pit in an oil field water injection pipe, (b) Welding decay, (c) Impingement failure, erosion-corrosion, (d) Galvanic corrosion of a magnesium shell that was cast around a steel core, (e) Crevice corrosion at the region that is covered by washers and (f) Stress-corrosion cracks in a pipe [80,81].

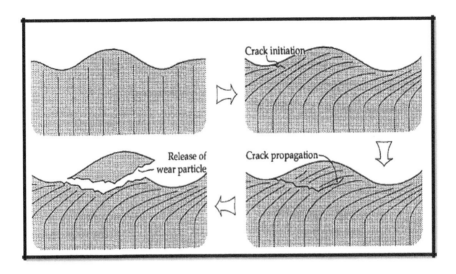

FIGURE 17.7 Formation of wear particles due to growth of surface-initiated cracks [83].

17.4.3.1.4 Fretting Wear
Fretting wear can be defined as surface destruction that is caused by small oscillatory sliding between the two contact surfaces, and it involves removal of substance from the contact surfaces during fretting effect [87].

17.4.3.1.5 Fatigue Wear
Fatigue wear is damaged substance surface under surface repeated shear stresses or strains and repeated loading that overdo the fatigue limit of this substance. This damaged is formed when the debris particle was separated by repeated cracks growing of microcracks (subsurface or superficial) and cracks on the substance surface. The five mechanisms are shown in Figure 17.8 [88].

17.4.4 Tribocorrosion

Tribocorrosion is characterized as the chemical – electrochemical – mechanical process prompting a damage of substances in rolling contact or sliding contact when immersed in destructive surroundings. The collective effects of mechanical loading and activity of corrosion lead to this wear or damage of substances [89]. In other words, tribocorrosion shields the study of surface changes that involve mechanical and chemical connections of body, counter-body, interfacial intermediate, and surroundings involving lubrication, friction, wear and tribologically activated electrochemical and chemical responses [90]. As a biomedical application, hip implant systems around 20–30% of damages can be ascribed to corrosion associated damage. Through wear and induced corrosion process where the implant surfaces are exposed to the relative motions which lead to wear loss with

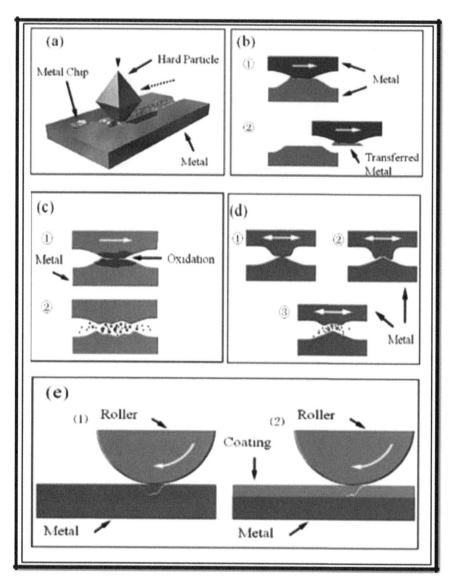

FIGURE 17.8 Wear mechanism representations in biomedical implants: (a) abrasive wear; adhesive wear; (c) oxidative wear; (d) fretting wear; and (e) fatigue wear in metal-on-metal mode (1) and metal-on-coating mode (2), respectively [88].

the corrosive attack of the body solutions [91]. It is shown in Figure 17.9 a Hip joint, cemented prosthesis components, indicates where fretting corrosion occurs in the hip implant assembly [92].

Fretting corrosion is a specific system of tribocorrosion that involve vibration and small amplitude relative movement between surfaces fixed to each other, common examples are orthopaedic implants where micro-movement occurs at fixed points

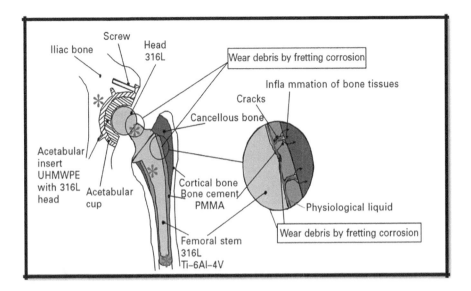

FIGURE 17.9 Hip joint, cemented prosthesis components, indicates where fretting corrosion occurs in the hip implant assembly [92].

and because the body solution which contain various inorganic and organic ions and molecules the action of corrosion occurs [93].

Tribocorrosion processes are a significant cause of occlusal abnormalities, and they can have harmful biological effects due to an increased release of metal ions and wear products. Moreover, abrasive wear products lead to decreased mechanical properties, particularly on the contact of a tribological pair [94].

In Uhlig's expression, the substances' damage is produced by two particular mechanisms:

- Mechanical wear mechanisms,
- Wear-accelerated corrosion mechanisms.

According to Uhlig's mechanisms, the total volume that is removed by mechano-electrochemical systems is assumed by below Eq. (17.2).

$$V_{tot} = V_{mech} + V_{chem} \qquad (17.2)$$

Where:

V_{tot}: The total removed volume

V_{mech}: The removed volume of metal by mechanical wear

V_{chem}: The removed volume of metal by wear-enhanced corrosion [95].

Figure 17.10 shows the degradation of substances that are caused by mechanical and chemical effects with different conditions. A common source of tribocorrosion is two or three bodies in contact sliding in a substance [96]. Pin-on-disc and simply immersing the contact in a specific electrolyte from conventional wear testers that used to

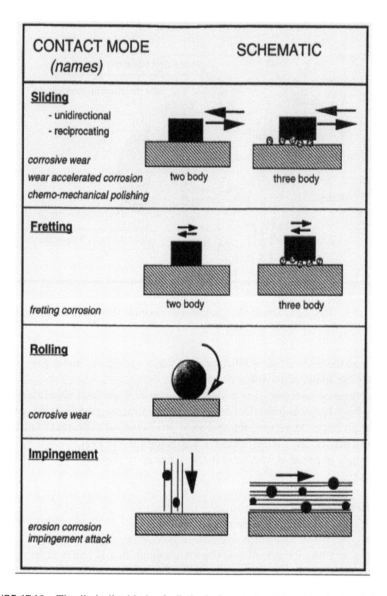

FIGURE 17.10 The dissimilar kinds of tribological contacts with mechanical and chemical influence [96].

evaluate material resistance to tribocorrosion [97]. Figure 17.11 shows schematic representation of the tribocorrosion experiments' tester [98].

17.4.4.1 Classification of Tribocorrosion

Tribocorrosion involves the communication between corrosion and the following: biological solutions, fretting, fatigue, abrasion and solid particle erosion. Figure 17.12 shows these interactions [99].

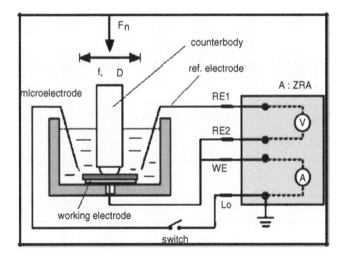

FIGURE 17.11   Schematic representation of corrosion-wear tests [98].

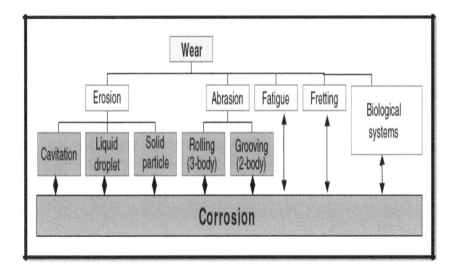

FIGURE 17.12   The interactions between the several wear mechanisms and corrosion [99].

17.4.4.2   Factors affecting Tribocorrosion

The performance of sliding contacts under electrochemical control of tribocorrosion depends on numerous variables, and Figure 17.13 listed the important of these factors schematically. According to this presentation, there are four types of factors of the behaviour of electrochemically controlled mechano-electrochemical systems:

- Electrochemical conditions usual at the rubbing surfaces' metal.
- Solution properties in the contact surfaces.

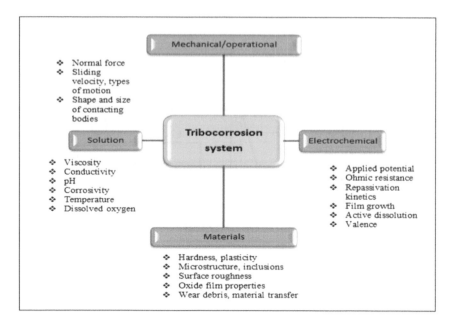

FIGURE 17.13   Factors influencing the tribocorrosion [100].

- Mechanical/ operational condition.
- Substances and surface characteristics of the specimen and the antagonist [100].

17.4.4.2.1   Materials

All material properties in the tribological contact are significant, for example ductility, rigidity, yield strength and hardness. The passivation performance of metals is mostly significant for tribocorrosion. Another important characteristics is the effect on the mechanical performance of substances such, the microstructure and presence the defects: non-metallic additions, phase distribution, dislocation density, grain size, grain orientation, segregations, etc [101].

17.4.4.2.2   Environment

The environmental factors play important roles in metal behaviours under tribocorrosion effect, such as solution viscosity, pH value, conductivity and temperature. Further factors that have effect on tribocorrosion behaviour and reactivity of metals are aggressive chloride ions and the concentration of dissolved oxygen [102].

17.4.4.2.3   Mechanical/Operational

For a given system, movement velocity, the type of contact surface (fretting or sliding) and loading forces from mechanical factors determine the degree of tribocorrosion while the factors that limit the arrangement of the rubbing surfaces and the area of contact region are the size and shape of contact bodies [103]. During tribocorrosion tests, the result of sliding velocity on the re-passivation process is an important factor. More active material is exposed to the electrolyte when

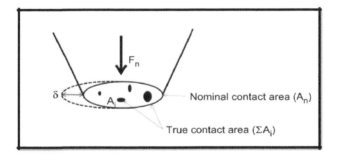

FIGURE 17.14   The pin-on-disc contact [105].

the ball slides on the surface [104]. Figure 17.14 showed the mechanics of fretting that involves two asperity-containing surfaces in contact under a normal load F_n and moving with cyclic displacement amplitude [105].

17.4.4.2.4   Electrochemical Factors

The electrochemical features are more significant in tribocorrosion, and it is considered chiefly since the phenomena of tribocorrosion were studied for several years by tribologists and electrochemists. Passive film growth, ohmic resistance, active dissolution and applied potential from basic electrochemical are factors that effect tribocorrosion rate. The tribologists have shown the effect of the surface oxidation on the mechanical rate during rubbing, whereas attention of electrochemists has been focused on the study re-passivation of the kinetics of the metal surfaces that activated with scratching, notably the mechanical and electrochemical mechanisms are dependent [106].

17.5   SUMMARY

It is evident biomedical materials and how they interact within the body is a complex process with many factors determining whether a biomaterial will be successful invitro. If these factors are not adequately appreciated, it could lead to a catastrophic failure which in these devices could lead to death or serious complications for those who rely on them. The tribological aspects of the biomedical materials have provided interesting results over the last few 70 years which moves from different sizes, configurations and materials used.

   In biomaterial research, there has been a huge growth in knowledge and research which has contributed to success that is enjoyed today, and it is no longer the case that one material will 'fit all', and consideration for the age, activity and even weight of the patient can have a bearing on the material used. The move towards bigger femoral heads to reduce the chance of dislocation provides complication regard to more wear but careful selection of materials and properties can allow the artificial hip to move through the various lubrication regimes without failure or the release of toxic particles throughout the body. The introduction of tribocorrosion does bring another interesting complication when it comes to material selection. Tribocorrosion can easily accelerate the wear process within articulating surfaces, and understanding the factors that aid this process can ensure that it can be mitigated as much as possible. But due to the requirements of the environment these materials are placed in this can prove difficult.

Titanium and titanium-based alloys have enjoyed success, being used in a variety of different applications such as hip replacements, heart values and tooth implants to name a few. Titanium is an outstanding material that is biocompatible within the body, but even this metal when subjected to the harsh environment of the human body can corrode. Recently, there has interest in using duplex stainless steel materials, and these materials are cheaper to manufacture but lack the biocompatibility and corrosion resistance that is inherent with titanium-based materials. To overcome this disadvantage, biomineralization of the surface can be achieved which not only improves the corrosion resistance but makes the surface bioactive improving the relationship between the bone and the implant, which is likely to lead to better long-term results.

Overall, tribology is an important field in biomedical devices, and without fully understanding the importance of all the factors, it is possible that as seen with early attempts at total hip arthroplasty failure can and will occur.

17.6   THE FUTURE OF BIOMEDICAL DEVICES

The advancements made over the past 100 years since the first recorded THR have been outstanding with huge advances made in materials used being one of the principle reasons. As advances in material science continue to develop, the focus will shift, moving away from being just bio-inert to a being highly bioactive, working with the body, at molecular, cellular and tissue levels [107]. These newest types of materials are termed 'fourth generation' and can even mimic the tissue that it has replaced. An example of a new generation material would be that of semiconductor silicon nanowires (SiNW) used to detect electrical signals in a single cell creating an in situ biosensor [108]. These biosensors can be placed inside synthetic tissue to feedback intracellular electrical signals, or even be used to support cell growth in cardiac tissue [109].

Other nanomaterials, such as graphene, have also recently been viewed as prospective materials to use in the next generation of devices due to the combined atomic thickness, chemical inertness and electronic properties that make it ideal for use in biosensors that can be used in combination with other biomaterials to work together at the molecular scale [110].

17.7   CONCLUSIONS

The materials used in biomedical devices can be considered to be one of the most important factors to predict how it will behave in a biotribological environment like the human body. Many of the early devices failed due to either biocompatibility issues or failure due to wear.

The biocompatibility of biomedical devices has improved, such that biomineralized coatings can be applied to materials so that they become bioactive, reducing the chances of loosening and thus the need to be replaced. The current standard of biomineralization applied to metallic implants is not perfect, with varying levels of thickness, poor adhesion to the substrate, high cost and low crystallinity which mean that newer methods using electrophoretic deposition could become the new standard. This would allow a low chance of devices failing due to poor anchoring when bone tissue is integrated into the bioactive coating.

When considering the wear of these devices, all three of the main tribo-pairs are suitable for use, with no current gold standard when it comes to material pairs alone. The choice of material should be determined on a by-case basis taking into account the likely lifetime needed, cost and any tribocorrosion issues which are often seen in metallic implants. As more and more coatings are developed for use in these devices, this may change but due to the highly corrosive nature that is inherent to these devices it will prove difficult.

REFERENCES

1. Elias CN, Lima JHC, Valiev R, Meyers MA. Biomedical applications of titanium and its alloys. *Jom.* 2008;(March):1–4.
2. Bártolo P, Bidanda B. *Bio-materials and Prototyping Applications in Medicine.* Springer; 2008.
3. Rack HJ, Qazi JI. Titanium alloys for biomedical applications. *Mater Sci Eng C.* 2006;26(8):1269–77.
4. Barão VAR, Mathew MT, Assunção WG, Yuan JCC, Wimmer MA, Sukotjo C. Stability of cp-Ti and Ti-6Al-4V alloy for dental implants as a function of saliva pH - an electrochemical study. *Clin Oral Implants Res.* 2012;23(9):1055–62.
5. de Viteri VS, Fuentes E. Titanium and titanium alloys as biomaterials. *Tribol Adv InTech.* 2013;155–81.
6. Roberge PR. *Handbook of Corrosion Engineering.* McGraw-Hill; 2000.
7. Ahmad Z. *Principles of Corrosion Engineering and Corrosion Control.* Elsevier; 2006.
8. Harsha AP, Tewari US, Venkatraman B. Three-body abrasive wear behaviour of polyaryletherketone composites. *Wear.* 2003;254(7–8):680–92.
9. Peterson DR, Bronzino JD. *Biomechanics: Principles and Applications.* CRC press; 2007.
10. Toptan F, Rocha LA. Tribocorrosion in metal matrix composites:149–67. DOI:10.4018/978-1-4666-7530-8.CH006.
11. Marques IdSV, Alfaro MF, Saito MT, da Cruz NC, Takoudis C, Landers R, et al. Biomimetic coatings enhance tribocorrosion behavior and cell responses of commercially pure titanium surfaces. *Biointerphases.* 2016;11(3):031008.
12. Ponthiaux P, Wenger F, Celis J. Tribocorrosion: Material behavior under combined conditions of corrosion and mechanical loading. *Corros Resist.* 2012:81–106.
13. Williams DF. Advanced applications for materials implanted within the human body. *Mater Sci Technol.* 1987;3(10):797–806.
14. Hench LL, Best SM. *Biomaterials Science.* 1996.
15. Brittain R. University of Leeds, Department of Mechanical Engineering taken from self produced lecture notes. Ali said to reference this way.
16. Bhamra TS, Tighe BJ. Mechanical properties of contact lenses: The contribution of measurement techniques and clinical feedback to 50 years of materials development. *Contact Lens Anterior Eye* [Internet]. 2017;40(2):70–81. Available from: http://www.sciencedirect.com/science/article/pii/S1367048416301709.
17. Cordeiro JM, Beline T, Ribeiro ALR, Rangel EC, da Cruz NC, Landers R, et al. Development of binary and ternary titanium alloys for dental implants. *Dent Mater* [Internet]. 2017;33(11):1244–57. Available from: http://www.sciencedirect.com/science/article/pii/S0109564117304906.

18. Tajirian AL, Goldberg DJ. A review of sutures and other skin closure materials. *J Cosmet Laser Ther* [Internet]. 2010 Dec 1;12(6):296–302. Available from: Doi: 10.3109/14764172.2010.538413.

19. Postlethwait RW, Willigan DA, Ulin AW. Human tissue reaction to sutures. *Ann Surg.* 1975;181(2):144.

20. Li RL, Russ J, Paschalides C, Ferrari G, Waisman H, Kysar JW, et al. Mechanical considerations for polymeric heart valve development: Biomechanics, materials, design and manufacturing. *Biomaterials* [Internet]. 2019;225:119493. Available from: http://www.sciencedirect.com/science/article/pii/S0142961219305927.

21. Yang Y, Franzen SF, Olin CL. In vivo comparison of hemocompatibility of materials used in mechanical heart valves. *J Heart Valve Dis.* 1996;5(5):532–7.

22. Zhang L, Chen L. A review on biomedical titanium alloys: Recent progress and prospect. *Adv Eng Mater.* 2019;21(4):1801215.

23. Pokorny G, Marchi L, Amaral R, Jensen R, Pimenta L. Lumbar total disc replacement by the lateral approach–up to 10 years follow-up. *World Neurosurg* [Internet]. 2019;122:e325–33. Available from: http://www.sciencedirect.com/science/article/pii/S1878875018323295.

24. Kettler A, Bushelow M, Wilke H-J. Influence of the loading frequency on the wear rate of a polyethylene-on-metal lumbar intervertebral disc replacement. *Eur Spine J.* 2012;21(5):709–16.

25. Katti KS. Biomaterials in total joint replacement. *Coll Surf B Biointerf.* [Internet]. 2004; 39(3):133–42. Available from: http://www.sciencedirect.com/science/article/pii/S0927776503003060.

26. Barakat A, Quayle J, Stott P, Gibbs J, Edmondson M. Results of hydroxyapatite ceramic coated primary femoral stem in revision total hip replacement. *Int Orthop.* 2020;1–6.

27. Mihov D, Katerska B. Some biocompatible materials used in medical practice. *Trakia J Sci.* 2010;8(2):119–25.

28. Hills BA, Butler BD. Surfactants identified in synovial fluid and their ability to act as boundary lubricants. *Ann Rheum Dis* [Internet]. 1984 Aug;43(4):641–8. Available from: https://pubmed.ncbi.nlm.nih.gov/6476922.

29. Ghosh S, Choudhury D, Das NS, Pingguan-Murphy B. Tribological role of synovial fluid compositions on artificial joints—a systematic review of the last 10 years. *Lubr Sci.* 2014;26(6):387–410.

30. Mirzajavadkhan A, Rafieian S, Hasan MH. Toxicity of metal implants and their interactions with stem cells: A review. *Int J Eng Mater Manuf.* 2020;5(1):2–11.

31. Raphel J, Holodniy M, Goodman SB, Heilshorn SC. Multifunctional coatings to simultaneously promote osseointegration and prevent infection of orthopaedic implants. *Biomaterials.* 2016;84:301–14.

32. Love BJ. *Biomaterials: A Systems Approach to Engineering Concepts.* Academic Press; 2017.

33. Kalai Selvan N, Shanmugarajan TS, Uppuluri VNVA. Hydrogel based scaffolding polymeric biomaterials: Approaches towards skin tissue regeneration. *J Drug Deliv Sci Technol.* [Internet]. 2020;55:101456. Available from: http://www.sciencedirect.com/science/article/pii/S1773224719308962.

34. Hammood AS, Thair L, Ali SH. Corrosion behavior evaluation in simulated body fluid of a modified Ti–6Al–4V alloy by DC glow plasma nitriding. *J Bio-and Tribo-Corrosion.* 2019;5(4):100.

35. Morks MF, Kobayashi A. Development of ZrO2/SiO2 bioinert ceramic coatings for biomedical application. *J Mech Behav Biomed Mater* [Internet]. 2008;1(2):165–71. Available from: http://www.sciencedirect.com/science/article/pii/S175161610700029X.

36. O'Hare P, Meenan BJ, Burke GA, Byrne G, Dowling D, Hunt JA. Biological responses to hydroxyapatite surfaces deposited via a co-incident microblasting technique. *Biomaterials.* 2010;31(3):515–22.

37. Wnek GE, Bowlin GL. *Encyclopedia of Biomaterials and Biomedical Engineering.* CRC Press; 2008.

38. Safronova T V, Kuznetsov A V, Korneychuk SA, Putlyaev VI, Shekhirev MA. Calcium phosphate powders synthesized from solutions with [Ca2+]/[PO4 3−]= 1 for bioresorbable ceramics. *Cent Eur J Chem.* 2009;7(2):184–91.

39. Teoh SH. *Engineering Materials for Biomedical Applications.* Vol. 1. World Scientific; 2004.

40. Pieniak D, Przystupa K, Walczak A, Niewczas AM, Krzyzak A, Bartnik G, et al. Hydro-thermal fatigue of polymer matrix composite biomaterials. *Materials (Basel).* 2019;12(22):3650.

41. Gepreel MA-H, Niinomi M. Biocompatibility of Ti-alloys for long-term implantation. *J Mech Behav Biomed Mater.* 2013;20:407–15.

42. Yang K, Ren Y. Nickel-free austenitic stainless steels for medical applications. *Sci Technol Adv Mater.* 2010;11(1):14105.

43. Hench LL, Wilson J. Surface-active biomaterials. *Science* (80-). 1984;226(4675):630–6.

44. Boss JH, Shajrawi I, Aunullah J, Mendes DG. The relativity of biocompatibility. A critique of the concept of biocompatibility. *Isr J Med Sci.* 1995;31(4):203–9.

45. Bocchiotti G, Verna G, Fracalvieri M, Fanton E, Datta G, Robotti E. Carbofilm-covered prostheses in plastic surgery: Preliminary observations. *Plast Reconstr Surg.* 1993;91(1):80–8.

46. Apelt D, Theiss F, El-Warrak AO, Zlinszky K, Bettschart-Wolfisberger R, Bohner M, et al. In vivo behavior of three different injectable hydraulic calcium phosphate cements. *Biomaterials.* 2004;25(7–8):1439–51.

47. Balamurugan A, Kannan S, Rajeswari S. Bioactive Sol-Gel Hydroxyapatite surface for biomedical application-invitro study. *Trends Biomater Artif Organs.* 2002;16(1):18–20.

48. Song YW, Shan DY, Han EH. Electrodeposition of hydroxyapatite coating on AZ91D magnesium alloy for biomaterial application. Mater Lett. 2008;62(17–18):3276–9.

49. Singh G, Singh S, Prakash S. Surface characterization of plasma sprayed pure and reinforced hydroxyapatite coating on Ti6Al4V alloy. *Surf Coatings Technol.* 2011;205(20):4814–20.

50. Hammood AS. Biomineralization of 2304 duplex stainless steel with surface modification by electrophoretic deposition. *J Appl Biomater Funct Mater.* 2020;18:2280800019896215.

51. Manam NS, Harun WSW, Shri DNA, Ghani SAC, Kurniawan T, Ismail MH, et al. Study of corrosion in biocompatible metals for implants: A review. *J Alloys Compd.* 2017;701:698–715.

52. Learmonth ID, Young C, Rorabeck C. The operation of the century: Total hip replacement. *Lancet.* 2007;370(9597):1508–19.

53. Hernigou P. Smith-Petersen and early development of hip arthroplasty. *Int Orthop.* 2014;38(1):193–8.

54. Judet J, Judet R. The use of an artificial femoral head for arthroplasty of the hip joint. *J Bone Joint Surg Br.* 1950;32(2):166–73.

55. August AC, Aldam CH, Pynsent PB. The McKee-Farrar hip arthroplasty. A long-term study. *J Bone Joint Surg Br.* 1986;68(4):520–7.

56. McKellop H, Park S-H, Chiesa R, Doorn P, Lu B, Normand P, et al. In vivo wear of 3 types of metal on metal hip prostheses during 2 decades of use. *Clin Orthop Relat Res.* 1996;329:S128–40.

57. Charnley J. Arthroplasty of the hip: A new operation. *Lancet* [Internet]. 1961 May [cited 2016 Oct 24];277(7187):1129–32. Available from: http://www.sciencedirect.com/science/article/pii/S0140673661920633

58. Dowson D, Jin Z-M. Metal-on-metal hip joint tribology. *Proc Inst Mech Eng Part H J Eng Med* [Internet]. 2006 Feb 1;220(2):107–18. Available from: Doi: 10.1243/095441105X69114.

59. Gao L, Dowson D, Hewson RW. A numerical study of non-Newtonian transient elasto-hydrodynamic lubrication of metal-on-metal hip prostheses. *Tribol Int* [Internet]. 2016; 93:486–94. Available from: http://www.sciencedirect.com/science/article/pii/S0301679X 15000900.

60. Dowson D, McNie CM, Goldsmith AAJ. Direct experimental evidence of lubrication in a metal-on-metal total hip replacement tested in a joint simulator. *Proc Inst Mech Eng Part C J Mech Eng Sci.* 2000;214(1):75–86.

61. Gao L, Wang F, Yang P, Jin Z. Effect of 3D physiological loading and motion on elasto-hydrodynamic lubrication of metal-on-metal total hip replacements. *Med Eng Phys.* 2009;31(6):720–9.

62. Law JI, Crawford DA, Adams JB, Lombardi Jr AV. Metal on metal total hip revisions: Pearls and pitfalls. *J Arthroplasty.* 2020;35:S68–S72.

63. Jarrett CA, Ranawat AS, Bruzzone M, Blum YC, Rodriguez JA, Ranawat CS. The squeaking hip: A phenomenon of ceramic-on-ceramic total hip arthroplasty. *J Bone Joint Surg.* 2009;91(6):1344–9.

64. Jeffers JRT, Walter WL. Ceramic-on-ceramic bearings in hip arthroplasty: State of the art and the future. *J Bone Jt Surg - Br* 2012;94–B(6):735–45.

65. Mattei L, Di Puccio F, Piccigallo B, Ciulli E. Lubrication and wear modelling of artificial hip joints: A review. *Tribol Int.* 2011;44(5):532–49.

66. Mochida Y, Boehler M, Salzer M, Bauer TW. Debris from failed ceramic-on-ceramic and ceramic-on-polyethylene hip prostheses. *Clin Orthop Relat Res.* 2001;389:113–25.

67. Jacobs CA, Christensen CP, Greenwald AS, McKellop H. Clinical performance of highly cross-linked polyethylenes in total hip arthroplasty. *J Bone Joint Surg.* 2007;89(12):2779–86.

68. Fisher J, Dowson D. Tribology of total artificial joints. *Proc Inst Mech Eng Part H J Eng Med.* 1991;205(2):73–9.

69. Liu A, Richards L, Bladen CL, Ingham E, Fisher J, Tipper JL. The biological response to nanometre-sized polymer particles. *Acta Biomater* [Internet]. 2015;23:38–51. Available from: http://www.sciencedirect.com/science/article/pii/S1742706115002408.

70. Baykal D, Siskey R, Underwood RJ, Briscoe A, Kurtz SM. Biotribology of PEEK bearings in multidirectional pin-on-disk testers. In: *PEEK Biomaterials Handbook.* Elsevier; 2019. p. 385–401.

71. Scholes SC, Unsworth A. Wear studies on the likely performance of CFR-PEEK/CoCrMo for use as artificial joint bearing materials. *J Mater Sci Mater Med.* 2009;20(1):163.

72. Wang A, Lin R, Stark C, Dumbleton JH. Suitability and limitations of carbon fiber reinforced PEEK composites as bearing surfaces for total joint replacements. *Wear* [Internet]. 1999;225–229:724–7. Available from: http://www.sciencedirect.com/science/article/pii/S0043164899000265.

73. Kandemir G, Smith S, Joyce TJ. Wear behaviour of CFR PEEK articulated against CoCr under varying contact stresses: Low wear of CFR PEEK negated by wear of the CoCr counterface. *J Mech Behav Biomed Mater* [Internet]. 2019;97:117–25. Available from: http://www.sciencedirect.com/science/article/pii/S1751616119303613.

74. Mathew MT, Ariza E, Rocha LA, Fernandes AC, Vaz F. TiCxOy thin films for decorative applications: Tribocorrosion mechanisms and synergism. *Tribol Int.* 2008;41(7):603–15.

75. Bryant M, Neville A. Fretting corrosion of CoCr alloy: Effect of load and displacement on the degradation mechanisms. *Proc Inst Mech Eng Part H J Eng Med.* 2017; 231(2):114–26.

76. Mischler S, Barril S, Landolt D. Fretting corrosion behaviour of Ti–6Al–4V/PMMA contact in simulated body fluid. *Tribol Surf Interf.* 2009;3(1):16–23.

77. Bryant M, Ward M, Farrar R, Freeman R, Brummitt K, Nolan J, et al. Characterisation of the surface topography, tomography and chemistry of fretting corrosion product found on retrieved polished femoral stems. *J Mech Behav Biomed Mater.* 2014;32:321–34.

78. Sin JR. *Investigation of the Corrosion and Tribocorrosion Behaviour of Metallic Biomaterials.* 2014.

79. Council NR. *Assessment of Corrosion Education.* National Academies Press; 2009.

80. Talbot DEJ, Talbot JDR. *Corrosion Science and Technology.* CRC Press; 2018.

81. Schweitzer PA. *Metallic Materials: Physical, Mechanical, and Corrosion Properties.* Vol. 19. CRC Press; 2003.

82. Vaughan J, Alfantazi A. Corrosion of titanium and its alloys in sulfuric acid in the presence of chlorides. *J Electrochem Soc.* 2006;153(1):B6–12.

83. Wandelt K. *Encyclopedia of Interfacial Chemistry: Surface Science and Electrochemistry.* Elsevier; 2018.

84. Sridhar TM, Rajeswari S. Biomaterials corrosion. *Corros Rev.* 2009;27(Supplement): 287–332.

85. Raju BR, Suresha B, Swamy RP, Kanthraju BSG. Investigations on mechanical and tribological behaviour of particulate filled glass fabric reinforced epoxy composites. 2013.

86. Stachowiak G, Batchelor AW. *Engineering Tribology* [Internet]. Elsevier Science; 2013. Available from: https://books.google.co.uk/books?id=_wVoTz1pDlwC.

87. Wright TM, Goodman SB. Implant wear in total joint replacement: Clinical and biologic issues, material and design considerations: *Symposium,* Oakbrook, Illinois, October 2000. American Academy of Orthopaedic Surgeons; 2001.

88. Dearnaley G. Adhesive and abrasive wear mechanisms in ion implanted metals. *Nucl Inst Meth Phys Res Sect B Beam Interact Mater Atoms.* 1985;7:158–65.

89. Davis JR. *Surface Engineering for Corrosion and Wear Resistance.* ASM International; 2001.

90. Bill RC. *Fretting Wear and Fretting Fatigue—How are They Related?*; 1983.

91. Yang Y. *Surface Treated cp-Titanium for Biomedical Applications: A Combined Corrosion, Tribocorrosion and Biological Approach.* Châtenay-Malabry, Ecole centrale de Paris; 2014.

92. Ponthiaux P, Wenger F, Drees D, Celis J-P. Electrochemical techniques for studying tribocorrosion processes. *Wear.* 2004;256(5):459–68.

93. Fischer A, Mischler S. Tribocorrosion: Fundamentals, materials and applications. *J Phys D Appl Phys.* 2006;39(15).

94. Doni Z, Alves AC, Toptan F, Pinto AM, Rocha LA, Buciumeanu M, et al. Tribocorrosion behaviour of hot pressed CoCrMo− Al$_2$O$_3$ composites for biomedical applications. *Tribol Surf Interf.* 2014;8(4):201–8.

95. Landolt D, Mischler S. *Tribocorrosion of Passive Metals and Coatings.* Elsevier; 2011.

96. Diomidis N, Mischler S, More NS, Roy M. Tribo-electrochemical characterization of metallic biomaterials for total joint replacement. *Acta Biomater.* 2012;8(2):852–9.

97. Walczak M, Drozd K. Tribological characteristics of dental metal biomaterials. *Curr Issues Pharm Med Sci.* 2016;29(4):158–62.

98. Mischler S. Triboelectrochemical techniques and interpretation methods in tribocorrosion: A comparative evaluation. *Tribol Int.* 2008;41(7):573–83.

99. Landolt D, Mischler S, Stemp M. Electrochemical methods in tribocorrosion: A critical appraisal. *Electrochim Acta.* 2001;46(24–25):3913–29.

100. Merl DK, Panjan P, Cekada M, Gselman P, Paskvale S. Tribocorrosion degradation of protective coatings on stainless steel. *Mater Tehnol.* 2013:47.

101. Vieira AC, Ribeiro AR, Rocha LA, Celis J-P. Influence of pH and corrosion inhibitors on the tribocorrosion of titanium in artificial saliva. *Wear.* 2006;261(9):994–1001.

102. Wood RJK. Tribo-corrosion of coatings: A review. *J Phys D Appl Phys.* 2007;40(18): 5502–21.

103. Landolt D. Electrochemical and materials aspects of tribocorrosion systems. *J Phys D Appl Phys.* 2006;39(15):3121–7.

104. Ramos LB, Simoni L, Mielczarski RG, Vega MRO, Schroeder RM, Malfatti C de F. Tribocorrosion and electrochemical behavior of DIN 1.4110 martensitic stainless steels after cryogenic heat treatment. *Mater Res.* 2017;20(2):460–8.
105. Swaminathan V, Gilbert JL. Fretting corrosion of CoCrMo and Ti6Al4V interfaces. *Biomaterials.* 2012;33(22):5487–503.
106. Mathew MT, Srinivasa Pai P, Pourzal R, Fischer A, Wimmer MA. Significance of tribocorrosion in biomedical applications: Overview and current status. *Adv Tribol.* 2009:2009.
107. Ning C, Zhou L, Tan G. Fourth-generation biomedical materials. *Mater Today.* 2016; 19(1):2–3.
108. Duan X, Gao R, Xie P, Cohen-Karni T, Qing Q, Choe HS, et al. Intracellular recordings of action potentials by an extracellular nanoscale field-effect transistor. *Nat Nanotechnol.* 2012;7(3):174–9.
109. Tian B, Liu J, Dvir T, Jin L, Tsui JH, Qing Q, et al. Macroporous nanowire nanoelectronic scaffolds for synthetic tissues. *Nat Mater.* 2012;11(11):986–94.
110. Ahammad AJS, Islam T, Hasan MM. Graphene-based electrochemical sensors for biomedical applications. In: *Biomedical Applications of Graphene and 2D Nanomaterials.* Elsevier; 2019. pp. 249–82.

18 Tribological Studies on Titanium Alloys for Biomedical Applications

*Vivudh Gupta, Balbir Singh, R.K. Mishra,
and Pawandeep Singh*
Shri Mata Vaishno Devi University

CONTENTS

18.1   INTRODUCTION

Over the past several decades, the use of titanium (Ti) alloys has increased manifolds in different industrial sectors such as transportation, aerospace, medical engineering, chemical industries, etc. Mechanical, physical and biological properties like low density, high strength, high resistance to corrosion, biocompatibility, etc. have led to their increased use in different areas (Pierret et al., 2014, Wu et al., 2017). However, Ti alloys exhibit poor tribological properties owing to their low resistance to plastic shearing, low work-hardening, low protection by surface oxides which are formed as a result of high temperature due to frictional heating during sliding operations. Other factors which contribute to poor tribological properties of Ti alloys include high adhesive wear, high frictional coefficients, low resistance to abrasion, etc (Chauhan and Dass, 2013, Alam and Haseeb, 2002). Due to high applicability of Ti alloys in different fields, efforts are continuously being made to improve their poor tribological

323

aspects. These include various coating and implantation techniques (Pierret et al., 2014). Surface treatment techniques have shown positive results in improving the tribological properties of Ti alloys. These techniques are further discussed in detail.

18.2   TITANIUM ALLOYS IN BIOMEDICAL FIELD

Nowadays, there are number of biocompatible (non-toxic) Ti alloys used as bone implants, dental implants and cardiac accessories (artificial vascular stents and artificial heart valve). Examples of such alloys are Ti6Al4V, Ti6Al7Nb, Ti-5Al-2.5Fe, Ti-5Al-3Mo-4Zr, Ti-15Mo, Ti-13Nb-13Zr, etc. These alloys have shown high resistance to corrosion and have low modulus and high strength. Moreover, they also possess good formability, machinability and good fatigue strength. The problems encountered in the use of Ti alloys as bone implants are high usage cost, high elastic modulus in comparison to bone and difficulty in attachment to bone due to non-reactivity with tissues (Oldani and Dominguez, 2012, Liu et al., 2004, 2017). Low wear resistance of these Ti alloys is one of the major reasons for implant failure. The debris generated due to wear accumulates in the tissues or bone marrow, which ultimately leads to inflammation and hypersensitivity. Such implications result in revision of surgery. Hence, tribological properties of biomedical Ti alloys need to be thoroughly investigated and worked upon to minimize these adverse effects occurring due to low wear resistance (Li et al., 2014).

18.3   TRIBOLOGICAL STUDIES ON
BIOMEDICAL TITANIUM ALLOYS

Tribological characteristics play a vital role in determining the performance of biocompatible Ti alloys. Choubey et al. (2004) conducted fretting wear experiments on biomedical Ti alloys such as Ti6Al4V, Ti-Al-2.5Fe and Ti-13Nb-13Zr. Tests were performed in Hank's solution (a simulated body fluid solution) at 10 N load for 10,000 cycles. It was reported that coefficient of friction (COF) was around 0.46–0.50 for different Ti alloys, except Ti-5Al-2.5Fe. COF in this exceptional case was found to be 0.3 while rubbing against bearing steel. Tribomechanical abrasion and cracking were reported as wear mechanisms in this tribological study of biomedical Ti alloys. Ming et al. (2006) studied the dry sliding tribological characteristics of Ti6Al4V against GCr15 steel at high speeds. The authors reported that when the temperature due to friction is greater than β-phase transformation temperature, there is a rapid growth of β crystal. As a result of which, there is a decrease in plasticity of Ti alloy and hence, resulting in poor tribological properties. Furthermore, oxides such as TiO, V_2O_3 and TiO_2 were also formed by temperature increase due to friction. These oxide layers are loose in nature, which further decrease wear resistance. Ramos-Saenz et al. (2010) studied the tribological properties of oxidized and non-oxidized Ti alloys like gamma-TiAl, Ti6Al4V and CP-Ti. Custom-made bone-implant pin arrangement was prepared to evaluate wear properties under dry and simulated biological conditions with the use of Ringer's solution. The researchers found high COF for oxidized gamma-TiAl surface and low COF for non-oxidized Ti alloys. Abrasion was the major wear mechanism in all the three Ti alloys for both dry and Ringer's solution media. Adhesion was also observed in case of dry medium during tribological study. Also, bone pins worn faster and debris generated due to adhesion got embedded on the surface of oxide. Similarly, Cvijović-Alagić et al. (2010) observed better wear resistance of Ti6Al4V than the orthopaedic Ti-13Nb-13Zr alloy in Ringer's solution at varying loads and sliding speeds. Mao et al. (2013) studied the behaviour of tribo-layer during dry sliding of Ti6Al4V under different loads (50–250 N) at different temperatures (25°C–500°C). It was found that the wear resistance of Ti6Al4V is higher at 400°C–500°C than that at 25°C–200°C. At high temperatures, more tribo-oxides were formed, thus giving more protection due to their higher hardness value. It was also observed that at lower temperatures, wear occurred due to adhesion, delamination and abrasion whereas oxidation occurred at high temperature which also kept the wear rate less. Tkachenko et al. (2014) found that the wear rate of biomedical Ti-Si-Zr alloys were 2–7 times lesser than Ti6Al4V but higher than CoCr alloy. In case of Ti6Al4V and Ti-Si-Zr alloys, there was very high material transfer to ceramic Si_3N_4 balls whereas CoCr alloy did not show any appreciable material transfer to Si_3N_4 balls.

18.4 SURFACE MODIFICATION METHODS TO IMPROVE TRIBOLOGICAL BEHAVIOUR OF BIOMEDICAL TI ALLOYS

Various surface engineering methods have been developed with a view to improving the tribological characteristics of Ti alloys. These methods are shown in Figure 18.1 and are discussed in detail as below.

18.4.1 USE OF PVD COATINGS AND LASER SURFACE ALLOYING

Physical vapour deposition (PVD) and laser surface alloying have been used for the purposes of tribological improvement of Ti alloys. Vadiraj and Kamaraj (2007) studied the effect of PVD TiN coating and plasma nitriding on the fretting behaviour of biomedical Ti6Al4V alloy. It was reported that both TiN coating and plasma nitriding offered improved resistance against fretting. Such methods also minimized friction and hence, delayed material failure. Thus, modification of surface layers proved effective in improving the fretting response. Martini and Ceschini

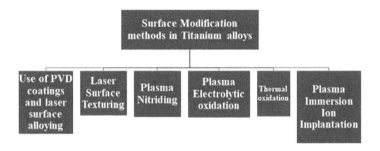

FIGURE 18.1 Surface modification methods to improve tribological properties in Ti alloys.

(2011) compared the dry sliding behaviour of three PVD coatings on Ti6Al4V to that of uncoated Ti6Al4V. These three PVD coatings were single-layered CrN (deposited by arc evaporation), alternate-layered CrN/NbN (also deposited by arc evaporation) and multi-layered WC/C (deposited by magnetron sputtering). In this study, the researchers identified the critical loads (i.e. loads at which coating life ends). Highest critical loads were found in case of WC/C- and CrN-coated Ti6Al4V alloy. Also, it was found that only WC/C was proven effective in minimizing COF and improving resistance to wear. This improved tribological performance was ascribed to the high H/E ratio of coating and matching of elastic modulus with Ti6Al4V substrate. Marin et al. (2016) used hybrid technology consisting of diffusive treatments (carburizing and nitriding) and use of PVD coatings (CrN and TiCN) on Ti6Al4V alloy. PVD-coated alloy showed limited resistance to wear, owing to the low hardness values of Ti6Al4V substrate during tribo-test, which ultimately led to fracture and delamination. However, the hybrid process comprising hardening and PVD coating on a substrate showed improved wear resistance. From this study, it was concluded that both carburized and nitrided PVD-coated Ti6Al4V (Grade 5) alloy can be used for different applications in biomedical industry. Jiang et al. (2000) produced TiN dendrites 'in situ' on Ti6Al4V substrate by laser surface alloying using gaseous nitrogen and found considerable enhancement of wear resistance under both dry sliding conditions and two-body abrasion. Similarly, Samuel et al. (2008) compared the wear behaviour of two orthopaedic Ti alloys namely boride-reinforced Ti-Nb-Zr-Ta (boride deposition via laser) and Ti6Al4V against two counterbodies – Si_3N_4 and stainless steel. Wear resistance of laser-deposited, boride-reinforced orthopaedic Ti-Nb-Zr-Ta alloy was better as compared to that of Ti6Al4V against softer counterbody of stainless steel. Presence of oxide layer on the surfaces of both the materials improved resistance to wear. However, when Si_3N_4 balls were used as a counterbody, TiB precipitates in boron-reinforced alloy tended to pull out, resulting in third-body abrasion and high COF.

18.4.2 LASER SURFACE TEXTURING

This method has been proven as one of the effective methods for improving wear resistance of Ti alloys. Hu et al. (2012) used laser surface texturing and application of a solid lubrication film to improve tribological properties of Ti6Al4V. In this study, regular micro-dimple pattern created on an alloy surface using laser micro-machining was analysed for tribological performance. Under dry sliding conditions, laser-textured surface having high dimple density exhibited low COF as compared to surface with no texturing at low loads and speed only. This improved frictional coefficient was ascribed to the effective capture of debris generated due to wear in surfaces with high dimple density. Moreover, laser-textured surfaces with burnished solid lubricant MoS_2 showed a reduction in friction and excellent wear resistant capability at all loads, which was ascertained to a high transferring capability of MoS_2 from the dimples to the load-bearing surface. Arenas et al. (2018) studied the tribological properties of laser surface–textured Ti6Al4V followed by coating with MoS_2 and graphene using cloth burnishing technique. Nd: Vanadate laser was employed to form crossed grooves having two different crossing angles (45° and 60°), thus forming a rhombus pattern. Their results suggested that the lifetime of both lubricants was longer on textured-coated surfaces as compared to that on polished surfaces. Also, graphene coating showed longer lifetime than MoS_2 coating in less than 40% textured areas. Similarly, Kümmel et al. (2019) improved the tribological performance of Ti6Al4V by texturing through nano-second pulsed laser. High friction and severe adhesive wear were observed during tribo-testing of non-textured surfaces whereas constant low friction and reduced wear were observed in case of laser-textured samples.

18.4.3 Plasma Nitriding

Plasma nitriding is a popular technique to improve the wear characteristics of biomedical Ti alloys. For example, Fu et al. (2000) carried out work on friction and wear characteristics of carbon nitride (CN_x) films deposited on plasma-nitrided Ti6Al4V alloy. Load-bearing capacity of CN_x-deposited plasma-nitrided Ti alloys was better than that of CN_x-deposited Ti alloys. This duplex treatment system was proven to be much efficient in maintaining stable and low COF and showed improved wear resistance than individual use of CN_x films and plasma nitriding on Ti6Al4V alloy. Lakshmi and Arivuoli (2004) evaluated the tribological properties of plasma-nitrided orthopaedic Ti-5Al-2Nb-1Ta alloy against ultra-high-molecular-weight polyethylene (UHMWPE) at different nitriding temperatures and nitriding time. Wear resistance of this alloy was considerably improved as compared to untreated Ti alloy. XRD analysis of wear debris evolved during tribo-test revealed formation of particles of TiO_2 and titanium oxynitride. Moreover, wear rate increased with increase in sliding speed and load. Tang et al. (2004) also observed a reduction in COF and decrease in wear during tribological study of plasma Mo-N-modified Ti6Al4V. Yetim et al. (2008) compared the tribological properties of two surface treatments on Ti6Al4V – one with plasma-nitrided samples and other with TiN film–deposited samples. Low COF and wear rates were observed in case of plasma-nitrided specimens than TiN film–deposited specimens. Similarly, Wang et al. (2011) studied COF and wear properties of plasma Ni-alloyed Ti6Al4V substrate wherein the authors reported that under dry sliding conditions, COF of this modified surface is 0.25 and it exhibits better tribological performance over Ti6Al4V due to high hardness of surface occurring because of Ti_2Ni intermetallic precipitates and excellent TiNi phase ductility. Also, wear mechanism in case of Ti6Al4V involved adhesion and abrasion whereas modified Ni-Ti6Al4V substrate exhibited wear due to micro-abrasion. Attabi et al. (2019), in their study, reported that worn regions in case of plasma-nitrided Ti6Al4V showed abrasion and their width is lesser as compared to that of untreated alloy which indicates improvement in tribological performance after plasma nitriding.

18.4.4 Plasma Electrolytic Oxidation (PEO)

This method is used for developing coatings on the alloy surface. Ceschini et al. (2008) studied the dry sliding friction and wear characteristics of Ti6Al4V treated with plasma electrolytic oxidation (PEO) technique and PVD coatings {TiN, (Ti, Al)N and CrN/NbN}. The results of this study revealed that TiN PVD coating on

Ti6Al4V gave the best tribo results up to 20 N only whereas PEO-treated Ti6Al4V showed reduced wear and COF up to higher load of 35 N. Mu et al. (2012) produced TiO_2/graphite composite on Ti6Al4V alloy surface using one-step PEO method. Composite coating exhibited a self-lubricating property due to the presence of graphite which ultimately reduced the COF to 0.15. Wear resistance property of this self-lubricating composite coating is also improved over TiO_2 coating and uncoated Ti6Al4V. Mu et al. (2013), in their other study, deposited TiO_2/MoS_2 composite coating on Ti6Al4V using one-step PEO. MoS_2, being a lubricant, resulted in improvement in tribological properties due to its transferring from the coating to the counterbody. Durdu and Usta (2014) successfully produced bioceramic coatings of hydroxyapatite and calcium apatite on Ti6Al4V using PEO technique. The researchers found that wear resistance of coatings produced at 90 min is lesser than the wear resistance of coatings produced at 60 min which may be ascribed to the fact that 90 min coatings were looser and porous than 60 min coatings. Tekin et al. (2016) compared the tribological properties of two PEO coatings on Ti6Al4V alloy. MoS_2 was impregnated to improve friction properties on PEO coatings. Less wear loss was observed in case of aluminate–phosphate electrolyte PEO coating as compared to silicate–phosphate electrolyte PEO coating. Impregnation of MoS_2 on PEO coatings decreased the COF as well.

18.4.5 THERMAL OXIDATION (TO)

Oxidation treatment is another technique for surface modification of Ti alloys. Dong and Bell (2000) compared the wear properties of Ti6Al4V treated by thermal oxidation (TO) to that of untreated Ti6Al4V. TO treatment was done in the atmosphere having 80% oxygen and 20% nitrogen, kept at 600°C for about 65 h. TO treatment improved wear resistance of Ti6Al4V due to the formation of tough and adherent rutile oxide. This rutile oxide eliminated the adhesive wear and improved boundary lubrication which ultimately led to low wear. Borgioli et al. (2005) also made an attempt to improve the wear resistance of Ti6Al4V alloy by TO method. In this case, TO was performed in furnace having proper air circulation at 1173 K for 2 h and at 10^5 Pa, followed by quenching in the compressed air to remove loose oxides on the alloy surface. TO treatment produced hardened oxide layers on surface which led to improvement in wear resistance of Ti6Al4V.

18.4.6 MISCELLANEOUS METHODS

Some other surface treatment methods have also been tried by the researchers to enhance tribological characteristics of Ti alloys. These are shown in Table 18.1.

18.5 LUBRICATION ASPECTS

Lubrication is one of the key aspects in tribology to reduce COF and wear. Amanov and Sasaki (2013) studied the tribological properties of polished, dimpled (by the use of Laser Surface Texturing (LST)), and dimpled Cr-doped diamond-like carbon film–coated (using unbalanced magnetron sputtering) Ti6Al4V alloy under Poly-Alpha-Olefin

TABLE 18.1

Other Surface Modification Methods

Researchers	Methods	Alloys	Remarks
Ueda et al. (2003)	Nitrogen plasma immersion ion implantation	Ti6Al4V	Technique was used to improve the tribological properties of specimens for artificial heart valves
Mello et al. (2009)	Plasma immersion ion implantation	Ti6Al4V	Attractive surface modification technique that improved wear resistance of Ti alloys
Luo et al. (2009)	Nitrogen ion implantation and carburization	Ti6Al4V	Both nitrogen ion implantation and carburization improved tribo properties of Ti alloys, and there was a considerable decrease in the size of wear debris
Amanov and Pyun (2017)	Local heat treatment (LHT) with and without ultrasonic nanocrystal surface modification (UNSM)	Ti6Al4V	Decrease in COF in LHT+ UNSM samples. Wear resistance of LHT +UNSM samples was higher than that of only LHT samples

(PAO) oil–lubricated sliding condition. Cr-plated steel pin acted as a counterbody. In this study, overcoated dimpled specimens showed lower COF and wear as compared to other polished and dimpled specimens due to its high hardness and better storage of debris generated due to wear. Luo, Yang and Tian (2013) studied the effect of bio-lubricants namely deionized water, bovine serum and physiological saline on the tribological characteristics of Ti6Al4V against silicon nitride balls. Wear resistance of Ti6Al4V under dry conditions was worst due to abrasion and oxidation wear. However, under three lubrication conditions, Ti6Al4V alloy exhibited a low COF and wear, and the major cause of wear was adhesion only. Lubricants were effective in preventing exposure to wear scar to atmosphere, which in turn reduced the oxidation rate of Ti alloy and maintained steady COF. Wang et al. (2014) studied the bio-tribological behaviour of Ti6Al4V treated by TO process under dry and lubricated conditions. Ti alloy was slidden against the UHMWPE pin to study its tribological behaviour for application in artificial cervical disc. Lubricants used in this study were distilled water and bovine serum (25 wt. %). In comparison with untreated samples under both dry and lubricated conditions, there was about a 50% reduction in COF and wear in TO-treated specimens. This improvement was attributed to the formation of hardened rutile coatings on alloy after TO treatment. Similarly, Luo et al. (2015) studied bio-tribological properties of TO-treated Ti6Al4V alloy under 25% bovine serum as a lubricant. Improvement of tribological properties by thermal oxidation proved that TO-treated Ti alloys can be used for artificial joints. Self-lubricating composite coatings on Ti6Al4V have also been proven effective in enhancing anti-wear properties (Ren et al., 2017). Self-emulsifying ester oil and aqueous solutions have also been used as lubricants by Yang et al. (2017) to study the tribological behaviour of Ti6Al4V-tungsten carbide tribo pair. Salguero et al. (2019), in their study, concluded that absorption of lubricant was promoted by the textured surfaces generated through laser process treatments. This absorption improved the friction and wear behaviour under sliding conditions as compared to surfaces which were not treated. Rathnam and Rathnam (2020) also found improved tribological properties of Ti6Al4V coated with Ni under lubrication condition in comparison to that of uncoated samples.

18.6 CONCLUSIONS

In nutshell, this chapter highlighted the tribological characteristics of Ti alloys used in the biomedical field. Various tribological problems in case of such alloys encountered during their usage are also elucidated in this chapter. Surface modification techniques to overcome those limitations including the use of coatings, surface texturing through laser, plasma nitriding, TO treatments, etc. which are already in use have also been incorporated. Lubrication aspects in the tribological study of biocompatible Ti alloys are also discussed here. The whole chapter summarizes some of the main findings of researchers working in this field. From a future point of view, it was observed that majority of the work is carried out in determining the tribological behaviour of Ti6Al4V only (with or without surface treatments). Other biomedical Ti alloys which are already in practice as orthopaedic implants or dental implants or cardiac accessories can be more explored for their tribological behaviour when subjected to different surface modification treatments. Hybrid technologies comprising the use of both surface treatments and lubrication aspects may also be explored for further improving friction and wear characteristics of Ti alloys used in the biomedical industry.

REFERENCES

Alam, M. O. & Haseeb, A., 2002. Response of Ti–6Al–4V and Ti–24Al–11Nb alloys to dry sliding wear against hardened steel. *Tribology International,* 35(6), pp. 357–362.

Amanov, A. & Pyun, Y.-S., 2017. Local heat treatment with and without ultrasonic nanocrystal surface modification of Ti-6Al-4V alloy: Mechanical and tribological properties. *Surface and Coatings Technology,* 326, pp. 343–354.

Amanov, A. & Sasaki, S., 2013. A study on the tribological characteristics of duplex-treated Ti–6Al–4V alloy under oil-lubricated sliding conditions. *Tribology International,* 64, pp. 155–163.

Arenas, M., Ahuir-Torres, J., García, I., Carvajal, H. & De Damborenea, J., 2018. Tribological behaviour of laser textured Ti6Al4V alloy coated with MoS_2 and graphene. *Tribology International,* 128, pp. 240–247.

Attabi, S., Mokhtari, M., Taibi, Y., Abdel-Rahman, I., Hafez, B. & Elmsellem, H., 2019. Electrochemical and Tribological Behavior of Surface-Treated Titanium Alloy Ti–6Al–4V. *Journal of Bio-and Tribo-Corrosion,* 5(1), pp. 2.

Borgioli, F., Galvanetto, E., Iozzelli, F. & Pradelli, G., 2005. Improvement of wear resistance of Ti–6Al–4V alloy by means of thermal oxidation. *Materials Letters,* 59(17), pp. 2159–2162.

Ceschini, L., Lanzoni, E., Martini, C., Prandstraller, D. & Sambogna, G., 2008. Comparison of dry sliding friction and wear of Ti6Al4V alloy treated by plasma electrolytic oxidation and PVD coating. *Wear,* 264(1–2), pp. 86–95.

Chauhan, S. & Dass, K., 2013. Dry sliding wear behaviour of titanium (Grade 5) alloy by using response surface methodology. *Advances in Tribology,* 2013.

Choubey, A., Basu, B. & Balasubramaniam, R., 2004. Tribological behaviour of Ti-based alloys in simulated body fluid solution at fretting contacts. *Materials Science and Engineering: A*, 379(1–2), pp. 234–239.

Cvijović-Alagić, I., Cvijović, Z., Mitrović, S., Rakin, M., Veljović, Đ. & Babić, M., 2010. Tribological behaviour of orthopaedic Ti-13Nb-13Zr and Ti-6Al-4V alloys. *Tribology Letters*, 40(1), pp. 59–70.

Dong, H. & Bell, T., 2000. Enhanced wear resistance of titanium surfaces by a new thermal oxidation treatment. *Wear*, 238(2), pp. 131–137.

Durdu, S. & Usta, M., 2014. The tribological properties of bioceramic coatings produced on Ti6Al4V alloy by plasma electrolytic oxidation. *Ceramics International*, 40(2), pp. 3627–3635.

Fu, Y., Loh, N. L., Wei, J., Yan, B. & Hing, P., 2000. Friction and wear behaviour of carbon nitride films deposited on plasma nitrided Ti–6Al–4V. *Wear*, 237(1), pp. 12–19.

Hu, T., Hu, L. & Ding, Q., 2012. Effective solution for the tribological problems of Ti-6Al-4V: Combination of laser surface texturing and solid lubricant film. *Surface and Coatings Technology*, 206(24), pp. 5060–5066.

Jiang, P., He, X., Li, X. A., Yu, L. & Wang, H., 2000. Wear resistance of a laser surface alloyed Ti–6Al–4V alloy. *Surface and Coatings Technology*, 130(1), pp. 24–28.

Kümmel, D., Hamann-Schroer, M., Hetzner, H. & Schneider, J., 2019. Tribological behavior of nanosecond-laser surface textured Ti6Al4V. *Wear*, 422, pp. 261–268.

Lakshmi, S. G. & Arivuoli, D., 2004. Tribological behaviour of plasma nitrided Ti-5Al-2Nb-1Ta alloy against UHMWPE. *Tribology International*, 37(8), pp. 627–631.

Li, Y., Yang, C., Zhao, H., Qu, S., Li, X. & Li, Y., 2014. New developments of Ti-based alloys for biomedical applications. *Materials*, 7(3), pp. 1709–1800.

Liu, X., Chen, S., Tsoi, J. K. & Matinlinna, J. P., 2017. Binary titanium alloys as dental implant materials—a review. *Regenerative Biomaterials*, 4(5), pp. 315–323.

Liu, X., Chu, P. K. & Ding, C., 2004. Surface modification of titanium, titanium alloys, and related materials for biomedical applications. *Materials Science and Engineering: R: Reports*, 47(3–4), pp. 49–121.

Luo, Y., Chen, W., Tian, M. & Teng, S., 2015. Thermal oxidation of Ti6Al4V alloy and its bio-tribological properties under serum lubrication. *Tribology International*, 89, pp. 67–71.

Luo, Y., Ge, S., Jin, Z. & Fisher, J., 2009. Effect of surface modification on surface properties and tribological behaviours of titanium alloys. *Proceedings of the Institution of Mechanical Engineers, Part J: Journal of Engineering Tribology*, 223(3), pp. 311–316.

Luo, Y., Yang, L. & Tian, M., 2013. Influence of bio-lubricants on the tribological properties of Ti6Al4V alloy. *Journal of Bionic Engineering*, 10(1), pp. 84–89.

Mao, Y., Wang, L., Chen, K., Wang, S. & Cui, X., 2013. Tribo-layer and its role in dry sliding wear of Ti–6Al–4V alloy. *Wear*, 297(1–2), pp. 1032–1039.

Marin, E., Offoiach, R., Regis, M., Fusi, S., Lanzutti, A. & Fedrizzi, L., 2016. Diffusive thermal treatments combined with PVD coatings for tribological protection of titanium alloys. *Materials & Design*, 89, pp. 314–322.

Martini, C. & Ceschini, L., 2011. A comparative study of the tribological behaviour of PVD coatings on the Ti-6Al-4V alloy. *Tribology International*, 44(3), pp. 297–308.

Mello, C., Ueda, M., Silva, M., Reuther, H., Pichon, L. & Lepienski, C., 2009. Tribological effects of plasma immersion ion implantation heating treatments on Ti–6Al–4V alloy. *Wear*, 267(5–8), pp. 867–873.

Ming, Q., Yong-Zhen, Z., Jian-Heng, Y. & Jun, Z., 2006. Microstructure and tribological characteristics of Ti–6Al–4V alloy against GCr15 under high speed and dry sliding. *Materials Science and Engineering: A*, 434(1–2), pp. 71–75.

Mu, M., Liang, J., Zhou, X. & Xiao, Q., 2013. One-step preparation of TiO2/MoS2 composite coating on Ti6Al4V alloy by plasma electrolytic oxidation and its tribological properties. *Surface and Coatings Technology*, 214, pp. 124–130.

Mu, M., Zhou, X., Xiao, Q., Liang, J. & Huo, X., 2012. Preparation and tribological properties of self-lubricating TiO2/graphite composite coating on Ti6Al4V alloy. *Applied Surface Science,* 258(22), pp. 8570–8576.

Oldani, C. & Dominguez, A., 2012. Titanium as a Biomaterial for Implants. *Recent Advances in Arthroplasty,* 218, pp. 149–162.

Pierret, C., Maunoury, L., Monnet, I., Bouffard, S., Benyagoub, A., Grygiel, C., Busardo, D., Muller, D. & Höche, D., 2014. Friction and wear properties modification of Ti–6Al–4V alloy surfaces by implantation of multi-charged carbon ions. *Wear,* 319(1–2), pp. 19–26.

Ramos-Saenz, C., Sundaram, P. & Diffoot-Carlo, N., 2010. Tribological properties of Ti-based alloys in a simulated bone–implant interface with Ringer's solution at fretting contacts. *Journal of the Mechanical Behavior of Biomedical Materials,* 3(8), pp. 549–558.

Rathnam, G. S. & Rathnam, C., 2020. Performance Assessment of Ni-Coated Titanium Alloys Under Wet Conditions. *Journal of Bio-and Tribo-Corrosion,* 6(2), pp. 1–8.

Ren, J., Liu, X.-B., Lu, X.-L., Yu, P.-C., Zhu, G.-X., Chen, Y. & Xu, D., 2017. Microstructure and tribological properties of self-lubricating antiwear composite coating on Ti6Al4V alloy. *Surface Engineering,* 33(1), pp. 20–26.

Salguero, J., Del Sol, I., Vazquez-Martinez, J., Schertzer, M. & Iglesias, P., 2019. Effect of laser parameters on the tribological behavior of Ti6Al4V titanium microtextures under lubricated conditions. *Wear,* 426, pp. 1272–1279.

Samuel, S., Nag, S., Scharf, T. W. & Banerjee, R., 2008. Wear resistance of laser-deposited boride reinforced Ti-Nb–Zr–Ta alloy composites for orthopedic implants. *Materials Science and Engineering: C,* 28(3), pp. 414–420.

Tang, B., Wu, P.-Q., Li, X.-Y., Fan, A.-L., Xu, Z. & Celis, J.-P., 2004. Tribological behavior of plasma Mo–N surface modified Ti–6Al–4V alloy. *Surface and Coatings Technology,* 179(2–3), pp. 333–339.

Tekin, K., Malayoglu, U. & Shrestha, S., 2016. Tribological behaviour of plasma electrolytic oxide coatings on Ti6Al4V and cp-Ti alloys. *Surface Engineering,* 32(6), pp. 435–442.

Tkachenko, S., Datskevich, O., Kulak, L., Jacobson, S., Engqvist, H. & Persson, C., 2014. Wear and friction properties of experimental Ti–Si–Zr alloys for biomedical applications. *Journal of the Mechanical Behavior of Biomedical Materials,* 39, pp. 61–72.

Ueda, M., Silva, M., Otani, C., Reuther, H., Yatsuzuka, M., Lepienski, C. & Berni, L., 2003. Improvement of tribological properties of Ti6Al4V by nitrogen plasma immersion ion implantation. *Surface and Coatings Technology,* 169, pp. 408–410.

Vadiraj, A. & Kamaraj, M., 2007. Effect of surface treatments on fretting fatigue damage of biomedical titanium alloys. *Tribology International,* 40(1), pp. 82–88.

Wang, Z., He, Z., Wang, Y., Liu, X. & Tang, B., 2011. Microstructure and tribological behaviors of Ti6Al4V alloy treated by plasma Ni alloying. *Applied Surface Science,* 257(23), pp. 10267–10272.

Wang, S., Liu, Y., Zhang, C., Liao, Z. & Liu, W., 2014. The improvement of wettability, bio-tribological behavior and corrosion resistance of titanium alloy pretreated by thermal oxidation. *Tribology International,* 79, pp. 174–182.

Wu, Z., Xing, Y., Huang, P. & Liu, L., 2017. Tribological properties of dimple-textured titanium alloys under dry sliding contact. *Surface and Coatings Technology,* 309, pp. 21–28.

Yang, Y., Zhang, C., Dai, Y. & Luo, J., 2017. Tribological properties of titanium alloys under lubrication of SEE oil and aqueous solutions. *Tribology International,* 109, pp. 40–47.

Yetim, A., Alsaran, A., Efeoglu, I. & Çelik, A., 2008. A comparative study: The effect of surface treatments on the tribological properties of Ti–6Al–4V alloy. *Surface and Coatings Technology,* 202(11), pp. 2428–2432.

19 Tribological Aspects of Artificial Joints

Ahmed Abdelbary
Egyptian Government

CONTENTS

19.1 INTRODUCTION

The start of introducing the word 'tribology' was in 1964. Presently, dictionaries define tribology as 'the study of friction, wear and lubrication, and the design of bearings' or 'the science of interacting surfaces in relative motion' (Dictionary; Jost, 1966; Oxford). Since the publication of the P. Jost report, the ability and understanding of modeling many tribology subjects has increased enormously.

Tribology was not limited to machinery, new areas of tribology have emerged such as space tribology, biotribology, green tribology, micro/nano-tribology and others. The term *Biotribology* is often used to mention the application of tribological principles between moving components in, artificial or natural, medical and biological systems.

Replacement of joint is one of the most crucial arthroplasties for many diseases of human joints. This treatment comprises the replacement of natural articulating joints, such as the hip, knee, shoulder and ankle with artificial joints. Presently, it is supposed that there are over 1,000,000 artificial joints implanted annually into patients

333

all over the world (Ellen, Forbush, & Groomes, 2020; Rizzo, 2020). Probably, by 2030, at least three million total knee arthroplasty and a half million total hip arthroplasty cases per year (Urban, Wolfe, Sanghavi, Fields, & Magid, 2017; Zawadzki et al., 2017). Several combinations of materials are now used in biomedical applications including metals, carbon ceramics and polymers.

Polymers, in particular, have shown a supreme achievement in this field due to its exceptional tribological properties. The dominant properties that suggested it to become a highly successful bearing material in these applications are (Lu et al., 2019; Shibata & Tomita, 2005; Sinha & Briscoe, 2009; Zhang, Sawae, Yamaguchi, Murakami, & Yang, 2015):

- Low coefficient of friction.
- High wear resistance.
- Ability of sterilization by γ-irradiation.
- Compatibility of wear debris with the human body.
- Long-time operation under free maintenance conditions.

In total joint replacements, polyethylene has remained one of the most suitable polymers of choice for the articulating surface (Atkinson, Dowson, Isaac, & Wroblewski, 1985; Dowson, Diab, Gillis, & Atkinson, 1985; Fisher & Dowson, 1991). Although polytetrafluoroethylene (PTFE) failed in these applications due to high wear rate, ultra-high-molecular-wight polyethene (UHMWPE) showed a superior wear resistance. First, UHMWPE was introduced by Atkinson, Charnley, Dowling, and Dowson (1976), Charnley (1972) as an acetabular cup sliding against stainless steel, alumina or ceramic femoral head. Recently, ultra-low-wear polyethylene (ULWPE) has been developed as a new type of polyethylene material due to its excellent wear properties compared to other polyethylene(s) (Bian et al., 2018).

Tribology of artificial joints plays an important role in its successful function. Nevertheless, it is unpredictable due to multifaceted biological and tribological manners and a relatively long-term wear (Jin, Zheng, Li, & Zhou, 2016). Currently, researches are conducted to evaluate the factors contributing to the total wear rate of the articulating surfaces, with the objective of minimizing wear. Nonetheless, one of the foremost challenges is the transformation of research data from *in vitro* to *in vivo* situations (Kashi & Saha, 2020). For instance, the wear debris generated from the test rig has a relatively larger size and broader distribution range related to those from the implanted artificial joint (Hongtao, Shirong, Shoufan, & Shibo, 2011). It was found that the typical diameter of the generated debris in test rig was about 7.5 μm. On the other hand, the typical diameter of the debris generated *in vivo* was about 1.3 μm. There is approximately 18% increasing of the debris generated *in vitro*.

In order to investigate the criteria of failure in artificial joints, a pioneer study was conducted by Atkinson et al. (1976) in the light of the examination of a number of worn hip joints of age, ranging from 1 to 12 years. It was observed that the deformation and wear processes of the acetabular cups appeared as follows:

1. At first, a running-in wear period where the entire cup surface wears in an abrasive manner.

2. During the increase in mobility, the upper half of the cup surface wears in an adhesive manner.
3. Finally, after several years of wear, fine fatigue surface cracks formed on the wear area.

These findings led to many questions which need to be answered:

- Are any of the detected wear features possible to improve early failure of the acetabular cup? Are there other suggested mechanisms?
- Are the fatigue cracks detected on the surface deleterious? Do they affect the serviceable life of the cup?
- Is there any effect of the fluctuating loading on the surface cracks and failure behaviour of the polymer particularly in fluid lubrication condition?
- Is there any effect of the size of wear particle on the clinical infection?

In conclusion, it was stated that no assumptions could be drawn and additional investigations are compulsory. To the best of our knowledge, hundreds of researches were established for more than five decades in order to answer those questions.

19.2   TRIBOLOGY OF ARTIFICIAL JOINTS

Generally, tribology has an important role in various medical components since they are usually elaborate with parts in relative motion. Regarding to artificial joints, tribology of contacting surfaces is often affecting their function. Now, it is extensively accepted that general basics of classical tribology could be applied to study the wear, friction and lubrication of natural and artificial joints in the human body (Stewart, 2010). Friction has a significant part in the Charnley arthroplasty design (Jin & Fisher, 2008). Likewise, wear has a dominant role in the prosthetic component. Wear debris can, also, be the reason for harmful biological reactions (Jin et al., 2016). Finally, proper lubrication could be a practical and effective solution to control both friction and wear.

19.2.1   FRICTION

At first, metal-on-metal hips were introduced by McKee–Farrar. Then, for its lowest frictional coefficient, PTFE cups were selected although an enormous wear was then found with this material. The impact of friction in the artificial joint design was introduced by Charnley (1972). Charnley utilized polymer–metal tribosystem to preserve low friction. He presented a cup made from UHMWPE and femoral heads made from stainless steel. Such systems are enormously effective in minimizing friction and wear rate of the polymer, as well as minor quantities of wear debris (Dowson, 1998; Dowson et al., 1985; Fisher & Dowson, 1991; Zietz et al., 2015).

In studying friction force (F) in artificial joints in terms of the applied load (W), sliding speed (V), apparent area of contact (Aap), the following laws of friction are defined:

1. (F) is directly proportional to (W).
2. (F) is independent of (Aap).
3. (F) is independent of (V).

The friction coefficient (f) is defined as the ratio between the magnitudes of the friction force F and the normal force N at the interface, therefore:

$$f = F/N \tag{19.1}$$

In studying the friction in artificial joints, we should consider the unusual contact behaviour of a revolute/spherical joint as illustrated in Figure 19.1 (Di Puccio & Mattei, 2015). Assuming a cylindrical pin in rotation, in a smooth contact, with a fixed angular speed (ω) within a collar, the applied load (W) acts at the centre of the pin itself. The normal force $N = -W$ functional at the point of contact K on the action line of W ensures the balance of the pin. If we cannot neglect the friction at the interface, the total contact force R, still with the same value of W, is the sum of tangent and perpendicular to the surface at the contact point N and F, respectively. Consequently, K is moved rearward in an angle $\varphi = \arctan(f)$, so that R restrains the motion and a torque M_f must be presented to preserve the rotation of the pin.

19.2.2   Wear Mechanisms

Wear and subsequent wear debris are considered the basic cause of prosthetic damage, producing osteolysis and adverse tissue reactions, which can be the cause of loosening of the prosthetic components. The basic mechanisms of wear in artificial joints are abrasion, adhesion, fatigue and creep. Nevertheless, scratching, pitting, delamination and burnishing also can be detected (Hood, Wright, & Burstein, 1983).

There are several aspects contributing to wear and, in many cases, it is not easy to define which mechanism will affect the contacting surfaces. In several cases,

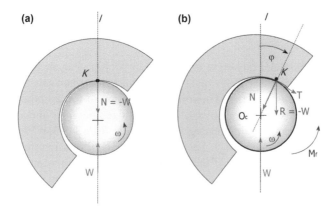

FIGURE 19.1   Forces in artificial joint: non-frictional contact (a) and frictional contact (b). (After Francesca Di Puccio and Mattei 2015 with permission.)

different types of wear are detected. The following mechanisms are introduced to designate the wear in artificial joints (Jin & Fisher, 2014):

Abrasive wear – is resulting from the motion of hard asperities on the harder counterface that plough the softer countersurface. The size of the generated wear debris is of a same value to the harder surface roughness.

Adhesive wear – when momentarily local sticking between asperities of the two surfaces occurs, subsequently broken in the movement.

Surface fatigue – repeated stress cycles at the material subsurface result in microcracks and debris detachment. This type of wear can also be detected on a macroscopic scale in the form of delamination. It was demonstrated that fatigue wear occurs when the mean value of cyclic stress applied to the material exceeds the material's fatigue strength (Abdelbary, 2014a). Adhesive and surface fatigue wear are, in many situations, working together, which over time have led to wear debris which are commonly larger than those of which produced in classical abrasive wear (Stewart, 2010).

Pitting – is usually related to surface fatigue wear and frequently has been detected in retrieved implants sterilized with gamma radiation in air, especially tibial plateaus (Bradford et al., 2004).

In studying the impact of contact stresses on the macroscopic and microscopic wear mechanisms, Cooper, Dowson, & Fisher, 1993; Cooper, Dowson, Fisher, Isaac, & Wroblewski, 1994 investigated the wear of UHMWPE sliding on ceramic and metallic counterparts under different tribological conditions. The experiments were performed in rotating and reciprocating tribometers and hip joint simulators and retrieve artificial joints taken from patients. They advocated that under fluctuating load, the asperity of polymer is cyclically stressed and deformed at the frequency of the load cycle. Such stress cycles can result in propagation of crack and fatigue at the polymer surface. It is suggested that cyclic loading is one of the key factors causing failure of artificial joints. Microscopic investigation of artificial joints indicated that under fluctuating load, subsurface cracks were detected in the highly stressed region. Propagation of these cracks results in accelerating failure and material removal from the polymer surface. Consequently, the wear processes, at macroscopic scale, greatly increased. It is remarkable to remember that the fatigue wear process is not only significant with respect to the increased volume wear debris generated. In fact, it may also generate relatively large debris particles which can result in adverse reactions in the body tissue. Recently, compared to polyethylene, highly cross-linked polyethylene has revealed significant enhancements in implant wear, with about 80% minimization in wear rates.

19.2.3 LUBRICATION

In general, lubrication refers to the existence of a substance between two surfaces in contact to minimize or avoid interaction of their asperities. In healthy natural joints, under applied load, the cartilage deforms and acts to dispense the load over a broader area, therefore minimizing the stress at contact. Synovial fluid is generally present as a fluid film lubrication that protects the cartilage from direct contact. After clinical joint replacements, lubrication has a vital role in the successful function of artificial joints. A pseudo-periprosthetic synovial fluid is found to be similar to those from patients with osteoarthritis (Yamada et al., 2000). To avoid artificial joint failure, it is imperative to reduce the contact stresses mainly for thin polymer cups. Effective fluid-film and boundary lubrication is the key factor in controlling wear and friction in artificial joints (Jin & Fisher, 2008). The relative motion between the articulating surfaces builds up a fluid pressure that separates the surfaces from direct contact. A natural hip during walking squeeze film lubrication arises when the articulating surfaces that are initially separated move together rapidly. Consequently, pools of lubricant can be trapped between the contact surfaces, slowly leaking out with time.

The regime of lubrication can be evaluated theoretically or experimentally. Theoretical calculation is based on the determination of the parameter (λ) which is defined as the ratio between the minimum film thickness h_{min} and the roughness of the composite surfaces ('7 Elastohydrodynamic Lubrication', 1993; Jalali-Vahid, Jagatia, Jin, & Dowson, 2001):

$$\lambda = \frac{h_{min}}{R_a} = \frac{h_{min}}{\left[\left(R_{a\_head}\right)^2 + \left(R_{a\_cup}\right)^2\right]^{1/2}} \tag{19.2}$$

where

R_a is mean surface roughness.

Consequently, if a typical h_{min} is assumed and the roughness values are measured, the parameter (λ) and the consistent lubrication regimes can be calculated, thus:

Boundary regime of lubrication, $\lambda < 1$
Mixed regime of lubrication, $1 < \lambda < 3$
Fluid film regime of lubrication, $\lambda > 3$

The relationship between coefficient of friction and film thickness of lubricant is usually presented in a *Stribeck curve*, characterizing the friction to the *Sommerfeld number S*, as shown in Figure 19.2.

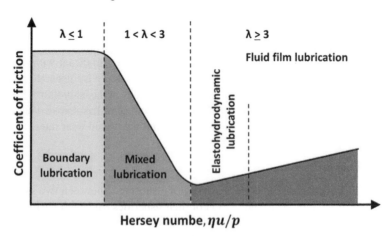

FIGURE 19.2 Stribeck diagram.

the Sommerfeld number is defined by:

$$S = \frac{\eta N}{P} \left(\frac{R}{C} \right)^2 \tag{19.3}$$

where
   R is the radius of the rotating element.
   C is the radial clearance between bearing and element.
   η is the viscosity of the fluid.
   P is the specific load given by:

$$P = \frac{W}{LD} \tag{19.4}$$

where
   W is the applied normal load.
   L and D are the length and the diameter of the bearing, respectively.

In artificial joint bearing types, soft polymer has a high surface roughness (about 1 μm), therefore metal-on polymer is considered to be a boundary lubrication. Accordingly, direct contact between surface asperities and wear are expected. Polyethylene does get burnished or polished *in vivo* which may minimize its roughness. However, the thickness of generated lubrication film is inadequate to provide substantial effect.

Metal-on-metal artificial prostheses are commonly designed to function in the mixed regime of lubrication. It is proposed that increasing the radii of bearing, while keeping the radial clearance low, resulted in increased predicted film thickness. Therefore, a complete transient elastohydrodynamic lubrication analysis under dynamic operating conditions is necessary in order to achieve a better understanding of the in vivo and in vitro lubrication mechanism in metal-on-metal hip implants (Liu et al., 2006).

Compared to polyethylene or metal, ceramic bearings are extensively hard and as such can be polished to a surface roughness $Ra = 0.004$ μm. A reduction in radial clearance (about 0.04 μm) of these bearings is realized by superior manufacturing tolerances and, as a result, leads to a fluid film regime during walk cycle (Jin, Dowson, & Fisher, 1997).

In contrast to the former findings, Abdelbary Abouelwafa, El Fahham and Hamdy (2013) found a dramatic increase in wear rate (up to about 3–6 times) of polyamide sliding in water lubrication condition against stainless steel counterface compared to dry condition, as shown in Figure 19.3. It is proposed that the high wear rates are attributed to the polymer plasticization which resulted from water absorption. Due to plasticization, diffusion of water molecules into the bulk polymer loosens the hydrogen bonds between polymer chains, consequently forming hydrogen bonds with amide groups. Accordingly, water reduces the intermolecular forces and deteriorates the mechanical properties of the polymer. The drop of attractive forces between polymer chains permits removal of material. As a result, high wear rates in water-lubricated sliding contacts are expected (Abdelbary, 2014b).

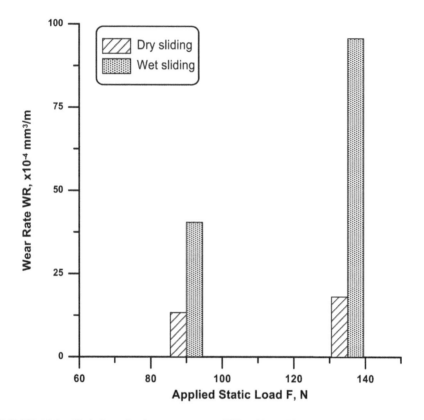

FIGURE 19.3 Variation of polymer wear rates WR with applied static load in dry and wet sliding conditions.

Another consideration to the consequence of water lubricant on the sliding inter-action is due to the inhabitation of the formation and transfer of polymer film on the counterpart. Water washes the metallic counterpart and removes the generated debris. This action makes the polymer fresh surface always exposed to the metal-lic counterpart. It is also important to consider that the running-in phase was not observed on wear curves in water lubrication tests. It is probable that transfer film of polymer, which is typically found in dry sliding, cannot be formed in lubricated sliding.

19.3   FAILURE CRITERIA

Typically, the implant's clinical life limit is when movement becomes painful, con-sequently affecting the quality of life of the patient (Burger, De Vaal, & Meyer, 2006; Kashi & Saha, 2020). It is crucial that we do not properly understand the ori-gin cause of mechanical failure in implants. Damage of articulating UHMWPE sur-faces is considered one of the major factors contributing to the failure of total joint replacements. Wright and Goodman (2001) investigated the influence of damaged

TABLE 19.1
The Common Defects Detected on the Inside of Retrieved Cups

Defect Noticed	ISO 12891-3 Items (12891-3, 2000)
Mechanical damage	A
Cracks	f, l
Scratches	a, d, j
Plastic flow	a, g, h
Adhesive wear	A
Embedded wear particles	a, c, e
Flaking	A

After Burger et al. (2006) with Permission.

joint components on patients. This damage not only affects the performance of the implant, but more significantly results in generation of debris particles to the fluids and surrounding tissues. Failure can be also related to adhesive wear of UHMWPE cup and fatigue fractures of the metal component (Rawal, Yadav, & Pare, 2016).

More than 100 retrievals were investigated by Burger et al. (2006, 2007) to characterize the implant failure criteria. They identified many common failures which are introduced in Table 19.1.

19.3.1 MECHANICAL DAMAGE

Mechanical damage is usually resulted from an acetabular cup not correctly aligned *in vivo*. Also, this impingement damage can arise after a severe wear when the femoral neck becomes in direct contact with the acetabular cup. Frequently, it results in pieces of cement or polymer being ripped from the edge of the cup, as shown in Figure 19.4a. The fragments of removed material, then, will result in rather large floating particles and probable loosening of the cup due to impact loading.

19.3.2 CRACKS

Due to localized stress, surface and subsurface cracks are frequently anticipated on the rim and in the high stress areas of the cup. Accumulation of plastic strain and cycling stress results in initiation of cracks at surface and subsurface. These cracks, with more cycling, deeply propagate into the substrate. Also, cracks join their neighbour cracks to create one large crack which can break from the bulk, producing pitting and spalling (Fisher & Dowson, 1991). As the process continues, a progressive loss of material from the worn surface is obtained. The cracks are highlighted by means of a colour dye penetrant, as shown in Figure 19.4b. Bradford et al. (2004) found similar observations in their study of *in vivo* wear mechanisms of highly cross-linked UHMWPE. In their study, almost all of the retrieved UHMWPE acetabular liners displayed a surface cracking, as shown in Figure 19.5.

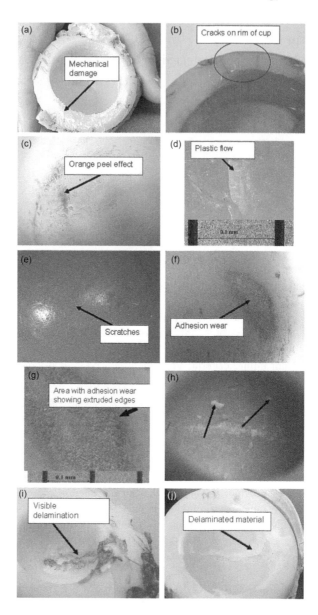

FIGURE 19.4 Acetabular cups illustrating different defects: (a) mechanical damage; (b) cracks on rim of cup; (c) orange peel effect; (d) plastic flow; (e) scratches; (f) adhesion wear; (g) area of adhesion wear associated with extruded edges; (h) cement particles embedded in cup surface; (i) and (j) examples of serious delamination. (After Burger et al. 2006 with permission.)

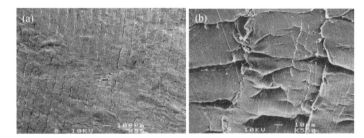

FIGURE 19.5 (a) Evolution of altered machining marks and surface cracks parallel to the marks, low-magnification and (b) Cracks perpendicular to the machining marks, high-magnification. (After Bradford et al. 2004 with permission.)

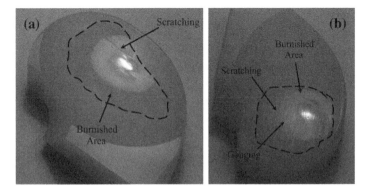

FIGURE 19.6 (a) and (b) worn and burnished areas and some slight scratching and gouging of UHMWPE tested in orthopaedic wear simulator. (After Flannery et al. 2008 with permission.)

19.3.3 SCRATCHES

The number of the inspected retrieved components confirmed indication of burnishing and scratching on the articulating surfaces (Flannery, McGloughlin, Jones, & Birkinshaw, 2008). These features were visible to the naked eyes, while microscale scratches were also detected (Burger et al., 2006). It is believed that the relatively large scratches are due to third-body wear no matter what caused the wear particles. The presence of such scratches can be formed during the time period shortly after implantation, as shown in Figure 19.4e. Also, the wear area looked shiny and very smooth, demonstrating that the surface had been burnished, besides some fine scratching in the sliding direction was detected, as shown in Figure 19.6.

Likewise, similar remarks were distinguished *in vitro* when testing the wear of polymer sliding on steel in water-lubricated conditions (Abdelbary et al., 2013). The wear surface of polymer specimens appeared relatively smooth and shiny, demonstrating that the surface had been burnished or lapped. At the end of wear tests, surface roughness (Ra) of worn polymer surfaces were found about 0.2–0.3 μm. On the other

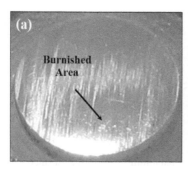

FIGURE 19.7   (a) and (b) scratching and burnished area on polyamide surface.

hand, surface roughness of the steel counterface was in order of 1.3–1.5 µm. It is postulated that, in the case of water lubricant condition, sliding wear of polyamide–steel tribosystems results in roughing the steel counterface and smoothing the polymer surface. Obviously, the former observation was due to the effect of rust on the steel surface, while the later observation seems to be due to the surface-lapping effect. Optical microscopic examination of the polymer-worn surfaces exhibited very fine scratches in the sliding directions, as shown in Figure 19.7.

19.3.4 PLASTIC FLOW

Optical investigation of acetabular cups displayed areas of plastic flow, as shown in Figure 19.4c and d. The affected area usually occurs just outside the region of high-contact stress. It seems as if the 'molten' polymer was softened satisfactorily to be extruded and was expelled from the area of high-contact stress and transferred to one where the stress was less (Burger et al., 2006). Moreover, if the compressive stress on the surface of the cup exceeded the maximum stress limits of the material, it will result in an outward flow and/or creep.

19.3.5 ADHESION WEAR

Generally, lack of lubrication and overheating cause abrasion and adhesion wear. These types of wear are usually taking place in the case of limited movement. In these situations, adhesion between asperities arises on two contact surfaces. Further movement results in elimination of the asperities of the softer surface. Optical microscopic investigation of affected areas exhibited rough patches, these patches are seen in the high-contact stress areas where lubrication was least, as shown in Figure 19.4f and g.

19.3.6 FLAKING

Flaking presents in areas where pieces separate from the base polymer material. It generates either as areas of delamination or craters. Figure 19.4i and j shows severe delamination of two cups. Such a type of defect is correlated with a defect within the material and occurs in the contact stress or high-stress areas (Burger et al., 2006).

19.3.7 EMBEDDED WEAR PARTICLES

Investigation of some of the retrieved cups exhibited wear debris embedded in the bulk material (Burger et al., 2006). The most common embedded particles found were cement used for the fixation of the implant as well as polymethacrylate (PMMA) wear particles. Figure 19.4h shows an acetabular cup with PMMA particles embedded in.

19.4 THE ROLE OF SURFACE AND COUNTERFACE DEFECTS

Dowson, Taheri and Wallbridge (1987) investigated the impact of imperfections upon metallic counterface on the wear behaviour of the polymer. They considered the effect of a single-imposed imperfection in the surface of stainless steel counterpart on the wear of UHMWPE in reciprocating wear test. The imposed imperfections are generated by diamond marks as longitudinal or transverse. It has been demonstrated that the polymer wear rate was increased to at least one order of magnitude due to a single transverse scratch, with the main factor being the piled-up steel along the scratch edge. After removing the piled-up polymer by lapping process, the rate of wear returned to its typical. On the other hand, longitudinal scratches produced a less effect in the polymer wear rate. The results supported the clarification of the alteration between *in vivo* and *in vitro* wear data of total replacement hip joints.

It was suggested that, under uniaxial loading and several frequencies, the fatigue behaviour of polymers is decreased due to the existence of a sharp notch (Crawford & Benham, 1975). Simultaneously, a reduction in the heat generated was also detected. Therefore, in the presence of a sharp notch, a typical fatigue failure was detected in polymers under the stresses which would have otherwise produced thermal softening.

19.5 GENERATION AND ANALYSIS OF WEAR DEBRIS

Wear is defined as 'the gradual removal of a material from contacting solid surfaces during their relative motion'. The generation of wear fragments in any tribosystem is one of the most substantial evidences of the wear. Meanwhile, if there are different mechanisms of wear, it should be expected that there is no typical or unique size or form of wear debris.

Usually, debris particles can be directly removed from open tribosystems as loosened debris. The detached debris can be elaborated in a three-body wear. Consequently, it can rush the deterioration of polymer-worn surface. In the case of closed tribosystems, wear debris remain circulating into the sliding contact and, therefore, form a transfer film of polymer.

During the last 50 years, nearly all prostheses which have been implemented clinically articulate with a boundary or mixed regime of lubrication (Di Puccio & Mattei,

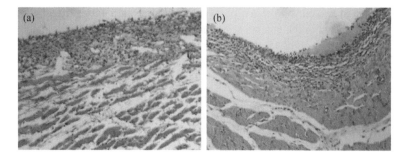

FIGURE 19.8 Accumulation of wear debris inside tissue. (a) ULWPE group and (b) UHMPWE group. (After Bian et al. 2018 with permission.)

2015; Fisher & Dowson, 1991; Zhang, Liu, Wang, Sheng, & Li, 2018). This permits the bearing surfaces to be in contact and consequently wear debris generation.

The main issues related to the surface of artificial joint bearing are wear and consequent debris which can be the source of antagonistic tissue reactions. It is widely recognized that the biocompatibility of wear debris has an imperative effect in bioapplication of artificial joints (Bian et al., 2018). Wear debris of UHMWPE and ULWPE which are generated at the joint surfaces result in granulomatous tissue, macrophage activity and necrosis of the bone surrounding the prosthesis (Fisher & Dowson, 1991), as shown in Figure 19.8 (Bian et al., 2018).

Experimental investigations of polymer wear in wet sliding media showed that the size of resulted wear debris is in the form of submicron (Abdelbary, 2014b). Distinguished wear debris was found suspended in the lubricating water. This scale of wear debris was detected in total hip joints; most UHMWPE wear debris have been shown to be of micron or submicron size (Flannery et al., 2008; Hatton et al., 2002; Kowandy, Mazouz, & Richard, 2006; Zhang et al., 2018). Also, wear debris of similar size were generated in hip simulators using bovine serum as a lubricant fluid.

Actually, the generation of *in vivo* wear debris takes place in a closed tribobiological system which is hard to be observed or tested directly (Slouf et al., 2008). Consequently, it is essential to perform the simulating wear test of artificial joints according to ISO standard. Simulation under situations that consider real dynamics and kinematics of applying a load to the components is significant to predict the tribological response of prosthesis. However, it is very problematic to simulate the wear performance of the artificial joint *in vivo* fully due to the complicated wear mechanism and movement process of the joint countersurfaces. For instance, the wear debris generated from the artificial joint simulator demonstrated a different morphology with that resulted *in vivo* although the wear loss of materials *in vitro* conforms to their wear loss under real circumstances.

In the artificial joint simulator, the wear debris of UHMWPE demonstrated a variation of morphologies due to the complex movements of the joint, as shown in Figure 19.9. It is clear that the debris have various shapes and sizes such as block, strip, spherical and plate. Correspondingly, it still can find that the shape of wear debris is more complicated and the size is larger. Small wear debris possibly take the spherical or near-spherical shapes (Hongtao et al., 2011).

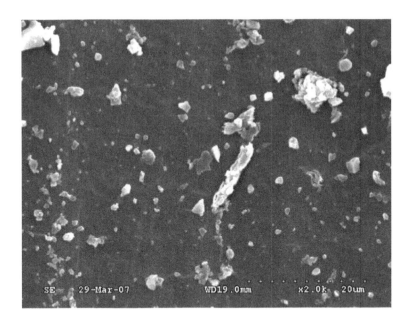

FIGURE 19.9   SEM micrograph of UHMWPE wear debris. (After Hongtao et al. 2011 with permission.)

Figure 19.10 shows typical morphologies of UHMWPE wear debris. The wear debris of spherical shape presented in Figure 19.10a has a diameter less than 50 μm, with distribution in quantity and wide size. Flat block or spindle shape of size ranging from ~1 to ~50 μm is mainly generated from the adhesive wear, as shown in Figure 19.10b. The flat block debris are produced from the cracked wear surface due to the fatigue stress and consequently separated from the worn surface at weak points by the force of adhesion. Generally, this type of debris has a size of 10–150 μm.

In almost all UHMWPE wear debris, irregular tearing debris are related to composite motion of friction pairs of artificial joints. This type of wear debris has irregular form, good third dimension, and rough surface, presented in Figure 19.10c. The dimension of such debris ranges from ~10 to ~150 μm. These debris are generated and grow following the friction direction, peeling off from worn surface progressively. For various operation conditions and various materials, wear debris has a wide diameter distribution and dissimilar sizes. A typical fatigue wear debris is sheet wear debris of a large plane size and a thin thickness, as shown in Figure 19.10d. This type of wear debris has a dimension of ~20 to ~100 μm. Furthermore, wear debris with a rod shape has a size of ~10 to ~100 μm, strong dimensional structure and an irregular shape, as presented in Figure 19.10e. It is significant to emphasize that the existence frequency of this type of wear debris is much less than other types. Finally, zonary debris is a rare form produced by kneading sheet wear debris into rope during the test process, as shown in Figure 19.10f. It is postulated that the *in vitro* wear debris present a different morphology compared to that from human body. Since the wear debris morphology is a significant factor to evaluate artificial joint materials, the wear debris should be considered one of the important parameters for joint simulator design (Hongtao et al., 2011).

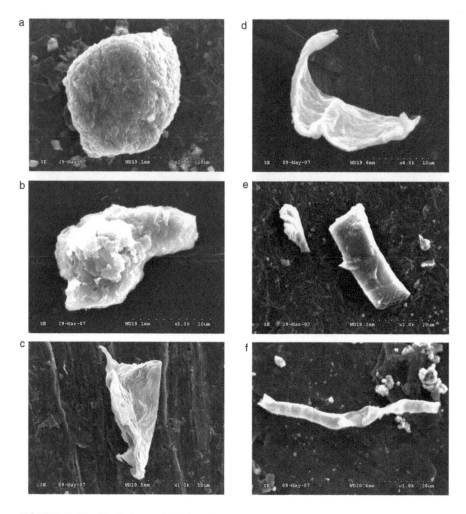

FIGURE 19.10 Typical morphologies of wear debris from periprosthetic tissue. (a) spherical wear debris of less than 50 μm diameter, (b) spindle or flat block shape, (c) tearing wear debris, (d) sheet wear debris, (e) rod wear debris, and (f) zonary wear debris. (After Hongtao et al. 2011 with permission.)

19.6 CONCLUSION

The tribology aspects of artificial joints are introduced and discussed in the light of the cutting-edge literatures. The conclusions are the following:

1. Friction, wear and lubrication of both natural and artificial joints can be considered using the basics of tribology.
2. In artificial joints, surface interaction and wear can be controlled through suitable lubrication.

3. In artificial joints, the common failure modes are wear, scratches, flaking, mechanical damage, plastic flow and cracks.
4. Wear fragments are sources for harmful tissue reactions and loosening of the prosthetic components.
5. The *in vitro* wear debris show a different morphology compared to that from human body.
6. The wear debris morphology is a significant feature to estimate the materials of artificial joint.

REFERENCES

Abdelbary, A. (2014a). 3- Fatigue wear of unfilled polymers. In A. Abdelbary (Ed.), *Wear of Polymers and Composites* (pp. 67–93). Oxford: Woodhead Publishing.

Abdelbary, A. (2014b). 4- Wear of polymers in wet conditions. In A. Abdelbary (Ed.), *Wear of Polymers and Composites* (pp. 95–112). Oxford: Woodhead Publishing.

Abdelbary, A., Abouelwafa, M. N., El Fahham, I. M., & Hamdy, A. H. (2013). The effect of surface defects on the wear of Nylon 66 under dry and water lubricated sliding. *Tribology International*, 59, 163–169. Doi: 10.1016/j.triboint.2012.06.004.

Atkinson, J. R., Charnley, J., Dowling, J. M., & Dowson, D. (1976). The wear of total replacement hip joints in the human body-A topgraphical survey of the surfaces of worn acetabular cups. *Paper Presented at the Third Leeds – Lyon Symposium on Tribology*, Leeds.

Atkinson, J. R., Dowson, D., Isaac, J. H., & Wroblewski, B. M. (1985). Laboratory wear tests and clinical observations of the penetration of femoral heads into acetabular cups in total replacement hip joints: III: The measurement of internal volume changes in explanted Charnley sockets after 2–16 years in vivo and the determination of wear factors. *Wear*, 104(3), 225–244. Doi: 10.1016/0043-1648(85)90050-X.

Bian, Y.-Y., Zhou, L., Zhou, G., Jin, Z.-M., Xin, S.-X., Hua, Z.-K., & Weng, X.-S. (2018). Study on biocompatibility, tribological property and wear debris characterization of ultra-low-wear polyethylene as artificial joint materials. *Journal of the Mechanical Behavior of Biomedical Materials*, 82, 87–94. Doi: 10.1016/j.jmbbm.2018.03.009.

Bradford, L., Baker, D. A., Graham, J., Chawan, A., Ries, M. D., & Pruitt, L. A. (2004). Wear and surface cracking in early retrieved highly cross-linked polyethylene acetabular liners. *Journal of Bone and Joint Surgery*, 86(6), 1271–1282.

Burger, N. D. L., De Vaal, P. L., & Meyer, J. P. (2006). Failure criteria for polyethylene acetabular cups. *South African Journal of Science*, 102(11–12), 572–574.

Burger, N. D. L., De Vaal, P. L., & Meyer, J. P. (2007). Failure analysis on retrieved ultra high molecular weight polyethylene (UHMWPE) acetabular cups. *Engineering Failure Analysis*, 14(7), 1329–1345. Doi: 10.1016/j.engfailanal.2006.11.005.

Charnley, J. W. (1972). The long-term results of low-friction arthroplasty of the hip performed as a primary intervention. *British Orthopedic Association in London*, 54(1): 61–76.

Cooper, J. R., Dowson, D., & Fisher, J. (1993). Macroscopic and microscopic wear mechanisms in ultra-high molecular weight polyethylene. *Wear*, 162–164, 378–384. Doi: 10.1016/0043-1648(93)90521-M.

Cooper, J. R., Dowson, D., Fisher, J., Isaac, G. H., & Wroblewski, B. M. (1994). Observations of residual sub-surface shear strain in the ultrahigh molecular weight polyethylene acetabular cups of hip prostheses. *Journal of Materials Science: Materials in Medicine*, 5(1), 52–57. Doi: 10.1007/BF00121154.

Crawford, R. J., & Benham, P. P. (1975). Some fatigue characteristics of thermoplastics. *Polymer*, 16(12), 908–914. Doi: 10.1016/0032-3861(75)90212-8.

Dictionary. Retrieved from http://www.dictionary.com.

Di Puccio, F., & Mattei, L. (2015). Biotribology of artificial hip joints. *World Journal of Orthopedics*, 6(1), 77–94. Doi: 10.5312/wjo.v6.i1.77.

Dowson, D. (1998). *History of Tribology* (Second edition). London: Institution of Mechanical Engineering.

Dowson, D., Diab, M. M. E.-H., Gillis, B. J., & Atkinson, J. R. (1985). Influence of counter-face topography on the wear of ultra high molecular weight polyethylene under wet or dry conditions. In *Polymer Wear and Its Control* (Vol. 287, pp. 171–187): American Chemical Society, St. Louis, Missouri.

Dowson, D., Taheri, S., & Wallbridge, N. C. (1987). The role of counterface imperfections in the wear of polyethylene. *Wear*, 119(3), 277–293. Doi: 10.1016/0043-1648(87)90036-6.

Ellen, M. I., Forbush, D. R., & Groomes, T. E. (2020). Chapter 80- Total Knee Arthroplasty. In W. R. Frontera, J. K. Silver, & T. D. Rizzo (Eds.), *Essentials of Physical Medicine and Rehabilitation* (Fourth edition) (pp. 443–450). Philadelphia, PA: Content Repository Only!

Fisher, J., & Dowson, D. (1991). Tribology of total artificial joints. *Proceedings of the Institution of Mechanical Engineers. Part H: Journal of Engineering in Medicine*, 205(2), 73–79. Doi: 10.1243/PIME_PROC_1991_205_271_02.

Flannery, M., McGloughlin, T., Jones, E., & Birkinshaw, C. (2008). Analysis of wear and friction of total knee replacements: Part I. Wear assessment on a three station wear simulator. *Wear*, 265(7), 999–1008. Doi: 10.1016/j.wear.2008.02.024.

Hatton, A., Nevelos, J. E., Nevelos, A. A., Banks, R. E., Fisher, J., & Ingham, E. (2002). Alumina–alumina artificial hip joints. Part I: A histological analysis and characterisation of wear debris by laser capture microdissection of tissues retrieved at revision. *Biomaterials*, 23(16), 3429–3440. Doi: 10.1016/S0142-9612(02)00047-9.

Hongtao, L., Shirong, G., Shoufan, C., & Shibo, W. (2011). Comparison of wear debris generated from ultra high molecular weight polyethylene in vivo and in artificial joint simulator. *Wear*, 271(5), 647–652. Doi: 10.1016/j.wear.2010.11.012.

Hood, R. W., Wright, T. M., & Burstein, A. H. (1983). Retrieval analysis of total knee prostheses: A method and its application to 48 total condylar prostheses. *Journal of Biomedical Materials Research*, 17(5), 829–842. Doi: 10.1002/jbm.820170510.

International Organization Standardization ISO 12891-3, (2000). Retrieval and analysis of surgical implants. Part 3, Analysis of retrieved polymeric surgical implants. In. Geneva.

Jalali-Vahid, D., Jagatia, M., Jin, Z. M., & Dowson, D. (2001). Prediction of lubricating film thickness in UHMWPE hip joint replacements. *Journal of Biomechanics*, 34(2), 261–266. Doi: 10.1016/S0021-9290(00)00181-0.

Jin, Z., Dowson, D., & Fisher, J. (1997). Analysis of fluid film lubrication in artificial hip joint replacements with surfaces of high elastic modulus. *Proceedings of the Institution of Mechanical Engineers, Part H: Journal of Engineering in Medicine*, 211(3), 247–256.

Jin, Z., & Fisher, J. (2008). 2- Tribology in joint replacement. In P. A. Revell (Ed.), *Joint Replacement Technology* (pp. 31–55): Woodhead Publishing, London.

Jin, Z., & Fisher, J. (2014). Tribology of hip joint replacement. In G. Bentley (Ed.), *European Surgical Orthopaedics and Traumatology: The EFORT Textbook* (pp. 2365–2377). Berlin, Heidelberg: Springer Berlin Heidelberg.

Jin, Z. M., Zheng, J., Li, W., & Zhou, Z. R. (2016). Tribology of medical devices. *Biosurface and Biotribology*, 2(4), 173–192. Doi: 10.1016/j.bsbt.2016.12.001.

Jost, P. (1966). Lubrication (Tribology) education and research, Technical report. Retrieved from

Jost HP (ed.). *Lubrication (Tribology) - A report on the present position and industry's needs*. Department of Education and Science, H. M. Stationary Office, London, UK, 1966.

Kashi, A., & Saha, S. (2020). 15- Failure mechanisms of medical implants and their effects on outcomes. In C. P. Sharma (Ed.), *Biointegration of Medical Implant Materials* (Second edition). (pp. 407–432): Woodhead Publishing, London.

Kowandy, C., Mazouz, H., & Richard, C. (2006). Isolation and analysis of articular joints wear debris generated in vitro. *Wear*, 261(9), 966–970. Doi: 10.1016/j.wear.2006.03.029.

Liu, F., Jin, Z. M., Hirt, F., Rieker, C., Roberts, P., & Grigoris, P. (2006). Transient elastohy-drodynamic lubrication analysis of metal-on-metal hip implant under simulated walking conditions. *Journal of Biomechanics*, 39(5), 905–914. Doi: 10.1016/j.jbiomech.2005.01.031.

Lu, K., Li, C., Wang, H.-z., Li, Y.-l., Zhu, Y., & Ouyang, Y. (2019). Effect of gamma irradiation on carbon dot decorated polyethylene-gold@ hydroxyapatite biocomposite on titanium implanted repair for shoulder joint arthroplasty. *Journal of Photochemistry and Photobiology B: Biology*, 197, 111504. Doi: 10.1016/j.jphotobiol.2019.05.001.

Oxford. Retrieved from https://en.oxforddictionaries.com.

Rawal, B. R., Yadav, A., & Pare, V. (2016). Life estimation of knee joint prosthesis by combined effect of fatigue and wear. *Procedia Technology*, 23, 60–67. Doi: 10.1016/j.protcy.2016.03.072.

Rizzo, T. D. (2020). Chapter 61- Total hip replacement**Based on a chapter in the third edition written by Juan A. Cabrera, MD and Alison L. Cabrera, MD. In W. R. Frontera, J. K. Silver, & T. D. Rizzo (Eds.), *Essentials of Physical Medicine and Rehabilitation* (Fourth Edition) (pp. 337–345). Philadelphia, PA: Content Repository Only!

Shibata, N., & Tomita, N. (2005). The anti-oxidative properties of α-tocopherol in γ-irradiated UHMWPE with respect to fatigue and oxidation resistance. *Biomaterials*, 26(29), 5755–5762. Doi: 10.1016/j.biomaterials.2005.02.035.

Sinha, S. K., & Briscoe, B. J. (2009). *Polymer Tribology*. London: Imperial College Press.

Slouf, M., Pokorny, D., Entlicher, G., Dybal, J., Synkova, H., Lapcikova, M., … Sosna, A. (2008). Quantification of UHMWPE wear in periprosthetic tissues of hip arthroplasty: Description of a new method based on IR and comparison with radiographic appearance. *Wear*, 265(5), 674–684. Doi: 10.1016/j.wear.2007.12.008.

Stewart, T. D. (2010). Tribology of artificial joints. *Orthopaedics and Trauma*, 24(6), 435–440. Doi: 10.1016/j.mporth.2010.08.002.

Urban, M. K., Wolfe, S. W., Sanghavi, N. M., Fields, K., & Magid, S. K. (2017). The incidence of perioperative cardiac events after orthopedic surgery: a single institutional experience of cases performed over one year. *HSS Journal*, 13(3), 248–254. Doi: 10.1007/s11420-017-9561-9569.

Wright, T. M., & Goodman, S. B. (2001). *Implant Wear in Total Joint Replacement: Clinical and Biologic Issues, Material and Design Considerations*. Rosemont, USA.

Yamada, H., Morita, M., Henmi, O., Miyauchi, S., Yoshida, Y., Kikuchi, T., … Seki, T. (2000). Hyaluronan in synovial fluid of patients with loose total hip prosthesis. *Archives of Orthopaedic and Trauma Surgery*, 120(9), 521–524. Doi: 10.1007/s004029900131.

Zawadzki, N., Wang, Y., Shao, H., Liu, E., Song, C., Schoonmaker, M., & Shi, L. (2017). Readmission due to infection following total hip and total knee procedures: A retrospective study. *Medicine*, 96(38), e7961. Doi: 10.1097/md.0000000000007961.

Zhang, D., Liu, H., Wang, J., Sheng, C., & Li, Z. (2018). Wear mechanism of artificial joint failure using wear debris analysis. *Journal of Nanoscience and Nanotechnology*, 18(10), 6805–6814. Doi: 10.1166/jnn.2018.15513.

Zhang, L., Sawae, Y., Yamaguchi, T., Murakami, T., & Yang, H. (2015). Effect of radiation dose on depth-dependent oxidation and wear of shelf-aged gamma-irradiated ultra-high molecular weight polyethylene (UHMWPE). *Tribology International*, 89, 78–85. Doi: 10.1016/j.triboint.2014.12.011.

Zietz, C., Reinders, J., Schwiesau, J., Paulus, A., Kretzer, J. P., Grupp, T., … Bader, R. (2015). Experimental testing of total knee replacements with UHMW-PE inserts: impact of severe wear test conditions. *Journal of Materials Science: Materials in Medicine*, 26(3), 134.

20 Biomedical Tribology
Wear of Polyethylene in Total Joint Replacement

Shahira Liza and Nur Hidayah Shahemi
Universiti Teknologi Malaysia

Tan Mean Yee
University Malaya

Sharifah Khadijah Syed Mud Puad
Universiti Teknologi Malaysia

CONTENTS

20.1   INTRODUCTION

A joint is formed by the ends of two or more bones that connect or make contact by connective tissue called cartilage. There are a total of 360 joints in the human body that included movable types. The articulating surfaces of the joint are covered by a thin layer of cartilage, which typically generates very minimum friction, 0.01 (Crawford and Murray, 1997; Hills and Buttler, 1984), and could last up to more than 80 years. This articular cartilage acts as a shock absorber and reduces friction in the joints during movement. However, when the cartilage is damaged by bone disease, osteoarthritis, the joint will become stiff, and a simple gesture can be severely painful.

Total joint replacement (TJR) is a surgery to replace or reconstruct problem or damaged joint with an artificial joint known as a prosthesis. Hip, knee, ankle, foot, shoulder, elbow, wrist and fingers replacements are the commonly performed joint replacement surgery. The prosthesis is made of plastic, metal or ceramic material. After the surgery, the prosthesis components can act as healthy natural joints. There are two widely used methods for fixing the joint with the bone. The surgeon can use polymethyl methacrylate bone cement for the fixation of joint prostheses (cemented implant), or the prostheses are press-fit into the bone (cementless implant). In some cases, the hybrid joint component known as a combination of cement and cementless elements is needed, depending on the components of a prosthesis and the patient's condition. TJR has been found to play a significant role in social rehabilitation in terms of relieving pain, joint functioning and improving patient's quality of life.

There has been a steady annual rise in the number of surgeries for joint replacement. It has been performed since the 1960s. The demand for primary TJRs in the United States is predicted growth of 673% for knee replacement and 174% for hip replacement between 2005 and 2030 (Kurtz et al., 2009). TJR of the hip and knee is among the most successful group of surgical procedures in the orthopaedic field. Over the years, the prosthesis components can fail for different reasons requiring repeat revision surgery. Considerable efforts have been made by manufacturers and researchers to increase the life span of TJR. New materials technologies and processes are developed to improve the existing prosthetic design that may help increase the mean life of the prosthesis.

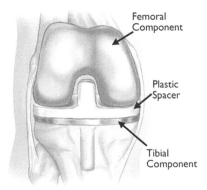

FIGURE 20.1 Total knee replacement. (https://orthoinfo.aaos.org/en/treatment/total-knee-replacement/.)

20.1.1 Total Knee Replacement (TKR)

A typical total knee replacement (TKR) prosthesis shown in Figure 20.1 consists of four components: a femoral component, a tibial insert, a patellar component and a tibial component. The tibial and femoral components can be made of a non-corrosive metal such as cobalt-chromium (CrCo) or titanium (Ti6Al4V) alloys. In contrast, tibial insert and patellar components can be made of UHMWPE, which is a medical-grade plastic. The patellar component glides up and down on the front femoral component in TKR.

20.1.2 Total Hip Replacement (THR)

The hip prosthesis is mainly composed of the acetabular component and femoral component. The bearing part in total hip replacement (THR) consisted of the acetabular liner against a femoral ball, and this part can be realized with a different combination of materials such as metal-on-metal (MoM), metal-on-ceramic (MoC) and metal-on-polymer (MoP). It has turned polymer-bearing components as the most favoured materials for this application (Figure 20.2), due to the limitation from metal bearing and ceramic bearing like the risk of elevated metal levels in the bloodstream (Sampson and Hart, 2012) and instantaneous catastrophic fracture (Drummond et al., 2015), respectively. Similar to the TKR, the CoCr/UHMWPE or Ti6Al4V/UHMWPE is commonly used as the bearing couple in THR.

20.1.3 Implant Failures

Currently, orthopaedic implants are designed to last a long time upwards of 15–20 years on average. Factors that affect the longevity of implants depend mainly on the type of prosthetic materials, type of implant fixation, socio-demographic factors, etc. Early failure that is less than two years after replacement surgery is related mostly to the operational failure, which causes infection and implant instability (Sharkey et al., 2002).

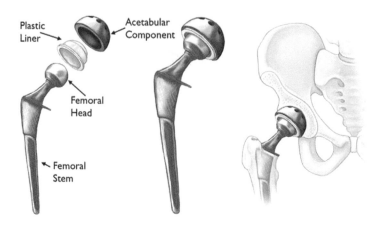

FIGURE 20.2   Total hip replacement. (https://orthoinfo.aaos.org/en/treatment/total-hip-replacement/.)

Meanwhile, damage or material deterioration is associated with long-term implant survival. According to the retrieval study by Wroblewski et al. (2002), the major cause for hip revision surgery under the age of 51-year-old patients was an aseptic loosening of the prosthesis, which is caused by osteolysis. Osteolysis or bone loss is a complex disease that occurs as a result of wearing out of the components in joint replacement called particulate debris. Over years of use, the particulate debris can irritate the tissues around the implant and begin to cause weakening of bone and may ultimately contribute to the failure of joint replacement.

20.2   UHMWPE AS BEARING COMPONENTS IN TJR

In the 1950s, when the first polymeric implant material (PTFE) was finally introduced in a clinical study, it was found that it has a very high wear rate, 0.5 mm per month (Oosthuizen and Snyckers, 2013). The voluminous mass released to the vast number of foreign-body giant cells (Oosthuizen and Snyckers, 2013) has caused an intense foreign-body reaction. Charnley verified it by injecting PTFE fine particulates into the thigh (Oosthuizen and Snyckers, 2013). The low cohesive strength of PTFE sliding with high cohesive strength of metal has caused high shear stress on PTFE's subsurface and causes subsurface crack (Myshkin and Kovalev, 2018). The subsurface crack was then propagating and forming delamination, which eventually removed and classified depending on its size as wear debris or wears particle production. Since the PTFE implant failure caused by loosening has increased by 57% higher than any other materials employed, the usage of PTFE as bearing materials was stopped before the 1960s.

As Dr Charnley is still hoping for the best outcome from PTFE, a modification was made by reinforcing fillers into the polymer matrix known as Fluorosint™ (Kurtz, 2015). It is expected that the filler could attribute in impeding subsurface cracks and increasing the bulk strength of the polymer. However, despite its excellent behaviour in in vitro test, Fluorosint™ has shown a weak performance in in vivo test.

It was due to the development of sticky surface, which could easily be worn away during the articulating activities.

In 1962, a new polymer, UHMWPE, was introduced as the bearing component in THR and has still been used ever since. UHMWPE has become a standard biomaterial for the loaded polymeric bearing component under the requirement of ISO 5834-5:2005 (Maksimkin et al., 2012). By the early 1990s, there were over ten different medical-grade UHMWPEs used for orthopaedic applications. These UHMWPE grades can be differentiated by their molecular weight, calcium stearate addition, method of sterilization and packaging (Li, 2001; Del Prever et al., 2009).

UHMWPE is a semicrystalline polymer; the microstructure consists of amorphous and crystalline regions. In the amorphous phase, the molecules were unorganized and highly flexible due to the random ordering of polymer chains (Hussain et al., 2020). The crystalline regions are composed of unit cells called lamellae. These can be further organized to form large spherical crystals (spherical semicrystalline regions) termed spherulites. The chain-folded lamellar spherulites of polyethylene consist of 150 CH2 groups in an orthorhombic shape arrangement. Lamellar spherulites are connected by tie molecules surrounding the crystalline regions. The tie molecules act as cross-links to promote a change and modify the polyethylene properties. Besides the cross-links, the mechanical properties of polyethylene depend on crystallinity, the number of tie molecules, amount and nature of chain entanglements, and chain orientation. The qualities that make UHMWPE suitable for implant applications are (Santavirta et al., 1998; Sobieraj and Rimnac, 2009):

- low friction coefficient
- high abrasion resistance
- high impact strength
- high ductility and biocompatibility
- high chemical resistance in vivo
- hydrophobic and resistant to aggressive media

In particular, UHMWPE possesses two essential properties that made it suitable for clinical use in orthopaedic. Firstly, UHMWPE exhibits tensile properties that are close to the human bone (Kurtz, 2015). Secondly, UHMWPE has the lowest friction compared to other polymers that contribute to the low wear degradation of the implant surface and reduce the risk of harm to the patient. Although the present UHMWPE wears out considerably less, the wear and damage of the UHMWPE components is one that limits the life span of the UHMWPE from 15 to 20 years (Shahemi et al., 2018). Therefore, some modifications are needed to improve UHMWPE performance.

20.3   WEAR OF UHMWPE

UHMWPE wear can be a significant factor limiting the long-term survivorship of the implant. Wear has been attributed to osteolysis and subsequent implant loosening. The polyethylene particles generated during the articulation are considered to be the most important inducers of osteolysis (Atkins et al., 2011). The wear debris formation

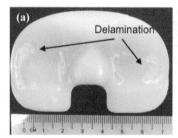
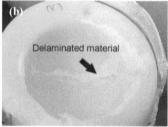

FIGURE 20.3 Worn surface of (a) UHMWPE tibial insert from TKR and (b) UHMWPE acetabular liner from THR (Burger et al., 2007.)

from UHMWPE during articulating activities is highly influenced in reducing the life expectancy of the implants. Wear will occur when two surfaces slide or roll against each other, and significant resistance to movement may arise due to frictional force. The frictional heating causes a temperature rise in sliding contact and then affects the wear resistance of the implant (Uddin and Majewski, 2013). Wear results in degradation of the implant surface and generally involves progressive loss of material. Figure 20.3 shows the example of surface damage on polyethylene components of knee and hip after revision surgery.

20.3.1 Wear Mechanism

Three primary wear mechanisms can occur in joint replacement, namely abrasive, adhesive and fatigue wear.

20.3.1.1 Abrasive Wear

Abrasive wear is the wear mechanism that occurs when hard rough material surface slides along the softer material under an applied load. This interaction results in the removal of softer material from the contact surface by asperities of the harder material surface during the sliding motion (Buford and Goswami, 2004). There are two standard basic modes of abrasive wear, namely two-body and three-body wear. Two-body wear results when abrasive particles slide on a soft surface, whereas in three-body wear, the abrasive particles are being embedded between two sliding surfaces. Loose bone cement, bone fragments or other particulate materials are examples of three-body abrasive wear, which can produce damage to bearing components (Bhatt and Goswami, 2008). Tan et al. (2020) reported that pitting and scratching on the UHMWPE insert are the result of hard particulates generated from the hard CrCoMo alloy-on-soft UHMWPE articulating surfaces. These particulates also caused excessive grooves formation and surface deformation of UHMWPE.

20.3.1.2 Adhesive Wear

In orthopaedic joint, adhesive wear usually occurs when two smooth bodies (polyethylene liners surface and metal femoral) are sliding against each other. Under the sliding motion, bond (junction) forms between contacting asperities and continues growing until they can support the applied load and developing adhesive bonds at the contact points. Then, when the applied force exceeds the adhesive strength, the junction bonds break. When the bond junction is broken, it creates wear particles from the weak material (polyethylene). The removal of polyethylene results in a small size of pits and voids on the bearing surface. The adhesive wears both UHMWPE acetabular and insert because the surface asperities undergo plastic deformation when it exceeds the transition limit (elastic limit). Liza et al. (2013) observed localized regions of plastic deformation on UHMWPE, which consist of shredded polyethylene layer and plastic flow of surface asperities.

20.3.1.3   Fatigue Wear

Fatigue wear occurs when surface contact between asperities is accompanied by repetitive local stresses or strains, which exceed the fatigue limit of that material. Propagated cracks generate wear particles due to fatigue. In polyethylene on the metal bearing, fatigue wear damage is dominant over the surface of polyethylene because of its weak fatigue strength than metal.

Under fatigue-loading conditions, both cracking and delamination at the subsurface level are present on the surface of polyethylene and contribute to orthopaedic joint failure by generating wear particles in a submicron scale (Buford and Goswami, 2004). Medel et al. (2011) reported that the fatigue was the predominant wear mechanism in in vivo degradation of polyethylene insert, and in vivo oxidation has been identified as one of the causes of surface fatigue damage. In this case, repetitive loading results in deterioration of mechanical properties of polyethylene such as loss of toughness and fatigue resistance (Kurtz, 2015).

20.3.2   Wear Process

Yamamoto et al. (2003) analysed the surface condition of the retrieved UHMWPE cups in THR. They reported that the polyethylene wear process consists of three steps. Firstly, the folding is generated when loading force is applied on the UHMWPE surface. Secondly, under sequential loading-shear force, ripples are formed. Thirdly, the formation of ripples decreases after exposure of UHMWPE cup to more severe wear conditions, which lead to fibril formation. Fibrils on the surface are generated by adhesion sites of UHMWPE and metal surface, which later torn off under articulating motion resulting in the creation of loose microwear particles (McKellop et al., 1995). The ripples and fibrils formation could also be seen on the wear scar of UHMWPE insert for TKR (Tamura et al., 2001).

20.3.3   Wear Features

Hood et al. (1983) classified articular surface damage in retrieved polyethylene components into seven wear damages modes, which are pitting, scratching, burnishing, embedded debris, abrasion, surface deformation and surface delamination. According to the previous analysis study, the standard wear features to be found on retrieved polyethylene components were pitting, scratching and delamination (Collier et al., 1996; Crowninshield et al., 2008; Diabb et al., 2009; Garcia et al., 2009; Hood et al., 1983; Liza et al., 2011). Pitting, scratching and delamination were classified

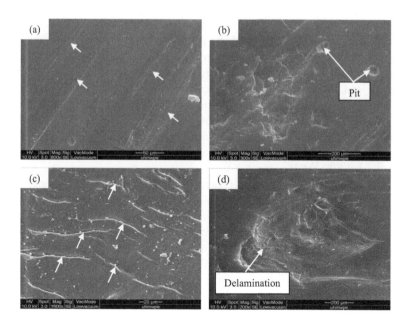

FIGURE 20.4   Wear damage features of retrieved UHMWPE tibial insert after 10 years of TKR: (a) scratches, (b) pits, (c) folding and (d) delamination (Liza et al., 2011.)

by Ho et al. (2007), as high-grade wear of retrieved polyethylene, while Wimmer et al. (2012) indicated that fatigue was the most common wear mechanism found on retrieved polyethylene, in which wear appearance formed was pitting and delamination. Figure 20.4 shows a combination of wear features in retrieved polyethylene after 10 years of implantation (Liza et al., 2011).

20.4   FACTORS INFLUENCING WEAR OF POLYETHYLENE

From the existing literature referenced in this chapter, multiple factors were found to influence wear behaviour and wear particles of UHMWPE. The factors known to affect polyethylene wear include the design of the implant, materials, processing method, surgical technique and patient factor.

20.4.1   DESIGN OF IMPLANT

Many researchers have reported design factors that influence the wear characteristics of polyethylene. These factors include the geometries of the implant component, surface topography contact stress, different type of articulation due to motion and body weight (Bhatt and Goswami, 2008; Goswami and Alhassan, 2008; Banchet et al., 2007). The following paragraph will discuss these factors in more detail.

20.4.1.1   Bearing Geometries

In TKR, less conforming and less constrained designs exhibit less wear and damage than the more conforming and more constrained (Abdelgaied et al., 2014).

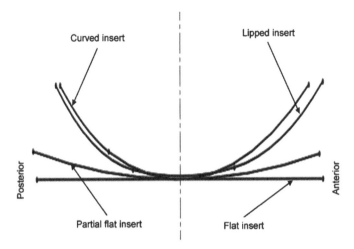

FIGURE 20.5 Schematic diagram for different polyethylene insert types with different conformities. (Abdelgaied et al., 2014.)

Figure 20.5 shows four different polyethylene insert designs, with varying levels of conformity: curved insert (most conformed) and flat insert (least conformed). More recently, Brockett et al. (2018) studied the effect of different insert designs: flat and lipped inserts on the wear performance of a fixed bearing TKR through a knee wear simulator study. This study highlighted the smallest wear scars observed on the flat inserts than the lipped inserts. The low wear of flat inserts is contributed by a reduction in the local contact area, and torque decreases as a less rotational motion at the insert–tray interface. These findings were supported by Galvin et al. (2009), and they proposed that the wear volume is associated with the contact area, thus suggesting that a low-conforming design reduces volumetric wear due to the lower contact area.

In THR, contact pressure and contact area for the cup can be differ depending on the cup geometry, abduction angle, thickness and gap between the cup and femoral head (Korhonen et al., 2005). For example, the cup abduction angle of more than 40° demonstrated high wear due to low contact area and high contact pressure in metal on polyethylene total hip articulations (Korduba et al., 2014). The large femoral head is often used in THR due to its design for hip stability and wide range of motion. However, Cho et al. (2016) suggested using femoral head sizes less than 40 mm to avoid fatigue fractures caused by oxidation during the implantation. Meanwhile, Daniel et al. (2016) demonstrated that the large femoral head (32 mm) could reduce the maximum contact pressure between the head and cup than the small femoral head (28 mm) and thus made it less susceptible to wear.

20.4.1.2 Surface Topography

The surface roughness of both metal and polyethylene is associated with the increased polyethylene wear in TJR. Numerous studies have consistently reported the effect of counterface roughness on the wear of conventional and cross-linked UHMWPE. Saikko et al. (2001) found that rough CoCr counterfaces (0.22–0.24

µm) produce larger polyethylene wear particles than polished CoCr counterfaces (0. 014–0.023 µm). Another study demonstrated the wear test using the hip simulator and linear reciprocating wear tester to examine the influence of CoCr femoral head surface roughness on the wear of UHMWPE. Increasing the surface roughness of the femoral head results in an increase in wear rate of UHMWPE, and the more substantial effect of counterface roughness is obtained by the reciprocating wear tester rather than from hip simulator (Wang et al., 1998). Marrs et al. (1999) studied the effect of CoCr metal roughness on the wear of acetylene-enhanced cross-linked UHMWPE using a multidirectional motion test. The results showed that the rough counterface, 0.09 µm, produces more wear than the smooth counterface, 0.01 µm. Previously, in 1992, Barrett et al. found optimum surface roughness (0.14–0.24 µm) where less wear can be found when stainless steel is tested against UHMWPE at low and medium sliding speeds (below 10 m/s). The effect of roughness is more dominant at the higher sliding speed, 10 m/s, and the increase of roughness can cause the abrasion of UHMWPE.

The surface roughness of polyethylene would change during sliding wear, as the surface roughness increased with the increased movement at a joint (Bahçe and Ender Emir, 2019). Copper et al. 1993 reported that polymer transfer films are responsible for the change in the wear resistance of UHMWPE with the surface roughness, Ra, of 0.02–0.04 µm. However, the use of protein-containing lubricants can effectively inhibit the transfer film to the articulating surface.

20.4.1.3   Contact Stress

Contact stress distribution at the bearing surfaces is affected by the design of the prosthesis and load applied and thus resulted in variation of wear. For hip prostheses, the size of the head and degree of conformity between the head and cup will affect the contact stress; for example, the smaller head diameter will increase the contact stress and reduce the wear rate. Besides, the contact stress distribution in UHMWPE can vary on the load applied, such as there was a small reduction in wear factor for the constant loads with increasing contact stress. The low wear is attributed to sufficient uniform stress distributions and high accumulation of plastic strain in the contact surface. On the other hand, the wear factor significantly increased by applying cyclic load over a constant UHMWPE contact area, and it increased more when the contact area was moving across the polymer surface. The contact stress in actual hip and knee prostheses varies during normal patient activities (stairs climbing, running or standing) (Barbour et al., 1995, 1997). For instance, in TKR, D'Lima et al. (2008) reported that walking is generating the lowest contact stresses (26 MPa), while stair climbing created about 32 MPa as the moderate contact stresses, and lunge activity was contributing the highest stresses (56 MPa).

20.4.1.4   Lubrication Condition

The biotribological system is more complicated. The durability of UHMWPE is not only relying on its mechanical and articulating activities. The reaction effect of macromolecules constituent in synovial fluid onto the UHMWPE bearing component should also be taken into account. Synovial fluid is a natural lubrication system and acting as a protective film that forms between joint surfaces. There are three major components of synovial fluid, including various types of proteins, several kinds of phospholipids and also hyaluronic acid. However, the interaction between the UHMWPE implant component and the synovial fluid's components which influenced the wear process is complicated.

Therefore, to understand an appropriate mechanism and reaction affecting the wear properties of UHMWPE, many studies were performed to reproduce the sliding activities under the physiological influences (Neogi and Wang, 2011; Rabe et al., 2011; Rahmati and Mozafari, 2018). As the joint is lubricated with synovial fluid, many investigations were carried out by modelling different types of synovial fluids as the lubricant for the sliding test. The international standard recommends using a synthetic synovial fluid; calf or bovine serum is pure or diluted with distilled or deionized water.

Several commercially available serum products contain protein concentration over a range of between 40 and 80 mg/ml (Rahmati and Mozafari, 2018). Some previous works suggested that the peroxide constituents of synovial fluid have caused an accelerated oxidative degradation of the UHMWPE (Murakami et al., 2007; Sawae et al., 2008). Meanwhile, Nečas et al. (2017) stated that the concentration of proteins had affected the wear behaviour of UHMWPE, which showed about two orders of wear rate increment when it was introduced in the lubricant. Moreover, Sawae et al. (2008) also indicated the effect of different protein types, and their interaction with the polyethylene surface contributes to accelerating the wear rate of the implant.

20.4.2 MATERIALS

There are two types of polyethylene that have been used as a bearing surface introduced by Charnley. High-density polyethylene (HDPE) was introduced first and now replaced by UHMWPE because of its poor performance due to resistance to wear. HDPE was reported to have 4.3 times wear rate compared with UHMWPE, which is significantly more abrasion resistance and wear resistance than HDPE (Kurtz, 2004). Different types of manufacturing processes produced by many forms of UHMWPE have recognized the potential that can affect the wear resistance of material such as heat pressing, polishing, gamma radiation, carbon-reinforced polymer chains and pressure crystallization (Bhatt and Goswami, 2008).

The structure of UHMWPE is a crucial factor that influences its friction, wear and mechanical properties. A higher degree of crystallinity results in higher hardness and elastic modulus as well as increases wear resistance of both conventional polyethylene and cross-linked polyethylene. The higher crystallinity of the polymer can be achieved by heating the polyethylene to a temperature above its melting point and by varying the hold time and cooling rate (Karuppiah et al., 2008; Affatato et al., 2008).

20.4.3 PROCESSING METHOD

20.4.3.1 Cross-Linking

The wear resistance can be achieved by increasing the cross-linking level of UHMWPE (Affatato et al., 2008; Lewis, 2001; Galvin et al., 2006; Kilgour and

Elfick, 2009). In previous studies, Lewis (2001) reviewed that there are reductions in wear rates with cross-linked UHMWPE. Furthermore, the study proposed four parameters used in the fabrication of cross-linked UHMWPE-bearing components that will affect the properties of cross-linked UHMWPE. The four parameters are starting type of resin, consolidation method, cross-linking process and optimum process variable.

On the other hand, Galvin et al. (2006) reported that highly cross-linked UHMWPE (10MRad) had 73% and 60% reduction in wear compared with non-cross-linked UHMWPE under smooth counterface and scratched counterface, respectively. The influence of cross-linking of polyethylene also was studied by Kilgour and Elfick (2009). 100kGy is an effective dose for cross-linked UHMWPE, as it had a 92% lower wear rate compared with non-cross-linked UHMWPE. The cross-linking method increased C-C covalent bonds between the polymer chains, which restrict the polymer chain movement. Consequently, it reduces plastic deformation and wear of polyethylene.

20.4.3.2   Sterilization Method

UHMWPE orthopaedic components must undergo sterilization before clinical use to prevent contamination. A different method of sterilization was reported to have influenced the wear outcome of polyethylene cups. The wear rate of polyethylene sterilized with gamma radiation in the air was dissimilar to another polyethylene sterilized in the absence of air (Stea et al., 2006). Sterilization in the air in the presence of oxygen is particularly harmful because it can promote free radical oxidation. Oxidation of free radical leads to chain scission and causes oxidative degradation in polyethylene. As a result, the wear resistance of polyethylene is reduced. Effect of sterilization method was studied by Affatato et al. (2002), and gamma-irradiated PE GUR 1020 (radiation sterilization) wore about 1.5 fold faster than that of ethylene oxide (EtO)-sterilized PE GUR1020 (gas sterilization). However, the retrieval study of 20 EtO-sterilized tibial inserts shows that the oxidation indices were low with no sign of delamination, which indicates that the polyethylene inserts remained stable for up to 10 years in implantation (MacDonald et al., 2012).

20.4.3.3   Shelf Storage

In the orthopaedic area, storage time or shelf life is defined as a period between polyethylene production and replacement operation. The shelf-ageing storage conditions are where polyethylene components were exposed to several environments before implantation. Generally, polyethylene components with long-term shelf storage tend to oxidize and are susceptible to severe wear. The high level of oxidation in the aged polyethylene increases as the shelf life increases and consequently reduces the degree of cross-linking (Lee and Lee, 1999). Wear of two groups acetabular liner was compared (Weimin et al., 2009), as a function of shelf life, which are 1 and 4 years on a hip simulator. The results show that the group of four-year-old liners has a higher wear rate compared to the one-year-old liner, which is 36.3 mg/million cycles and 23.1 mg/million cycles, respectively. Thus, the extended shelf life reduces the wear resistance of polyethylene.

20.4.4 SURGICAL TECHNIQUE

To date, not many reports about the influence of surgical technique are available. An improved surgical procedure during joint replacement could result in the slow progress of wear. In vivo study by Gallo et al. (2010) found that variables related to surgical techniques such as cup positioning during THR can affect the dislocation rate and led to polyethylene wear. However, patient positioning during surgery is the primary factor that can determine the orientation of the acetabular cup.

20.4.5 PATIENT FACTOR

Patient factors such as age, body mass index and activity level were expected to play an essential role in polyethylene wear. A comparative study by Devane et al. (1999) reported the correlation between patient and wear rate of polyethylene in THR. Younger and more active patients are contributed to the higher polyethylene wear rate. The hip simulator studies have demonstrated that the wear behaviour can be significantly changed when applying various simulated activities such as normal walking, fast walking and jogging sequences at varying cycle speeds and loads. Fast walking contributes to higher wear of cross-linked polyethylene than jogging. This result indicates that jogging has a relatively small effect on polyethylene wear under short periods of increased load and speed (Bowsher and Shelton, 2001).

20.5 DEVELOPMENT OF POLYETHYLENE TO IMPROVE WEAR RESISTANCE

Considerable efforts have been made by manufacturers and researchers to improve the wear resistance of UHMWPE and substantially extend the longevity of hip and knee prosthesis, which reduce the frequent need for a revision procedure. It is perhaps attained by the improvement of its performances, to provide durable implants in young and active patients. It is possible to improve the wear resistance of UHMWPE, including developing a new type of UHMWPE, manufacturing technique, method of sterilization as well as surface modification. A further explanation for each method will be described below to fully understand how the modification can contribute to reducing the wear of UHMWPE material.

20.5.1 MODIFIED UHMWPE FOR TJR

Various modifications of UHMWPE have been studied, and many experiments have also been carried out to reduce the wear debris production of UHMWPE. Modification is done as the conventional UHMWPE has exposed the component to failure from the aseptic loosening, which accounts for nearly 30% of all performed revisions surgery (Fawsitt et al., 2019).

20.5.1.1 Hylamer

In the early 1990s, a new form of UHMWPE was introduced as an option to conventional polyethylene with high crystallinity to improve wear resistance. This type of polyethylene has different physical properties from conventional UHMWPE such as higher density, yield and ultimate strength (Li, 2001). The higher crystallinity was achieved by controlling the morphology of the polymer using high pressures, high temperatures and very slow cooling rates during the manufacturing process (Kurtz et al., 1999). To date, fewer works were reported on the average wear rate of Hylamer acetabular liners compared with conventional UHMWPE. However, a study by Huddleston et al. in 2010 found a similar mean linear wear rate between Hylamer (0.21 mm/year) and conventional liners (0.20 mm/year) from the clinical records but patients with Hylamer liner were found to have a risk of pelvic osteolysis. This is because the Hylamer produces smaller wear particles than conventional polyethylene, and the high amount of small size particles are expected to result in a more intense biological response.

20.5.1.2   Highly Cross-Linked UHMWPE

More recently, cross-linked polyethylene is preferred in joint replacement because of its resistance to wear better than conventional UHMWPE. Polyethylene can be cross-linked by peroxide cross-linking, radiation cross-linking and silane cross-linking, but most of the medical device manufacturers use only radiation cross-linking commercially (Lewis, 2001). Cross-linking by radiation is used to decrease the large-scale deformation ability of the polymer to increase wear resistance. Radiation (both gamma and electron beam) has been recently used to form highly cross-linked UHMWPE that has better adhesive and abrasive wear resistance than non-cross-linked UHMWPE (Santavirta et al., 1998; Kurtz et al., 2008; Affatato et al., 2005). The degree of cross-linking depends on the radiation dose, atmosphere and heat treatment (Rimnac and Kurtz, 2005). Current clinically available cross-linked UHMWPE is usually irradiated at doses ranging from 40 to 100 kGy.

Gamma-ray irradiation was introduced to allow cross-linking and increasing the mechanical properties of polyethylene. The gamma ray would trigger the long polymer's chain to break, which is also known as chain scission. The broken chains would tend to form bonding with the parallel backbone polymer chain to replace the missing bond. Thus, creating interconnecting links with the other polymers' backbone chain is known as cross-linking.

Cross-linking is beneficial in improving the polymer mechanical properties by restricting the polymer chain movement. This has proven to increase the wear resistivity of the polymer as well. However, free radicals that are the by-product of the chain scission enable the reaction with the molecules that existed in the environment, which frequently lead to an oxidation effect. As a consequence, wear resistance will be reduced but so does the toughness of the polymer. Post-irradiation thermal is usually used to minimize the occurrence of oxidation, and this process decreases the concentration of residual free radical by encouraging additional cross-linking (Affatato et al., 2005).

20.5.1.3   Vitamin E with UHMWPE

On the other hand, the vitamin E infusion method was also being studied by various researchers to reduce the oxidation effect in the polymer. The infusion method led to the interruption of the oxidation cycle by decreasing the reactivity of the radical species and retards the oxidation process (Affatato et al., 2018; Baena et al., 2015; Bracco et al., 2017). There are two types of methods for the vitamin E infusion. The first method is by diffusing vitamin E into UHMWPE after the irradiation cross-linking (Baena et al., 2015; Markut-Kohl et al., 2009).

This method will not affect the efficiency of the UHMWPE cross-linking as the irradiation does not occur when vitamin E is incorporated. However, this method has a two-step process that makes it more challenging to execute. Moreover, this method also exposes UHMWPE to oxidation during irradiation and in storage until vitamin E is incorporated (Baena et al., 2015). The second method of the vitamin E infusion is executed by blending a liquid antioxidant into the UHMWPE powder before the material fabrication. The radiation cross-linking will then be performed as the finishing process. Although the process is much easier than the first method, the presence of vitamin E during the irradiation process reduces the efficiency of UHMWPE cross-linking due to the scavenging effect of the infused antioxidant.

20.5.1.4 UHMWPE Composites

UHMWPE-based composite is known to improve not only their mechanical properties but also their tribological properties. Different types of fillers can be incorporated into polyethylene depending on their material's properties and compatibility in functioning on the properties enhancement or at a particular application. Since polyethylene-bearing component in TJR is a soft material, it is expected that relatively it would be affected the most, especially in resisting the shear stress from the hard materials (metal). However, the effective use of composites relies on the compatibility and dispersion of the filler throughout the matrix. The efficient interaction between the matrix and fillers can increase the load transfer capacity, which resulted in high mechanical properties. On the other hand, the extremely long chains of polyethylene have made the reinforcement process more difficult and led to filler agglomerations. The agglomerated fillers potentially act as localized stress concentration points (Ashraf et al., 2018; Garcia-Seisdedos et al., 2019) and cause crystal defect. Therefore, choosing the optimum filler type and suitable fabrication process can help in minimizing or eliminating the side effects of the reinforcing fillers on the UHMWPE's wear properties.

In the 1970s, a carbon-reinforced UHMWPE, known as Poly II™, was created for orthopaedic bearing materials. The incorporation of carbon fibres did improve the mechanical properties of UHMWPE. However, Poly II™ has displayed a catastrophic short-term TKR clinical failure, as many patients presented with osteolysis (Busanelli et al., 1996; Wright et al., 1988). Poor crack propagation resistance of Poly II, resulting from the poor interaction between carbon fibres and UHMWPE matrix, causes the failure after TKR (Connelly et al., 1984).

On the other hand, the use of natural biocompatible particles such as natural coral (NC) can be added as a reinforcement filler into UHMWPE to increase the wear resistance of UHMWPE composite. The enhancement of microhardness and scratching resistance were contributed by the increased content of NC particles as well as it was noted on the variation of the size distribution of wear debris (Ge et al., 2009). Another study was done by Plumlee and Schwartz (2009), and the result showed a significant reduction in wear by UHMWPE reinforced with up to 20 wt. % of micro-sized zirconium particle when compared to unfilled UHMWPE without scarifying impact toughness. Instead of having excellent corrosion resistance and biocompat-ibility, zirconia particles also may have a severe effect on the wear of metal counter-face, and it needs further investigation.

However, modifying UHMWPE by filling nano-hydroxyapatite (n-HA) together with gamma radiation has a different result from other types of UHMWPE compos-ite. There are not only enhancements of tribological properties but also changes in the contact angle and hardness of UHMWPE-nHA composite. There is only abrasive wear of N-HA material that was characterized at worn surface differing from the unfilled material which has both abrasive wear and adhesive wear.

20.5.2 Manufacturing Technique

Currently, polyethylene tibial and acetabular components are manufactured either by machining or by direct compression moulding (DCM). The literature suggests that the DCM process can produce better articular-sided wear characteristics (smooth sur-face finish) of compression-moulded inserts than machined components (Currier et al., 2000). However, in 2007, Lancin et al. studied the wear performance of machined and moulded tibial inserts consolidated from the same resin (GUR 1020 and Basell 1900, respectively) under knee simulator testing. They found similar surface wear character-istics of machined and moulded tibial inserts, which suggest that the fabrication method does not play an important role in controlling the wear behaviour of UHMWPE.

Meanwhile, pressure, temperature and time are the most influential parameters in the consolidation process for UHMWPE, which can adversely affect the mechanical performance of the entire UHMWPE implants. For example, Wang and Ge (2007) found that the pressure in the compression moulding has a significant influence on the mechanical properties and tribological behaviour of UHMWPE sample. The UHMWPE samples moulded at 15 MPa indicate the good mechanical properties and higher wear resistance when compared to others that designate compression pressure which is 10 and 20 MPa. Besides, compression pressure affected the microstruc-tures of UHMWPE. For the UHMWPE samples moulded at 10 MPa, results in grain defects decrease the bonding strength between grain boundaries and at 20 MPa, exhibit stress-induced orientation microstructure in UHMWPE.

20.5.3 Sterilization Method

There are a number of sterilization treatments that can be employed, including gamma radiation, non-irradiation-based methods such as EtO and gas plasma. In orthopaedic, the most common sterilization is gamma radiation. Gas sterilization method, EtO and gas plasma do not generate cross-linking, which are beneficial to wear (Affatato et al., 2002). The higher dosage of gamma radiation during steriliza-tion is believed to work well for reducing wear (Hamilton and Hopper 2005). In gen-eral, UHMWPE orthopaedic components are sterilized by the manufacturer from 25 to 40 kGy of 60Co gamma rays.

Oxidation degradation can occur on the sterilized polyethylene, especially when performing ionizing radiation in the presence of oxygen. In order to mitigate oxidative degradation, several new methods have been developed, and one of them is gamma sterilization in the inert environment during which argon, nitrogen, vacuum, etc., were used to minimize oxidation during sterilization (Fisher et al., 2004). These methods are very effective as they can sustain the advantage of the low level of oxidation and high level of cross-linking during gamma irradiation sterilization, but they are ineffective in preventing in vivo oxidative degradation in polyethylene.

20.5.4　Surface Modification

Other than that, another attempt of UHMWPE wear properties improvement is by surface modifications which is likely performed to improve the lubrication condition and to limit wear – several techniques that involved including surface treatment, and surface texturing by different sources. There are many different surface modifications that have been reported to improve the wear reduction benefitted from the geometrical aspect, and lubrication effect from it (Baena et al., 2015).

One of the most widely used techniques in surface treatment is known as plasma treatment. Plasma was used to lightly cross-link the subsurface of UHMWPE and graft or deposit CFx groups on the surface. Klapperich et al. in 2011 used Ar/CsFg. The cross-linked surface was expected to inhibit the alignment of the crystalline regions along the sliding direction and enhance the wear resistance, while the low-surface energy CFx groups were expected to reduce the surface adhesion (friction). This treatment was intended for TKRs to avoid fully cross-linking the bulk UHMWPE, which lowers the resistance to crack propagation, making the TKRs subject to fatigue and delamination failures. The results indicated that the coefficient of friction was only marginally affected by the plasma treatment, but a significantly improved wear resistance was observed with the appropriately tailored treatments. It can be observed by changing some surface properties such as surface energy and surface chemistry through surface modification, which helps a lot in reducing the wear effect of UHMWPE. This is due to the surface energies of material, which are said to play an important role in frictional in terms of in vivo environment. Widmer et al. (2001) reported oxygen-plasma treated surface demonstrated low wear and friction, which attributed by the adsorbed protein layer formed on surface contact and potentially improve boundary lubrication.

Surface texturing also has been considered as a feasible way to improve the wear properties of UHMWPE. The microcavities (also known as dimples) formed on the surface act as lubricant reservoirs that build up a load-carrying hydrodynamic pressure that separates the surfaces of the counterparts and reduces the bearing surface contact. Moreover, dimples trap wear debris to minimize further abrasive wear on the bearing surfaces (Ippolito et al., 2017; Zhang et al., 2012, 2013, 2015). Hence, many studies on improving the wear resistance of UHMWPE by developing texture or pattern on the surface of UHMWPE had been researched. Surface texturing has first been used to improve the wear performance of implant in the study of Clarke (1971). Clarke has stated that the surface of the natural joint was not smooth in the last decade and studies have shown that the textured surface has better tribological properties compared to an untextured surface (Zhang et al., 2013). Various techniques have been used to fabricate surface texture (Ippolito et al., 2017; Kustandi et al., 2010). The effect of surface texturing on improving the tribological properties depends on the dimple's shape, dimples area density, dimples depth and pattern of dimples. There are many types of dimples pattern that have been fabricated and studied, such as nano-wells textured, mesh texture, microgroove texture and microdimple texture. According to the study of Zhang et al. (2015), the numerical results showed that the geometrical characteristics of dimples such as diameter size, area density and depth are important factors that will affect the load-carrying capacity of the dimpled texture (Zhang et al., 2015). Dimple textured surface can improve the wear resistance during a sliding motion due to the ability to trap wear particles where the dimple diameter size and area density play an important role in it.

20.6   CONCLUSIONS

In summary, it is important to understand the wear mechanism and the factor influencing wear as an attempt to reduce wear in TJR. Fatigue wear is the most predominant mechanism in contributing to surface failures in the UHMWPE bearing component. Therefore, the significant findings that acquire from this chapter would be meant to enhance the durability and improve the UHMWPE life span to reduce the frequent need for revision procedure. It is perhaps attained by the improvement of wear properties, expanding knowledge on the new generation of UHMWPE, namely irradiated vitamin E-doped, highly cross-linked, composite UHMWPE.

ACKNOWLEDGEMENTS

The research study was not funded by any grants.

REFERENCES

Abdelgaied, A., Brockett, C.L., Liu, F., Jennings, L.M., Jin, Z., & Fisher, J., 2014. The effect of insert conformity and material on total knee replacement wear. *Proceedings of the Institution of Mechanical Engineers, Part H*, 228(1), pp.98–106. Available at: Doi: 10.1177/0954411913513251.

Affatato, S., Bersaglia, G., & Rocchi, M., 2005. Wear behavior of crosslinked polyethylene assessed in vitro under severe conditions. *Biomaterials*, 26, pp.3259–3267. Available at: Doi: 10.1016/j.biomaterials.2004.07.070.

Affatato, S., Bordini, B., Fagnano, C., Taddei, P., Tinti, A., & Toni, A., 2002. Effect of the sterilisation method on the wear of UHMWPE acetabular cups tested in a hip joint simulator. *Biomaterials*, 23(6), pp.1439–1446. Doi: 10.1016/S0142-9612(01)00265-4.

Affatato, S., Ruggiero, A., Jaber, S. A., Merola, M., & Bracco, P., 2018. Wear behaviours and oxidation effects on different uhmwpe acetabular cups using a hip joint simulator. *Materials*, 11(3), pp.1–13. Available at: Doi: 10.3390/ma11030433.

Affatato, S., Zavalloni, M., Taddei, P., Foggia, M.D., Fagnano, C. & Viceconti, M., 2008. Comperative study on the wear behavior of different conventional and crosslinked polyethylenes for total hip replacement. *Tribology International*, 41(8), pp. 813–822. Available at: Doi: 10.1016/j.triboint.2008.02.006.

Ashraf, M.A., Peng, W., Zare, Y., & Rhee, K.Y., 2018. effects of size and aggregation/agglomeration of nanoparticles on the interfacial/interphase properties and tensile strength of polymer nanocomposites. *Nanoscale Research Letters*, 13. pp.1–7. Available at: Doi: 10.1186/s11671-018-2624-0.

Atkins, G.J., Haynes, D.R., Howie, D.W., & Findlay, D.M., 2011. Role of polyethylene particles in peri-prosthetic osteolysis: A review. *World Journal of Orthopedics*, 2(10), pp.93–101. Available at: Doi: 10.5312/wjo.v2.i10.93.

Baena, J.C., Wu, J., & Peng, Z., 2015. Wear performance of UHMWPE and reinforced UHMWPE composites in arthroplasty applications: A review. *Lubricants*, 3(2), pp.413–436. Available at: Doi: 10.3390/lubricants3020413.

Bahçe, E., & Emir, E., 2019. Investigation of wear of ultra high molecular weight polyethylene in a soft tissue behaviour knee joint prosthesis wear test simulator. *Journal of Materials Research and Technology*, 8(5), pp.4642–4650. Available at: Doi: 10.1016/j.jmrt.2019.08.008.

Barbour, P.S.M., Barton D.C., & Fisher, J., 1995. The influence of contact stress on the wear of UHMWPE for total replacement hip prostheses, *Wear*, 181–183, pp.250–257. Available at: Doi: 10.1016/0043-1648(95)90031-4.

Barbour, P.S.M., Barton D.C., & Fisher, J., 1997.The influence of stress conditions on the wear of UHMWPE for total joint replacements. *Journal of Materials Science: Materials in Medicine*, 8(10), pp. 603–611. Available at: Doi: 10.1023/a:1018515318630.

Barrett, T.S., Stachowiak, G.W., & Batchelor, A.W., 1992. Effect of roughness and sliding speed on the wear and friction of ultra-high molecular weight polyethylene. *Wear*, 153 (2), pp.331–350. Available at: Doi: 10.1016/0043-1648(92)90174-7.

Banchet, V., Fridrici, V., Abry, J.C., & Kapsa, P., 2007. Wear and friction characterization of materials for hip prosthesis. *Wear*, 263(7–12), pp. 1066–1071. Available at: Doi: 10.1016/j.wear.2007.01.085.

Bhatt, H., & Goswami, T. 2008. Implant wear mechanisms-basic approach. *Biomedical Materials*, 3(4), pp.1–8. Available at: Doi: 10.1088/1748-6041/3/4/042001.

Bowsher, J.G., & Shelton, J.C., 2001. A hip simulator study of the influence of patient activity level on the wear of crosslinked polyethylene under smooth and roughened femoral conditions. *Wear*, 250(1–12), pp.167–179. Available at: Doi: 10.1016/S0043-1648(01)00619–6.

Bracco, P., Bellare, A., Bistolfi, A., & Affatato, S., 2017. Ultra-high molecular weight polyethylene: Influence of the chemical, physical and mechanical properties on thewear behavior. A review. *Materials*, 10(7), pp.1–22. Available at: Doi: 10.3390/ma10070791.

Brockett, C.L., Carbone, S., Fisher, J., & Jennings, L.M., 2018. Influence of conformity on the wear of total knee replacement: An experimental study. *Proceedings of the Institution of Mechanical Engineers, Part H: Journal of Engineering in Medicine*, 232(2), pp.127–134. Available at: Doi: 10.1177/0954411917746433.

Buford, A., & Goswami, T., 2004. Review of wear mechanism in hip implants: Paper I-General. *Material and Design*, 25(5), pp.385–393. Available at: Doi: 10.1016/j.matdes.2003.11.010.

Burger, N.D.L., de Vaal, P.L., & Meyer, J.P., 2007. Failure analysis on retrieved ultra-high molecular weight polyethylene (UHMWPE) acetabular cups. *Engineering Failure Analysis*, 14(7), pp.1329–1345. Available at: Doi: 10.1016/j.engfailanal.2006.11.005.

Busanelli, L., Squarzoni, S., Brizio, L., Tigani, D., & Sudanese, A., 1996. Wear in carbon fiber-reinforced polyethylene (poly-two) knee prostheses. *La Chirurgia Degli Organi Di Movimento,* 81(3), pp.263–267. Available at: https://pubmed.ncbi.nlm.nih.gov/9009408/.

Cho, M-R., Choi, W.K., and Kim, J.J., 2016. Current concepts of using large femoral heads in total hip arthroplasty. *Hip Pelvis* 28(3): pp.134–141. Available at: http://dx.doi.org/10.5371/hp.2016.28.3.134.

Clarke, I.C., 1971. Human articular surface contours and related surface depression frequency studies. *Annals of the Rheumatic Diseases*, 30(1), pp.15–23. Available at: Doi: 10.1136/ard.30.1.15.

Collier, J.P., Sperling, D.K., Currier, J.H., Sutula, L.C., Saum, K.A., & Mayor, M.B. 1996. Impact of gamma sterilization on clinical performace of polyethylene in the knee. *The Journal of Arthroplasty*, 11(4), pp.377–389. Available at: Doi: 10.1016/S0883-5403(96)80026-X.

Connelly, G.M,. Rimnac, C.M., Wright, T.M., Hertzberg, R.W., & Manson, J.A., 1984. Fatigue crack propagation behavior of ultrahigh molecular weight polyethylene. *Journal of Orthopaedic Research*, 2(2), pp.119–125. Available at: Doi: 10.1002/jor.1100020202.

Crawford, R.W., & Murray, D.W., 1997. Total hip replacement: Indications for surgery and risk factors for failure. *Annals of the Rheumatic Diseases*, 56(8), pp.455–457. Available at: Doi: 10.1136/ard.56.8.455.

Crowninshield, R.D, Wimmer, M.A., Jacobs, J.J., & Rosenberg, A.G., 2008. Clinical performance of contemporary tibial polyethylene components. *The Journal of Arthroplasty*, 21(5), pp.754–761. Available at: Doi: 10.1016/j.arth.2005.10.012.

Currier, B.H., Currier, J.H., Collier, J.P., Mayor M.B., 2000. Effect of fabrication method and resin type on performance of tibial bearings. *Journal of Biomedical Materials Research*, 53, pp.143–151. Available at: Doi: 10.1002/(sici)1097-4636(2000)53:2<143::aid-jbm3>3.0.co;2-5.

Daniel, M., Rijavec, B., Dolinar, D., Pokorný, D, Iglič, A., Kralj-Igli, V., 2016. Patient-specific hip geometry has greater effect on THA wear than femoral head size. *Journal of Biomechanics*, 49(16), pp.3996–4001. Available at: Doi: 10.1016/j.jbiomech.2016.10.030.

Devane, P.A., & Horne, J.G., 1999. Assessment of polyethylene wear in total hip replacement. *Clinical Orthopaedics and Related Research*, 369, pp.59–72. Available at: Doi: 10.1097/00003086-199912000-00007.

Diabb, J., Juarez-Hernandez, A., Reyes, A., Gonzalez-Rivera, C., Hernandez-Rodriguez, M.A. L. 2009. Failure analysis for degradation of a polyethylene knee prosthesis component. *Engineering Failure Analysis*, 16(5), pp.1770–1773. Available at: Doi: 10.1016/j.engfailanal.2009.02.007.

D'Lima, D.D., Steklov, N., Fregly, B.J., Banks, S.A., & Colwell, C.W. Jr., 2008. In vivo contact stresses during activities of daily living after knee arthroplasty. *Journal of Orthopaedic Research*, 26(12), 1549–1555. Available at: Doi: 10.1002/jor.20670.

Drummond, J., Tran, P. & Fary, C., 2015. Metal-on-metal hip arthroplasty: A review of adverse reactions and patient management. *Journal of Functional Biomaterials*, 6(3), pp.486–499. Available at: Doi: 10.3390/jfb6030486.

Fawsitt, C.G., Thom, H.H.Z., Hunt, L.P., Nemes, S., Blom, A.W., Welton, N.J., Hollingworth, W., López-López, J.A., Beswick, A.D., Burston, A., Rolfson, O., Garellick, G., Marques, E.M.R., 2019. Choice of prosthetic implant combinations in total hip replacement: Cost-effectiveness analysis using UK and Swedish hip joint registries data. *Value in Health*, 22(3), pp.303–312. Available at: Doi: 10.1016/j.jval.2018.08.013.

Fisher, J., McEwen, H.M.J., Barnett, P.I., Bell, C., Stone, M.H., & Ingham, E., 2004. Influences of sterilising techniques on polyethylene wear. *The Knee*, 11(3), pp.173–176. Available at: Doi: 10.1016/j.knee.2003.10.002.

Gallo, J., Havranek, V., & Zapletalova, J., 2010. Risk factors for accelerated polyethylene wear and osteolysis in ABG I total hip arthoplasty. *International Orthopaedics*, 34(1), pp.19–26. Available at: Doi: 10.1007/s00264-009-0731-3.

Galvin, A.L., Kang, L., Tipper, J., Stone, M., Ingham, E., Jin, Z., & Fisher, J., 2006. Wear of crosslinked polyethylene under different tribological conditions. *Journal of Materials Science: Materials in Medicine*, 17(3), pp.235–243. Available at: Doi: 10.1007/s10856-006-7309-z.

Galvin, A.L., Kang, L., Udofia, I., Jennings, L.M., McEwen, H.M.J., Jin, Z., & Fisher, J., 2009. Effect of conformity and contact stress on wear in fixed-bearing total knee prostheses. *Journal of Biomechanics*, 42(12), pp.1898–1902. Available at: Doi: 10.1016/j. jbiomech.2009.05.010.

Garcia, R.M., Kraay, M.J., Messerschmitt, P.J., Goldberg, V.M. & Rimnac, C.M., 2009. Analysis of retrieved ultra-high-molecular-weight polyethylene tibial components from rotating-platform total knee arthroplasty. *The Journal of Arthroplasty*, 24(1), pp.131–138. Available at: Doi: 10.1016/j.arth.2008.01.003.

Garcia-Seisdedos, H., Villegas, J.A., & Levy, E.D., 2019. infinite assembly of folded proteins in evolution, disease, and engineering. *Angewandte Chemie - International Edition*, 58(17), pp.5514–5531. Available at: Doi: 10.1002/anie.201806092.

Ge, S., Wang, S., & Huang, X., 2009. Increasing the wear resistance of UHMWPE acetabular cups by adding natural biocompatible particles. *Wear*, 267(5–8), pp.770–776. Available at: Doi: 10.1016/j.wear.2009.01.057.

Goswami, T., & Alhassan, S., 2008. Wear rate model for UHMWPE in total hip and knee arthroplasty. *Materials and Design*, 29, pp. 389–296, Available at: Doi: 10.1016/j. matdes.2007.02.015.

Hamilton, W.G. & Hopper, R.H., 2005. The effect of total hip arthroplasty cup design on polyethylene wear rate. *Journal of Arthroplasty*, 20(3), pp. 63–72. Available at: Doi: 10.1016/j.arth.2005.05.007.

Hills, B.A., & Buttler, B.D., 1984. Surfactants identified in synovial fluid and their ability to act as boundary lubricants. *Annals of the Rheumatic Diseases*, 43(4), pp.641–648. Available at: Doi: 10.1136/ard.43.4.641.

Ho, F.-Y., Ma, H.-M., Liau, J.-J., Yeh, C.-R., Huang, C.-H., 2007. Mobile-bearing knees reduce rotational asymmetric wear. *Clinical Orthopaedics and Related Research*, 462, pp. 143–149. Available at: Doi: 10.1097/BLO.0b013e31806dba05.

Hood, R.W., Wright, T.M., & Burstein, A.H., 1983. Retrieval analysis of total knee protheses: a method and its application to 48 total condylar protheses. *Journal of Biomedical Materials Research*, 17(5), pp.829–842. Available at: Doi: 10.1002/jbm.820170510.

Huddleston, J.I., Harris, A.H.S., Atienza, C.A., & Woolson, S.T., Hylamer vs conventional polyethylene in primary total hip arthoplasty: A long term case control study of wear rates and osteolysis. *Journal of Arthoplasty*, 25(2), pp.203–207. Available at: Doi: 10.1016/j.arth.2009.02.006.

Hussain, M., Naqvi, R.A., Abbas, N., Khan, S.M., Nawaz, S., Hussain, A., Zahra, N., & Khalid, M.W., 2020. Ultra-high-molecular-weight-polyethylene (UHMWPE) as a promising polymer material for biomedical applications: a concise review. *Polymers*, 12(2), 323, pp.1–28. Available at: Doi: 10.3390/polym12020323.

Ippolito, C., Yu, S., Lai, Y.J., & Bryant, T., 2017. Process parameter optimization for hot embossing uniformly textured UHMWPE surfaces for orthopedic bearings. *Procedia CIRP*, 65, pp.163–167. Available at: Doi: 10.1016/j.procir.2017.04.039.

Karuppiah, K.S.K., Bruck, A.L., Sundararajan, S., Wang, J., Lin, Z., Xu, Z-H., Li, X., 2008. Friction and wear behavior of ultra high molecular weight polyethylene as a function of polymer crystallinity. *Acta Biomaterilia*, 4, pp. 1401–1410. Available at: Doi: 10.1016/j. actbio.2008.02.022.

Kilgour, A., & Elfick, A., 2009. Influence of crosslinked polyethylene structure on wear of joint replacements. *Tribology International*, 42(11–12), pp.1582–1594. Available at: Doi: 10.1016/j.triboint.2008.11.011.

Klapperich, C.M., Komvopoulos, K., Pruitt, L., 2011. Plasma surface modification of medical-grade ultra-high molecular weight polyethylene for improved tribological properties. *MRS Online Proceedings Library*, 550, pp. 331–336. Available at: Doi: 10.1557/PROC–550-331.

Korduba, L.A., Essner, A., Pivec, R., Lancin, P., Mont, M.A., Wang, A., Delanois, R.E., 2014. Effect of acetabular cup abduction angle on wear of ultrahigh-molecular-weight polyethylene in hip simulator testing. *American Journal of Orthopedics*, 43(10), pp.466–471. Available at: https://pubmed.ncbi.nlm.nih.gov/25303445/.

Korhonen, R.K., Koistinen, A., Konttinen, Y.T., Santavirta, S.S., & Lappalainen, R., 2005. The effect of geometry and abduction angle on the stresses in cemented UHMWPE acetabular cups – finite element simulations and experimental tests. *BioMedical Engineering OnLine*, 4(32), pp.1–14. Available at: Doi: 10.1186/1475-925X-4-32.

Kustandi, T. S., Choo, J.H., Low, H.Y., Sinha, S.K., 2010. Texturing of UHMWPE surface via NIL for low friction and wear properties. *Journal of Physics D: Applied Physics*, 43(1), pp. 1–6. Available at: Doi: 10.1088/0022-3727/43/1/015301.

Kurtz, S.M., 2015. *UHMWPE Biomaterials Handbook: Ultra High Molecular Weight Polyethylene in Total Joint Replacement and Medical Devices*: Third Edition. Available at: Doi: 10.1016/C2013-0-16083-7.

Kurtz, S.M., Lau, E., Ong, K., Zhao, K., Kelly, M., & Bozic, K.J., 2009. Future young patient demand for primary and revision joint replacement: National projections from 2010 to 2030. *Clinical Orthopaedics and Related Research*, 467(10), pp.2606–2612. Available at: Doi: 10.1007/s11999-009-0834-6.

Kurtz, S.M., Medel, F.J. & Manley, M., 2008. Wear in highly crosslinked polyethylenes. *Current Orthopaedics*, 22(6), pp. 392–399. Available at: Doi: 10.1016/j.cuor.2008.10.011.

Kurtz, S.M., Muratoglu, O.K., Evans, M., & Edidin, A.A., 1999. Advances in the processing, sterilization and crosslinking of ultra-high molecular polyethylene for total joint arthroplasty. *Biomaterials*, 20, pp. 1659–1688. Available at: Doi: 10.1016/S0142-9612(99)00053-8.

Lancin, P., Essner, A., Yau, S.S., Wang, A.G., 2007. Wear performance of 1900 direct compression molded, 1020 direct compression molded, and 1020 sheet compression molded UHMWPE under knee simulator testing. *Wear*. 263, pp.1030–1033. Available at: Doi: 10.1016/j.wear.2007.01.120.

Lee, K.-Y., & Lee, K.H., 1999. Wear of shelf-aged UHMWPE acetabular liners. *Wear*, 225–229(2), pp. 728–733. Available at: Doi: 10.1016/S0043-1648(99)00027-7.

Lewis, G., 2001. Properties of crosslinked ultra-high-molecular-weight polyethylene. *Biomaterials*, 22(4), pp.371–401. Doi: 10.1016/s0142-9612(00)00195-2.

Li, S., 2001. Ultra high molecular weight polyethylene: From charnley to cross-linked. *Operative Techniques in Orthopaedics*, 11(4), pp.288–295. Available at: Doi: 10.1016/S1048-6666(01)80044-6.

Liza, S., Haseeb, A.S.M.A. & Abbas, A.A., 2013. The wear behaviour of cross-linked UHMWPE under dry and bovine calf serum-lubricated conditions. *Tribology Transactions*, 56 (1), pp.130–140. Available at: Doi: 10.1080/10402004.2012.732199.

Liza, S., Haseeb, A.S.M.A., Abbas, A.A., & Masjuki, H.H., 2011. Failure analysis of retrieved UHMWPE tibial insert in total knee replacement. *Engineering Failure Analysis*, 18(6), pp.1415–1423. Available at: Doi: 10.1016/j.engfailanal.2011.04.00.

MacDonald, D., Hanzlik, J., Sharkey, P., Parvizi. J., & Kurtz, S.M., 2012. In vivo oxidation and surface damage in retrieved ethylene oxide-sterilized total knee arthroplasties. *Clinical Orthopaedics and Related Research*, 470, pp.1826–1833. Available at: Doi: 10.1007%2Fs11999-011-2184-4.

Maksimkin, A.V., Kaloshkin, S.D., Tcherdyntsev, V.V., Senatov, F.S., & Danilov, V.D., 2012. Structure and properties of ultra-high molecular weight polyethylene filled with disperse hydroxyapatite. *Inorganic Materials: Applied Research* 3(4), pp.288–295. Available at: Doi: 10.1134/S2075113312040132.

Markut-Kohl, R., Archodoulaki, V. M., Seidler, S., & Skrbensky, G., 2009. PE-UHMW in Hip implants: Properties of conventional and crosslinked prosthetic components. *Advanced Engineering Materials*, 11(10), pp.148–154. Available at: Doi: 10.1002/adem.200900050.

Marrs, H., Barton, D.C., Jones, R.A., Ward, I.M., Fisher, J., & Doyle, C., 1999. Comparative wear under four different tribological conditions of acetylene enhanced cross-linked ultra high molecular weight polyethylene. *Journal of Materials Science: Materials in Medicine*, 10, pp.333–342. Available at: Doi: 10.1023/a:1026469522868.

McKellop, H.A., Campbell, P., Park, S.H., Schmalzried, T.P., Grigoris, P., Amstutz, H.C., Sarmiento, A., 1995. The origin of submicron polyethylene wear debris in total hip arthroplasty. *Clinical Orthopaedics and Related Research*, 311, pp.3–20. Available at: https://pubmed.ncbi.nlm.nih.gov/7634588/.

Medel, F.J., Kurtz, S.M., Sharkey, P., Parvizi, J., Klein, G., Hartzband, M., Kraay, M., & Rimnac, C.M., 2011. In vivo oxidation contributes to delamination but not pitting in polyethylene components for total knee arthroplasty. *Journal of Arthroplasty*, 26(5), pp.802–810. Available at: Doi: 10.1016/j.arth.2010.07.010.

Murakami, T., Sawae, Y., Nakashima, K., Yarimitsu, S., & Sato, T. (2007). Micro- and nanoscopic biotribological behaviours in natural synovial joints and artificial joints. *Proceedings of the Institution of Mechanical Engineers, Part J: Journal of Engineering Tribology*, 221(3), 237–245. Doi: 10.1243/13506501JET245.

Myshkin, N., & Kovalev, A., 2018. Adhesion and surface forces in polymer tribology—A review. *Friction*, 6(2), pp.143–155. Available at: Doi: 10.1007/s40544-018-0203-0.

Nečas, D., Sawae, Y., Fujisawa, T., Nakashima, K., Morita, T., Yamaguchi, T., Vrbka, M., Křupka, I., & Hartl, M., 2017. The influence of proteins and speed on friction and adsorption of metal/uhmwpe contact pair. *Biotribology*, 11, pp.51–59. Available at: Doi: 10.1016/j.biotri.2017.03.003.

Neogi, P., & Wang, J. C., 2011. Stability of two-destimensional growth of a packed body of proteins on a solid surface. *Langmuir*, 27(9), pp.5347–5353. Available at: Doi: 10.1021/la104616b.

Oosthuizen, P., & Snyckers, C., 2013. Controversies around modern bearing surfaces in total joint replacement surgery. *SA Orthopaedic Journal*, 12(1), pp. 44–50. Available at: http://www.scielo.org.za/pdf/saoj/v12n1/09.pdf.

Plumlee, K., & Schwartz, C. J., 2009. Improved wear resistance of orthopaedic UHMWPE by reinforcement with zirconium particles. *Wear*, 267(5–8), pp.710–717. Available at: Doi: 10.1016/j.wear.2008.11.028.

Rabe, M., Verdes, D., & Seeger, S., 2011. Understanding protein adsorption phenomena at solid surfaces. *Advances in Colloid and Interface Science*, 162(1–2), pp.87–106. Available at: Doi: 10.1016/j.cis.2010.12.007.

Rahmati, M., & Mozafari, M., 2018. Protein adsorption on polymers. *Materials Today Communications*, 17, pp.527–540. Available at: Doi: 10.1016/j.mtcomm.2018.10.024.

Rimnac, C.M., & Kurtz, S.M., 2005. Ionizing radiation and orthopaedic prostheses. *Nuclear Instrument and Methods in Physics Research*, 236(1–4), pp. 30–37. Available at: Doi: 10.1016/j.nimb.2005.03.245.

Saikko, V., Calonius, O., & Keränen, J., 2001. Effect of counterface roughness on the wear of conventional and crosslinked ultrahigh molecular weight polyethylene studied with a multidirectional motion pin on disk device. *Journal of Biomedical Materials Research*, 57(4), pp.506–512. Available at: Doi: 10.1002/1097-4636(20011215)57:4<506::aid-jbm 1196>3.0.co;2-h.

Sampson, B., & Hart, A., 2012. Clinical usefulness of blood metal measurements to assess the failure of metal-on-metal hip implants. *Annals of Clinical Biochemistry*, 49, pp.118–131. Available at: Doi: 10.1258/acb.2011.011141.

Santavirta, S., Konttinen, Y., Lappalainen, R., Anttila, A., Goodman, S., Lind, M., Smith, L., Takagi, M., Gomez-Barrena, E. & Nordsletten, L., 1998. Materials in total joint replacement. *Current Orthopaedics*, 12(1), pp.51–57. Available at: Doi: 10.1016/S0268-0890 (98)90008-1.

Sawae, Y., Yamamoto, A., & Murakami, T., 2008. Influence of protein and lipid concentration of the test lubricant on the wear of ultra high molecular weight polyethylene. *Tribology International*, 41(7), pp.648–656. Available at: Doi: 10.1016/j.triboint.2007.11.010.

Shahemi, N., Liza, S., Abbas, A. A., & Merican, A. M., 2018. Long-term wear failure analysis of uhmwpe acetabular cup in total hip replacement. *Journal of the Mechanical Behavior of Biomedical Materials*, 87, pp.1–9. Available at: Doi: 10.1016/j.jmbbm.2018.07.017.

Sharkey, P.F., Hozack, W.J., Rothman, R.H., Shastri, S., Jacoby, S.M., 2002. Why Are Total Knee Arthroplasties Failing Today? *Clinical Orthopaedics and Related Research*, 404, pp.7–13. Available at: Doi: 10.1097/01.blo.0000036002.13841.32.

Sobieraj, M. & Rimnac, C., 2009. Ultra high molecular weight polyethylene: mechanics, morphology, and clinical behavior. *Journal of the Mechanical Behavior of Biomedical Materials*, 2(5), pp.433–443. Available at: Doi: 10.1016/j.jmbbm.2008.12.006.

Stea, S., Antonietti, B., Baruffaldi, F., Visentin, M., Bordini, B., Sudanese, A. & Toni, A., 2006. Behavior of Hylamer polyethylene in hip arthroplasty: Comparison of two gamma sterilization techniques. *International Orthopaedics*, 30(1), pp.35–38. Available at: Doi: 10.1007/s00264-005-0022-6.

Tamura, J., Clarke, I.C., Kawanabe, K., Akagi, M., Good, V.D., Williams, P.A., Masaoka, T., Schroeder, D., Oonishi, H., 2001. Micro-wear patterns on UHMWPE tibial inserts in total knee joint simulation. *Journal of Biomedical Materials Research* 61(2), pp.218–225. Available at: Doi: 10.1002/jbm.10027.

Tan, M.Y., Liza, S., Khadijah S.M.P., Abbas, A.A., Merican, A.M., Ayob, K.A., Zulkifli, N.W.M., Masjuki, H.H., 2020. Surface analysis of early retrieved polyethylene tibial inserts for both knees in total knee replacement. *Engineering Failure Analysis*, 109, p. Available at: Doi: 10.1016/j.engfailanal.2019.104279.

Uddin, M.S., & Majewski, P., 2013. Frictional heating in hip implants – a review, *Procedia Engineering*, 56, pp.725–730. Available at: Doi: 10.1016/j.proeng.2013.03.185.

Wang, S., & Ge, S., 2007. The mechanical property of tribological behavior of UHMWPE; Effect of molding pressure, *Wear*, 263(7–12), pp. 949–956. Available at: Doi: 10.1016/j.wear.2006.12.070.

Wang, A., Polineni, V.K., Stark, C., & Dumbleton, J.H., 1998. Effect of femoral head surface roughness on the wear of ultrahigh molecular weight polyethylene acetabular cups. *Journal of Arthroplasty*, 13(6), pp.615–620. Available at: Doi: 10.1016/s0883-5403 (98)80002-8.

Weimin, F., Huanghe, S., Xiang, L., Feng, L., & Qing, W., 2009. The impact of storage time on the wear rates of ultra high molecular weight polyethylene acetabular liners in hip simulators. *Journal of Arthoplasty*, 24 (4), pp.543–547. Available at: Doi: 10.1016/j.arth.2008.01.308.

Widmer, M.R., Heuberger, M., Vörös, J., & Spencer, N.D., 2001. Influence of polymer surface chemistry on frictional properties under protein-lubrication conditions: Implications for hip-implant design. *Tribology Letters*, 10, pp.111–116. Available at: Doi: 10.1023/A:1009074228662.

Wimmer, M.A., Laurent, M.P., Haman, J.D., Joshua, J. Jacobs, J.J., & Galante, J.O., 2012. Surface damage versus tibial polyethylene insert conformity: A retrieval study. 470(7), pp.1814–1825. Available at: Doi: 10.1007/s11999-012-2274-y.

Wright, T.M., Faris, P.M., & Bansal, M., 1988. Analysis of surface damage in retrieved carbon fiber-reinforced and plain polyethylene tibial components from posterior stabilized total knee replacements. *Journal of Bone and Joint Surgery*, 70(9), pp.1312–1319. Available at: https://pubmed.ncbi.nlm.nih.gov/3053722/.

Wroblewski, B.M., Siney, P.D., & Fleming, A., 2002. Charnley low-frictional torque arthroplasty in patients under the age of 51 years. Follow-up to 33 years. *Journal of Bone and Joint Surgery*, 84(4), pp.540–543. Available at: Doi: 10.1302/0301-620x.84b4.10293.

Yamamoto, K., Imakiire, A., Masaoka, T., Shishido, T., Mizoue, T., Clarke, I.C., Shoji, H., Kawanabe, K., & Tamura, J., 2003. Wear mode and wear mechanism of retrieved acetabular cups. *International Orthopaedics*, 27, pp.286–290. Available at: Doi: 10.1007/s00264-003-0477-2.

Zhang, B., Huang, W., Wang, X., 2012. Biomimetic surface design for ultrahigh molecular weight polyethylene to improve the tribological properties. *Proceedings of the Institution of Mechanical Engineers, Part J: Journal of Engineering Tribology*, 226 (8), pp.705–713. Available at: Doi: 10.1177/1350650112437829.

Zhang, B., Huang, W., Wang, J., Wang, X., 2013. Comparison of the effects of surface texture on the surfaces of steel and UHMWPE. *Tribology International*, 65, pp.138–145. Available at: Doi: 10.1016/j.triboint.2013.01.004.

Zhang, Y.L., Zhang, X.G., Matsoukas, G., 2015. Numerical study of surface texturing for improving tribological properties of ultra-high molecular weight polyethylene. *Biosurface and Biotribology*, 1(4), pp.270–277. Available at: Doi: 10.1016/j.bsbt.2015.11.003.

21 Tribological Review of Medical Implants Manufactured by Additive Manufacturing

Mohd Javaid
Jamia Millia Islamia

Suresh Babu
Indraprastha Apollo Hospital

Shanay Rab
Jamia Millia Islamia

Raju Vaishya
Indraprastha Apollo Hospital

Abid Haleem
Jamia Millia Islamia

CONTENTS

21.1   INTRODUCTION

Additive manufacturing (AM) provides better product design and development capability as per our requirements. This manufacturing platform consists of 3D scanners, 3D printing, designing, scanning and printing software. In this technology, materials are added layer by layer in the built platform with the input of the computer-aided design (CAD) digital model. AM is popular due to its capability of customization and lesser wastage. This technological platform quickly provides design modification and other significant technological changes (Javaid & Haleem, 2018; Panin et al., 2019).

In the medical field, there is an essential requirement of customization, which can be made possible at a lower cost by the application of AM. This technology has lesser usage of energy as compared to the traditional manufacturing process. The flexibility in design and manufacturing is easily possible by the implementation of this technology. The research is easily carried out for the design and development of products. This technology provides unlimited possibilities in the medical field by producing patient-specific implants at a lesser time and cost (Conner et al., 2014; Bustillos et al., 2018; Ju et al., 2020).

Many product development/ manufacturing companies are implementing additive manufacturing to reduce cost and time. For this, part accuracy and data clarity are required in the digital 3D model. Software like Catia and Solidworks are used to create a computer-aided 3D digital model. Software like mimics and 3D doctors are used to extract the 3D models from magnetic resonance (MR) scans and computed tomography scans and print prototypes of required implants, scaffolds and internal nerves using 3D printing technology (Bartolomeu et al., 2017; Haleem & Javaid, 2019a). In orthopaedics, missing bones and other hard tissue of the body are efficiently manufactured by the application of this technology. This opens new opportunities to sort out complex surgical problems. In medical education, AM is used for the better training of medical students (Khun et al., 2018; Van Eijnatten et al., 2018; Haleem & Javaid, 2019b).

Tribology is defined as the science and engineering of interacting surfaces in relative motion. Tribology is a highly interdisciplinary field, including physics, chemistry, materials science, mathematics, biology and engineering (Jin et al., 2016; Popov et al., 2018). Its study includes friction, wear and lubrication involved at moving contacts.

In this chapter, we discuss additive manufacturing technologies along with the tribological behaviour of medical implants. 3D printing technology can easily incorporate biomaterial aptly and produce a bone-like structure. The significant benefits of this technology are reduction in pre-planning costs and surgery time to improve patient safety. 3D printed models give an added perfectionism and increase surgeon efficiency during operation. Soon, doctors can easily create any required body part, especially related to medical sciences such as the exact map of cranial nerves, ankle bone repairs and the lumbar spine.

21.2   NEED FOR ADDITIVE MANUFACTURING

In ongoing patient treatment and surgery, there is a requirement to design and develop and customize medical parts in less time. So, the application of this technology is introduced in the medical field for customization of products because all patient data are not the same. It provides a flexible solution for the research and development path in the medical field (Salmi et al., 2013; Tan et al., 2020). This study analyses ongoing additive manufacturing capabilities in different fields. Different capabilities of AM in providing direction in development are studied. By the proper implementation of this technology, researchers observed that there is a lesser rejection rate, lesser waste generation, yet a better quality of design and manufacturing. Medical product innovation can be done without restrictions, which was not previously possible with traditional manufacturing technologies.

21.3   MAJOR STEPS FOLLOWED BY ADDITIVE MANUFACTURING

The step followed by additive manufacturing for the design and manufacturing of parts are as under:

- Designing of 3D CAD models using various designing software and other scanning technologies
- Editing/ post-processing of CAD data and changes as per requirement
- Conversion of 3D CAD model into Standard Triangulate Language format
- Required parts printing by using appropriate 3D printing technologies with appropriate material
- Removing printed part from the built tray
- Post-processing of the 3D printed part to increase accuracy and strength

21.4   HISTORY AND BACKGROUND OF TRIBOLOGY

Tribology in Greek means the science of rubbing surfaces and is defined as the science and technology of interacting surfaces in relative motion. Humankind has been aware and has used the elements of tribology for its advancement, like the cavemen used the friction between two stones to lite a fire and shape hunting gear. Similarly, Egyptians and Sumerians (3500–35BC) used water, bitumen, oil, and fat as lubricants; they discovered the use of a leather belt to reduce friction between the wheel and axle carriages (Dowson, 1975; Charnley, 1979; Jost, 1990). The first scientific approach to explain the interaction between interacting surfaces in recorded history was by Italian genius Leonardo Davinci, who explained the friction between sliding horizontal and inclined surfaces. He postulated that the frictional force between interacting surfaces depended on the normal loads and was independent of the surface area in contact. Frenchman Guillaume Amontons (1663–1705) later corroborated this observation . Amontons's observations along with his French countryman

Charles Augustin Coulomb's (1736–1806) theory on dry friction between solid bodies are commonly referred to as Coulomb's laws. Englishman Sir Isaac Newton's (1646–1727) contribution to the branch of fluid mechanics is significant.

Two centuries later, German Richard Stribeck (1861–1950) described adhesion, deformation and lubrication as the principal elements of friction. The British Department of Education and Science in 1966 constituted an interdisciplinary commission for the evaluation of friction, wear and lubrication and recommend tools to tackle them. It was headed by Jost and came to be known as the Jost Report. This commission defined the term 'tribology'. Sir John Charnley, through his low-friction hip arthroplasty designs, introduced the concept of tribology in the design and survivorship of total hip systems (Czichos, 1978; ZumGahr, 1987; Czichos & Habig, 2003).

21.5 TRIBOLOGICAL SYSTEM

Figure 21.1 shows the pictographic view of the tribological system. The whole tribological system is affected due to various influencing factors. The interface element is considered as lubricants, and these lubricants are a part of tribology, but in some cases, the lubrication can be built into material of tribo elements. The operational environment contains several types of input, such as type of motion, load, and velocity temperature.

21.6 COMPONENTS AND DYNAMICS OF A TRIBOSYSTEM

There is always a force between two interacting surfaces that keeps them either together or in relative motion: this force is known as friction. It is static or dynamic, respectively. Since we are interested in the interaction between the articulating surfaces of the artificial joint, henceforth, friction will mean dynamic friction, and all discussion will be in the context of surfaces in relative motion. Force is needed to initiate motion between two interacting surfaces, which should overcome the frictional

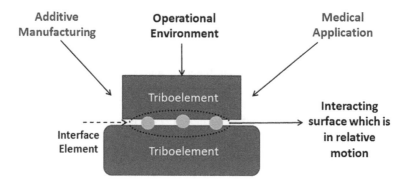

FIGURE 21.1 Pictographical view of a tribological system.

force. Thus, during this process, energy is dissipated in the form of heat from the system and results in the loss of material from the articulating surfaces. The introduction of an interfacial medium in the form of a lubricant between the interacting surfaces helps reduce the force required, minimizes wear and absorbs the heat generated (Berrien, 1999; Ashby et al., 2002; Bucholz et al., 2006; Callister, 2007).

Therefore, friction, wear and lubrication are interrelated and influence the functioning of the joint. This interaction occurs in an environment, and in the case of artificial joints, it is the human body. It constitutes a 'tribosystem' and consists of a body, a counter body, interfacial medium (lubricant) and the environment in which they operate. The operating inputs are variable loads, relative speed, loading time and ambient temperature and the outputs, motion and work. As there cannot be 100% efficiency, there will be losses in terms of wear. In joint arthroplasty, the wear of the articulating surfaces determines the joint's longevity, other factors notwithstanding. Wear of the articulating surfaces can cause mechanical failure by malalignment and subsequent loosening or, more commonly, by osteolysis secondary to tissue reaction to wear debris, which can be metallic, polyethylene or less commonly ceramic (Craig Jr, 2011; Statistics, 2010).

21.6.1 FRICTION

It is the force that resists motion between articulating surfaces. It occurs as there are no smooth surfaces of medical implants, and all surfaces have irregularities at the microscopic level called asperities, consisting of peaks and valleys. Friction is directly proportional to the load applied to the surface.

$$\text{Friction} \propto \text{Load}, \text{Friction} = \mu \times \text{Load}$$
$$\mu = \text{Friction/Load}$$

where μ is the co-efficient of friction and has no units.

It is to be noted that the 'coefficient of friction' depends on the smoothness of the articulating surfaces and is different for different bearing surfaces. Different interfaces have different coefficient of friction (COF). Metal-on-metal (MoM) articulations have different COF as compared to metal-on-ceramic (MoC), ceramic-on-ceramic (CoC), ceramic-on- polyethylene (CoP) or metal-on-polyethylene (MoP). Contact between the surfaces can be increased to reduce unit loads without affecting the friction, as in larger heads. Table 21.1 shows the coefficient of friction on different surfaces.

Linear motion between surfaces leads to friction, whereas angular motion or rotational motion causes frictional torque.

$$\text{Since Moment} = \text{Force} \times \text{Distance}$$

Frictional Torque = Fiction × Distance from the centre of motion, which is the radius

Therefore, increasing head diameters increase the frictional torque.

TABLE 21.1

Coefficient of Friction on Different Surfaces

SNo	Contacting Surfaces	Typical Coefficient of Friction
1	Metal on metal (dry)	0.41
2	Metal on metal (lubricated)	0.06
3	Teflon on Teflon	0.04
4	Ice on ice	0.03
5	Cartilage on cartilage	0.005

21.6.2 WEAR

It is the progressive loss of material from the articulating surfaces. It can be linear (loss of thickness) or volumetric, which is more representative. It is directly proportional to the applied loads and sliding distance between the surfaces.

$$\text{The volume of wear } \alpha \text{ Load } \times \text{Sliding distance}$$

$$\text{The volume of wear } = \kappa \times \text{Load} \times \text{sliding distance}, \; k \text{ is the co-efficient of wear}$$

The wearability of a bearing surface is given by k (coefficient of wear). As seen from the equation, it depends on the applied loads and sliding distance and is not the intrinsic property of a bearing surface.

It is to be noted that both friction and wear depend on the applied loads and, therefore, on the body weight, which can have a cumulative effect and is a function of time. It can be increased in obese patients, young patients with an increased level and a prolonged period of activity, high loading activities, and greater gait cycles.

Wear can result from different modes, and at any given point of time, the different modes may be acting simultaneously or sequentially. The different modes of wear are adhesive, abrasive, fatigue, erosive and corrosive. According to McKellop, wear in artificial joints can be of four modes. Mode-1 is a normal functioning joint, Mode-2 between bearing and non-bearing surfaces, Mode-3 in the presence of a third body and Mode-4 occurs between non-bearing surfaces as in a dysfunctional joint.

It is seen that corrosive wear acting in consonance with other types of wear can hasten wear rates acting in a vicious cycle. It is important to note that wear patterns leading to failure are different in hip and knee arthroplasties. In the hip, adhesive and abrasive wear are the main modes of wear, whereas in the knee, fatigue wear is the chief mode of failure.

21.6.3 TRIBOCORROSION AND TRIBOCHEMICAL REACTION

An artificial joint does not function in an inert environment but is bathed in the synovial fluid. Corrosion is a chemical phenomenon and not a wearing process. However, when the articulating surfaces are functioning in a chemically active environment with simultaneous chemical processes, wear can no longer be studied in isolation and should be evaluated through a wider window of 'tribocorrosion' (Hutchings & Shipwa, 2007; National Joint Registry, 2013; Sudeep et al., 2013).

In situations where the tribochemical reactions dominate, as in the human body, a tribofilm or tribochemical reaction layer is formed over the articulating surfaces and consists of a matrix of metallic and polyethylene nanoparticles and protein complexes from the interfacial medium (synovial fluid), and the composition of this layer varies broadly across patients (Hertz, 1982; Bartel et al., 1985).

Wear particles (especially, roughened surfaces and fibular shapes elicit stronger reaction) in the nanometre range are phagocytosed by macrophages in the periarticular space and elicit an inflammatory reaction by the release of Prostaglandin E_2, Interleukin6, tumour necrosis factor – alpha and Matrix Metalloproteins. These act to upregulate the functioning of the osteoclasts, which causes osteolysis at the bone-implant interface (Li et al., 1997; Zienkiewicz et al., 2013).

21.6.4 LUBRICATION

A lubricating medium is an interface that helps separate the articulating surfaces and, thereby, decreases wear rates. The lubricating property of any medium depends on its dynamic viscosity, which is the intrinsic property of the lubricating medium (Greenwood and Williamson, 1966; Mattei et al., 2013).

Lubricating mechanisms can be one of the three types depending on the 'distance between the asperities' expressed by the parameter λ, which is the ratio between the minimum fluid thickness and the combined height of the asperities' two opposing surfaces.

In Regime 1 ($\lambda > 3$), fluid film [hydrodynamic or elastohydrodynamic] lubrication exists, and there is complete separation of the asperities. If the fluid pressure can cause elastic deformation of the asperities, it is elastohydrodynamic; otherwise, if the fluid film is of sufficient thickness to cause the surfaces to float, it is called hydrodynamic lubrication. This is the ideal mode of lubrication in minimizing wear.

In Regimen 2 ($1 < \lambda < 3$), mixed film lubrication occurs, where the asperities are partly in contact, and the asperities and lubricant film share the load.

In Regimen 3 ($1 > \lambda$), boundary lubrication occurs, and the asperities are in contact with each other, leading to high friction and wear.

The different regimes are not mutually exclusive and can operate in the same tribosystem at different times and in different areas of the articulating surface. The mode of lubrication achieved is determined by the viscosity of the lubricant, speed of relative motion and loads. Sommerfeld's number summarizes this relationship, as shown in the equation.

$$S \propto \mu \upsilon / W$$

where S is the Sommerfeld number, μ is viscosity of the lubricant, u, the speed and W, the magnitude of the load. Viscosity, in the case of synovial fluid, depends on the concentration of hyaluronic Acid.

It can be derived from the above equation that as the viscosity of the synovial fluid and the sliding speed (diameter) increases, lubrication moves to a more favourable fluid film regime. Other factors responsible for fluid film lubrication are hard bearing surfaces, larger diameter heads and optimum radial clearance (the difference in diameter of the inner surface of the acetabular liner and that of the femoral head) that leads to polar articulation (Wang, 2001; Van Citters et al., 2013).

21.6.5 ANALYSIS OF WEAR

Quantifying wear in the real world is not always ideal, as a recreation of the actual tribosystem that exists in the body is hard to replicate. Nevertheless, estimation of wear is an important exercise in the quest to develop long-lasting joints and minimize wear rates (Willert, 1996; Williams et al., 2007; Virtanen et al., 2008).

Theoretical and physical methods can analyse wear. Theoretical methods are mechanical finite element analysis or chemical molecular dynamics computer simulation. Physical methods can be in vitro or in vivo studies. In vitro studies are carried out in physical laboratories using simulators' various models, whereas in vivo studies are done by radiological assessment, patient feedback and retrieval studies. There are several confounders in any method used, and interpretation of results should consider these variables before making broad assumptions (Utzschneider et al., 2009; Atwood et al., 2011; Teeter et al., 2012).

21.7 MATERIALS COMMONLY USED FOR MEDICAL IMPLANTS

The development of materials for implants in medicine is directly determined by the characteristics and nature of the tissue, organs and body systems that are being replaced. Mechanical integrity and wear resistance of an implant material play a critical role in joint replacement's long-term stability. The evaluation of the implant material's durability is carried out by tribological testing.

The measurement of friction force and the friction coefficient are of great importance for tribological behaviours of material in use, and for some, it is especially critical, like for human joint implants. Another major challenge is anticipating the type of wear to which components will be subjected according to the external loading behaviour. The low wear rate within the implant is the required characteristic of the material in use for hip and knee replacement systems. In the case of total hip replacement, wear debris released in the joint system produces highly unwanted consequences, such as osteolysis leading to lowered implant performance and further revision surgeries (Ferraris & Spriano, 2016; Ratner et al., 2020).

In general, all implant materials belong to one of the following three categories: metals, polymers and ceramics. The major concern in the medical implant is related to the aseptic loosening of the implants due to wear and tear. For this, a thorough knowledge of the tribological behaviour of the implant material is paramount. The tribological behaviour of the implant materials include a high wear resistance and a low friction coefficient when sliding occurs against body tissues. Generally, a high friction coefficient or a decrease in the wear resistance can cause the implant to loosen. Moreover, the wear debris generated can cause inflammation destructive to the bone supporting the implant (Distler & Boccaccini, 2020; Kuo et al., 2020).

The material selection must also take into account better biocompatibility along with the surrounding tissues and the tribological behaviours of the materials. Although many researchers/technologists/scientists have made noticeable contributions towards the advancement and development of various implant materials and their associated tribological behaviours, some of such important materials are compiled in a tabulated form, as shown in Table 21.2. Admittedly, the list of materials included is not comprehensive, but a short description would be highly useful for the readers to understand the extensively used materials in a concise form.

21.8   BIOTRIBOLOGY OF TOTAL HIP ARTHROPLASTY

The total hip arthroplasty (THA) is a true tribological system, where the femoral head is the body, the acetabular liner is the counter body and the synovial fluid is the interfacial medium. The system input is in the form of variable loads involved in daily activities, and the moment of friction, motion and generation of wear particles and metallic ions constitutes the output (Argenson and O'Connor, 1992; Thorwarth et al., 2010).

The wear mode in the THA is a combination of reciprocating and multi-directional sliding wear. Tribofilms are not significantly seen in polyethylene and ceramic bearings. The problem with MoM bearings is that they violate the basic tribological principle of 'not' mating identical materials of high ductility. Despite this discrepancy, it is ironic to note that the wear generated in MoM bearings is the lowest, bettered only by CoC bearings. Failures in MoM appear more a function of the large-size heads (Williams et al., 2004; Atwood et al., 2008).

Unique to the artificial hips is the second articulation at the taper junction between the femoral neck and the femoral head. The fretting, crevice and galvanic wear appear to be the chief mode of wear at the junction.

For a ball/cup system, a tribologist's most important task is to prevent catastrophic adhesive wear by using non-mating couples as in MoP or CoP. Tribological characteristics of articulating surfaces can be improved with the introduction of newer materials like Polyether-ether-ketone (PEEK) and new coupling of ceramic-on-metal. Another strategy is to improve the surface properties by surface coating with Diamond-like-Carbon (DLC) and Chromium nitride.

21.9   BIOTRIBOLOGY OF TOTAL KNEE ARTHROPLASTY

Wear in total knee arthroplasty (TKA) poses difficult problems since quantitative measurement of wear is difficult due to the absence of good reference data, design and system selection as important factors as are the materials selected for bearing surfaces.

The chief mode of wear in a TKA is the fatigue failure of the polyethylene bearing. Fatigue failure occurs in all types of gamma sterilized tibial inserts, regardless of design, material and fabrication (moulded or machined). The incidence of fatigue failure increases with time *in vivo* as the levels of oxidation increase with the number of cycles. Higher initial oxidation means fewer cycles before failure (Atwood et al.,

TABLE 21.2

Description of Commonly Used Materials in Medical Implant

Material	Application	Major Properties and Descriptions	References
Metals			
Titanium-based alloys	Artificial hip joints, artificial knee joints, bone plates, screws of fracture fixation, cardiac valve prostheses, pacemakers jaw bone, etc	High strength, good biocompatibility, excellent mechanical properties, etc	Brånemark et al. (1982), Morais et al. (2007), Elias et al. (2008), Sidambe (2014), Koike et al. (2011), Geetha et al. (2009)
Cobalt–chromium-based alloys	Artificial hip joint, knee joints artificial bone replacement, dental implants	Good wear resistance, good castability, excellent mechanical properties, etc	Arvidson et al. (1987), Kereiakes et al. (2003), Cheng et al. (2010), Hermawan et al. (2011)
Stainless steel (commonly 316L)	Fracture fixation, stents, surgical instruments etc	High wear resistance, higher creep, etc	Fellah et al. (2004), Yao et al. (2017)
Ceramics			
Zirconia	Dentistry implants, artificial bone fillers, hip and knee prostheses, hip joint heads temporary supports, tibial plates, and dental crowns, etc	High tensile strength, high hardness, corrosion resistance, good biocompatibility	Denry et al. (2008), Kollar et al. (2008)
Hydroxyapatite	Bone grafts, spinal fusion, bone repairs, bone fillers, maxillofacial reconstruction, etc	Excellent biocompatibility, osteogenic properties (interaction with osteoblasts/ osteoclasts), good bioactivity, etc	Petit (1999), Ioku et al. (2006), Okuda et al. (2008), Sinha et al. (2008)
Silicon nitride	Prosthetic hip and knee joints, dental implants, etc	High strength, high fracture resistance, biocompatible and stable *in vivo*, etc	Kue et al (1999), Neumann et al. (2004), Dante and Kajdas (2012), Websteret et al. (2012)
Silicon carbide	Bone prosthetics, dental implants, stents, membranes, orthopaedic implant, surface modification of biomaterials and tissue engineering, etc	Very high bond strength, high wear resistance, low thermal expansion coefficient, lightweight and porosity, etc	Carter (2000), Coletti (2007)
Bioglass	Dental prosthetics, bone substitution and tissue engineering, dental roots, etc	Strong interfacial bonding with the bone, mechanically strong bond, etc	Holand and Beall (2019), Rashwan et al. (2019)

(Continued)

TABLE 21.2 (*Continued*)
Description of Commonly Used Materials in Medical Implant

Material	Application	Major Properties and Descriptions	References
		Polymers	
Polyether Ether Ketone (PEEK)	Orthopaedics surgery, spinal implants, bone screws and pins, dental implants, etc	Excellent biocompatibility, good mechanical properties, low moisture absorption, high flexible	Lu and Friedrich (1995), Najeeb et al. (2016), Honigmann et al. (2018), Haleem et al. (2018)
Silicone polymers	Medical device lubrication excipients for topical formulations, ophthalmologic applications, fibrous capsule formation, intraarticular implants	Good biocompatibility good bio durability, hydrophobicity, chemical stability, and thermal stability, etc	Steiert et al. (2013), Pugliese et al. (2009)
Polyethylene	Implant for facial and cranial reconstruction, tissue in-growth	Good biocompatibility, non-toxic, non-antigenic	Onate et al. (2001), Affatato et al. (2016), Ardestani, Amenábar and Edwards (2017)

2006; Asano et al., 2007). The tibial inserts that were gamma sterilized and were on the shelf for a long time fail earlier. Even the gamma barrier-sterilized inserts eventually fail with increasing oxidation, and it is just a matter of time. Though the time to failure cannot be predicted, the threshold oxidation level following which the insert fails can be ascertained.

The design factors that influence the wear patterns in a TKA are

I. **Insert conformity** – Flat inserts show that the flat designs provide better tolerance and lesser wear rates than more conforming inserts because of the high stresses generated as it offers more freedom for torsional moments.

II. **Backside wear and tribal insert locking mechanisms** – In a posterior stabilized fixed bearing knee, the locking mechanism is insufficient in resisting the rotational torque transferred from the femur. Therefore, fixed bearings don't truly remain fixed with time; there is a relative motion between the tibial insert and tibial plate, causing abrasive wear and further loosening of the locking mechanism.

III. **Mobile bearings** – The wear with mobile bearings, as seen in rotating platform tibial trays, is lower owing to the more conforming kinematics.

IV. **Cement** – Wear rates are reported to be less in uncemented, polished trays and porous coated stems when compared to cemented, rough and uncoated stems.

The wear rate at least for the initial period is reported to be low with highly cross-linked polyethylene inserts than the conventional ultra-high-molecular-weight polyethylene (UHMWPE) owing to the increased strength of highly cross-linked polyethylene (HXP).

21.10 LIMITATIONS AND FUTURE SCOPE

There is a requirement for precise data either from an MRI or 3D scan of the patient's injury. The cost of 3D printing and software is high, which is not affordable for all. Skilled manpower is required for designing and printing required parts through additive manufacturing. Limited materials option is available in this technology, and the part is printed by powder material having lesser strength. There is an extra cost of 3D printed parts during pre-planning of the surgery.

In the future, AM will play an important role in developing various medical and industrial parts with better efficiency. This will become an emerging technology to solve complex surgical problems and to manufacture patient-specific medical implants and other medical devices. It provides new opportunities to improve patient outcomes with appropriate patient implants. Many scientists have been working on these processes to build up a strong uphold of the prevalence of AM in medical sciences, especially in surgical planning and prosthetics. With a wide variety of tasks performed by AM, it is clear that its contributions will disrupt the medical industry. The prototype of the patient's damaged part can be created by using AM processes, which can be useful for surgery and practices. This study provides great significance for the upcoming researchers and for future innovations.

The future of biotribology rests on developing more inert, strong and corrosion-resistant bearing surfaces. A more efficient system of in vitro simulators must be developed, and retrieval analysis of implants should cover a suitable number of retrieved joints. HXP and agents to prevent oxidation like Vit E incorporation in the polyethylene to reduce in vivo oxidation, DLC, and chromium nitride coating to increase the surface strength of bearing surfaces goes a long way in developing better and efficient tribosystems so that the goal of 'fifty at 2050' becomes a reality.

21.11 CONCLUSION

This technology is a take on designing a new product with better geometric freedom. The prototype manufactured by this technology can be successfully applied to check the product feasibility before starting a full production system. AM simplified the design process, which can be helpful to upgrade tools as per requirement. This plays a vital role in concept generation for various engineering parts. Nowadays, by the applications of this technology, it is feasible to convert first design ideas into physical products. AM reduces the design and manufacturing steps to increase the efficiency and reliability of the product. Many developments and improvements of existing materials have been introduced into clinical practice over the last few years. More stable materials have been investigated by many laboratories across the world with low reactivity to oxygen, better biocompatibility, lower density, increased durability, decreased reactions in human fluids environment, etc.

REFERENCES

Affatato, S., Freccero, N., & Taddei, P. (2016). The biomaterials challenge: A comparison of polyethylene wear using a hip joint simulator. *Journal of the Mechanical Behaviour of Biomedical Materials*, *53*, 40–48.

Ardestani, M. M., Edwards, P. P. A., & Wimmer, M. A. (2017). Prediction of polyethylene wear rates from gait biomechanics and implant positioning in total hip replacement. *Clinical Orthopaedics and Related Research®*, *475*(8), 2027–2042.

Argenson, J. N., & O'Connor, J. J. (1992). Polyethylene wear in meniscal knee replacement. A one to nine-year retrieval analysis of the Oxford knee. *The Journal of Bone and Joint Surgery. British Volume*, *74*(2), 228–232.

Arvidson K., Cottler-Fox M., Hammarlund E., & Friberg U. (1987). Cytotoxic effects of cobalt-chromium alloys on fibroblasts derived from human gingiva. *European Journal of Oral Sciences*, *95*(4), 356–363.

Asano, T., Akagi, M., Clarke, I. C., Masuda, S., Ishii, T., & Nakamura, T. (2007). Dose effects of cross-linking polyethylene for total knee arthroplasty on wear performance and mechanical properties. *Journal of Biomedical Materials Research Part B: Applied Biomaterials: An Official Journal of The Society for Biomaterials, The Japanese Society for Biomaterials, and The Australian Society for Biomaterials and the Korean Society for Biomaterials*, *83*(2), 615–622.

Ashby, M. F., Messler, R. W., Asthana, R., Furlani, E. P., Smallman, R. E., Ngan, A. H. W., … Mills, N. (2009). *Engineering Materials and Processes Desk Reference*. Oxford: Elsevier. 55–56.

Atwood, S. A., Currier, J. H., Mayor, M. B., Collier, J. P., Van Citters, D. W., & Kennedy, F. E. (2008). Clinical wear measurement on low contact stress rotating platform knee bearings. *The Journal of Arthroplasty*, *23*(3), 431–440.

Atwood, S. A., Kennedy, F. E., Currier, J. H., Van Citters, D. W., & Collier, J. P. (2006). *In vitro* study of backside wear mechanisms on mobile knee-bearing components. *Journal of Tribology - ASME Journal* 128, 275–281,

Atwood, S. A., Van Citters, D. W., Patten, E. W., Furmanski, J., Ries, M. D., & Pruitt, L. A. (2011). Tradeoffs amongst fatigue, wear, and oxidation resistance of cross-linked ultra-high molecular weight polyethylene. *Journal of the Mechanical Behavior of Biomedical Materials*, *4*(7), 1033–1045.

Bartel, D. L., Burstein, A. H., Toda, M. D., & Edwards, D. L. (1985). The effect of conformity and plastic thickness on contact stresses in metal-backed plastic implants. 107: 193–199.

Bartolomeu, F., Buciumeanu, M., Pinto, E., Alves, N., Carvalho, O., Silva, F. S., & Miranda, G. (2017). 316L stainless steel mechanical and tribological behavior—A comparison between selective laser melting, hot pressing and conventional casting. *Additive Manufacturing*, *16*, 81–89.

Berrien, L. S. J. (1999). *Biotribology: Studies of the Effects of Biochemical Environments on the Wear and Damage of Articular Cartilage* (Doctoral dissertation, Virginia Tech). Virginia Polytechnic Institute and State University.

Brånemark, P. I., Adell, R., Albrektsson, T., Lekholm, U., Lundkvist, S., & Rockler, B. (1983). Osseointegrated titanium fixtures in the treatment of edentulousness. *Biomaterials*, *4*(1), 25–28.

Bucholz, R.W., Heckman, J.D., & Court-Brown, C.M (eds) (2006). *Rockwood and Green's Fractures in Adults*. 6th edn. Philadelphia: Lippincott, Williams, and Wilkins.

Bustillos, J., Montero, D., Nautiyal, P., Loganathan, A., Boesl, B., & Agarwal, A. (2018). Integration of graphene in poly (lactic) acid by 3D printing to develop creep and wear-resistant hierarchical nanocomposites. *Polymer Composites*, *39*(11), 3877–3888.

Callister, W. D. (2007). *Material Science and Engineering: An Introduction*. 7th edn. New York: Wiley.

Carter, G. E., Casady, J. B., Bonds, J., Okhuysen, M. E., Scofield, J. D., & Saddow, S. E. (2000). Preliminary investigation of SiC on Silicon for biomedical applications. In *Materials Science Forum* (Vol. 338, pp. 1149–1152). Trans Tech Publications Ltd., Zurich-Uetikon, Switzerland.

Charnley, J. (1979). *Low Friction Arthroplasty of the Hip*, New York.

Cheng, H., Xu, M., Zhang, H., Wu, W., Zheng, M., & Li, X. (2010). Cyclic fatigue properties of cobalt-chromium alloy clasps for partial removable dental prostheses. *The Journal of Prosthetic Dentistry*, *104*(6), 389–396.

Coletti, C. (2007). *Silicon Carbide Biocompatibility, Surface Control and Electronic Cellular Interaction for Biosensing Applications*. Ph.D., Electrical Engineering, University of South Florida, Tampa, FL USA.

Conner, B. P., Manogharan, G. P., Martof, A. N., Rodomsky, L. M., Rodomsky, C. M., Jordan, D. C., & Limperos, J. W. (2014). Making sense of 3-D printing: Creating a map of additive manufacturing products and services. *Additive Manufacturing*, *1*, 64–76.

Craig Jr, R. R. (2011). *Mechanics of Materials*. 3rd ed. New York: John Wiley & Sons, 237–275.

Czichos, H. (2009). *Tribology: A Systems Approach to the Science and Technology of Friction, Lubrication, and Wear* (Vol. 1). Elsevier.

Czichos, H., &Habig K-H. (2003). *Triblogie-Handbuch*. 2nd ed. Munich: Carl Hanser Verlag.

Dante, R. C., & Kajdas, C. K. (2012). A review and a fundamental theory of silicon nitride tribochemistry. *Wear*, *288*, 27–38.

Denry, I., & Kelly, J. R. (2008). State of the art of zirconia for dental applications. *Dental Materials*, *24*(3), 299–307.

Distler, T., &Boccaccini, A. R. (2020). 3D printing of electrically conductive hydrogels for tissue engineering and biosensors–A review. *Acta Biomaterialia*, *101*, 1–13.

Dowson, D. (1975). *History of Tribology*. London: Longman;

Elias, C. N., Lima, J. H. C., Valiev, R., & Meyers, M. A. (2008). Biomedical applications of titanium and its alloys. *Jom*, *60*(3), 46–49.

Fellah, M., Labaïz, M., Assala, O., Iost, A., & Dekhil, L. (2013). Tribological behaviour of AISI 316L stainless steel for biomedical applications. *Tribology-Materials, Surfaces & Interfaces*, *7*(3), 135–149.

Ferraris, S., & Spriano, S. (2016). Antibacterial titanium surfaces for medical implants. *Materials Science and Engineering: C*, *61*, 965–978.

Geetha, M., Singh, A. K., Asokamani, R., & Gogia, A. K. (2009). Ti based biomaterials, the ultimate choice for orthopaedic implants–a review. *Progress in Materials Science*, *54*(3), 397–425.

Greenwood, J.A., & Williamson, J.B.P. (1966). Contact of nominally flat surfaces. *Proceedings, Mathematical, Physical, and Engineering Sciences, Proceedings of the Royal Society of London*, 295: 300–319.

Haleem, A., & Javaid, M. (2019a). 3D scanning applications in medical field: A literature-based review. *Clinical Epidemiology and Global Health*, *7*(2), 199–210.

Haleem, A., & Javaid, M. (2019b). Polyether ether ketone (PEEK) and its manufacturing of customised 3D printed dentistry parts using additive manufacturing. *Clinical Epidemiology and Global Health*, *7*(4), 654–660.

Haleem, A., Javaid, M., Vaish, A., & Vaishya, R. (2019). Three-dimensional-printed polyether ether ketone implants for orthopedics. *Indian Journal of Orthopaedics*, *53*, 377–379.

Hermawan, H., Ramdan, D., & Djuansjah, J. R. (2011). Metals for biomedical applications. *Biomedical Engineering-from Theory to Applications*, 411–430.

Hertz, H. R. (1882). *On Contact between Elastic Bodies*. Germany, Leipzig: Gesammelte Werke (Collected Works).

Holand, W., & Beall, G. H. (2019). *Glass-Ceramic Technology*. John Wiley & Sons, pp. 381–413.

Honigmann, P., Sharma, N., Okolo, B., Popp, U., Msallem, B., &Thieringer, F. M. (2018). Patient-specific surgical implants made of 3D printed PEEK: Material, technology, and scope of surgical application. *BioMed Research International*, *7*, 184–191.

Hutchings, I., & Shipwa, P. (2007). *Tribology*. 2nd ed. Butterworth-Heinemann Ltd.

Ioku, K., Kawachi, G., Sasaki, S., Fujimori, H., & Goto, S. (2006). Hydrothermal preparation of tailored hydroxyapatite. *Journal of Materials Science*, *41*(5), 1341–1344.

Javaid, M., & Haleem, A. (2018). Additive manufacturing applications in medical cases: A literature-based review. *Alexandria Journal of Medicine*, *54*(4), 411–422.

Jin, Z. M., Zheng, J., Li, W., & Zhou, Z. R. (2016). Tribology of medical devices. *Biosurface and Biotribology*, *2*(4), 173–192.

Jost, H. P. (1990). Tribology—origin and future. *Wear*, *136*(1), 1–17.

Ju, J., Zhou, Y., Wang, K., Liu, Y., Li, J., Kang, M., & Wang, J. (2020). Tribological investigation of additive manufacturing medical Ti6Al4V alloys against Al_2O_3 ceramic balls in artificial saliva. *Journal of the Mechanical Behavior of Biomedical Materials*, *104*, 103602.

Kereiakes, D. J., Cox, D. A., Hermiller, J. B., Midei, M. G., Bachinsky, W. B., Nukta, E. D., … Guidant Multi-Link Vision Stent Registry Investigators. (2003). Usefulness of a cobalt chromium coronary stent alloy. *The American Journal of Cardiology*, *92*(4), 463–466.

Khun, N. W., Toh, W. Q., Tan, X. P., Liu, E., & Tor, S. B. (2018). Tribological properties of three-dimensionally printed Ti–6Al–4V material via electron beam melting process tested against 100Cr6 steel without and with Hank's solution. *Journal of Tribology*, *140*(6).

Koike, M., Greer, P., Owen, K., Lilly, G., Murr, L. E., Gaytan, S. M., … Okabe, T. (2011). Evaluation of titanium alloys fabricated using rapid prototyping technologies—electron beam melting and laser beam melting. *Materials*, *4*(10), 1776–1792.

Kollar, A., Huber, S., Mericske, E., & Mericske-Stern, R. (2008). Zirconia for teeth and implants: A case series. *International Journal of Periodontics & Restorative Dentistry*, *28*(5), 479–87.

Kue, R., Sohrabi, A., Nagle, D., Frondoza, C., & Hungerford, D. (1999). Enhanced proliferation and osteocalcin production by human osteoblast-like MG63 cells on silicon nitride ceramic discs. *Biomaterials*, *20*(13), 1195–1201.

Kuo, C. N., Chua, C. K., Peng, P. C., Chen, Y. W., Sing, S. L., Huang, S., & Su, Y. L. (2020). Microstructure evolution and mechanical property response via 3D printing parameter development of Al–Sc alloy. *Virtual and Physical Prototyping*, *15*(1), 120–129.

Li, G., Sakamoto, M., & Chao, E. Y. (1997). A comparison of different methods in predicting static pressure distribution in articulating joints. *Journal of Biomechanics*, *30*(6), 635–638.

Lu, Z. P. & Friedrich, K. (1995). On sliding friction and wear of PEEK and its composites. *Wear*, *181*, 624–631.

Mattei, L. Di Puccio, F, & Ciulli, E. (2013). A comparative study of wear laws for soft-on-hard hip implants using a mathematical wear model. *Tribology International*, *63*, 66–77.

Morais, L. S., Serra, G. G., Muller, C. A., Andrade, L. R., Palermo, E. F., Elias, C. N., & Meyers, M. (2007). Titanium alloy mini-implants for orthodontic anchorage: Immediate loading and metal ion release. *Acta Biomaterialia*, *3*(3), 331–339.

Najeeb, S., Zafar, M. S., Khurshid, Z., & Siddiqui, F. (2016). Applications of polyetheretherketone (PEEK) in oral implantology and prosthodontics. *Journal of Prosthodontic Research*, *60*(1), 12–19.

National Joint Registry. (2013). 10th Annual Report 2013. Available from: https://www.njr-centre.org.uk.

Neumann, A., Reske, T., Held, M., Jahnke, K., Ragoss, C., & Maier, H. R. (2004). Comparative investigation of the biocompatibility of various silicon nitride ceramic qualities *in vitro*. *Journal of Materials Science: Materials in Medicine*, *15*(10), 1135–1140.

Okuda, T., Ioku, K., Yonezawa, I., Minagi, H., Gonda, Y., Kawachi, G., & Ikeda, T. (2008). The slow resorption with replacement by bone of a hydrothermally synthesised pure calcium-deficient hydroxyapatite. *Biomaterials, 29*(18), 2719–2728.

Onate, J. I., Comin, M., Braceras, I., Garcia, A., Viviente, J. L., Brizuela, M., & Alava, J. I. (2001). Wear reduction effect on ultra-high-molecular-weight polyethylene by application of hard coatings and ion implantation on cobalt-chromium alloy, as measured in a knee wear simulation machine. *Surface and Coatings Technology, 142*, 1056–1062.

Panin, S. V., Buslovich, D. G., Kornienko, L. A., Alexenko, V. O., Dontsov, Y. V. & Shil'ko, S. V. (2019). Structure, as well as the tribological and mechanical properties, of extrudable polymer-polymeric UHMWPE composites for 3D printing. *Journal of Friction and Wear, 40*(2), 107–115.

Petit, R. (1999). The use of hydroxyapatite in orthopaedic surgery: A ten-year review. *European Journal of Orthopaedic Surgery & Traumatology, 9*(2), 71–74.

Popov, V. V., Muller-Kamskii, G., Kovalevsky, A., Dzhenzhera, G., Strokin, E., Kolomiets, A., & Ramon, J. (2018). Design and 3D-printing of titanium bone implants: Brief review of approach and clinical cases. *Biomedical engineering letters, 8*(4), 337–344.

Pugliese, D., Bush, D., & Harrington, T. (2009). Silicone synovitis: Longer term outcome data and review of the literature. *JCR: Journal of Clinical Rheumatology, 15*(1), 8–11.

Rashwan, M., Cattell, M. J., & Hill, R. G. (2019). The effect of barium content on the crystallisation and microhardness of barium fluormica glass-ceramics. *Journal of the European Ceramic Society, 39*(7), 2559–2565.

Ratner, B. D., Hoffman, A. S., & McArthur, S. L. (2020). Physicochemical surface modification of materials used in medicine. In *Biomaterials Science* (pp. 487–505). Academic Press.

Salmi, M., Paloheimo, K. S., Tuomi, J., Wolff, J., & Mäkitie, A. (2013). Accuracy of medical models made by additive manufacturing (rapid manufacturing). *Journal of Cranio-Maxillofacial Surgery, 41*(7), 603–609.

Sidambe, A. T. (2014). Biocompatibility of advanced manufactured titanium implants—A review. *Materials, 7*(12), 8168–8188.

Sinha, A., Mishra, T., & Ravishankar, N. (2008). Polymer assisted hydroxyapatite microspheres suitable for biomedical application. *Journal of Materials Science: Materials in Medicine, 19*(5), 2009–2013.

Statistics NCHS. Number of all-listed procedures for discharges from short-stay hospitals, by procedure category and age: The United States, 2010. Available from: https://www.cdc.gov/nchs/data/nhds/4procedures/ 2010pro4_numberprocedureage.pdf.

Steiert, A. E., Boyce, M., & Sorg, H. (2013). Capsular contracture by silicone breast implants: Possible causes, biocompatibility, and prophylactic strategies. *Medical Devices (Auckland, NZ), 6*, 211–218.

Sudeep, I., Nosonovsky, M., Satish Vasu, K., Michael, R. L., Pradeep, L. M. (2013). *Tribology for Scientists and Engineers*. Springer-Verlag,

Tan, J. H. K., Sing, S. L., & Yeong, W. Y. (2020). Microstructure modelling for metallic additive manufacturing: A review. *Virtual and Physical Prototyping, 15*(1), 87–105.

Teeter, M. G., Brandt, J. M., Naudie, D. D., Bohm, E. R., McCalden, R. W., & Holdsworth, D. W. (2012). Measurements of surface and subsurface damage in retrieved polyethylene tibial inserts of a contemporary design. *Journal of Long-Term Effects of Medical Implants, 22*(1), 21–31,

Thorwarth, G., Falub, C. V., Müller, U., Weisse, B., Voisard, C., Tobler, M., & Hauert, R. (2010). Tribological behavior of DLC-coated articulating joint implants. *Acta Biomaterialia, 6*(6), 2335–2341.

Utzschneider, S., Harrasser, N., Schroeder, C., Mazoochian, F., & Jansson, V. (2009). Wear of contemporary total knee replacements–a knee simulator study of six current designs. *Clinical Biomechanics, 24*(7), 583–588.

Van Citters, D. W., Currier, J. H., Mayor, M. B., et al. (2013). 1000 Knee retrievals by the numbers: Important lessons learned; questions still to be answered. *Annual Meeting of the American Academy of Orthopaedic Surgeons*, Chicago.

Van Eijnatten, M., van Dijk, R., Dobbe, J., Streekstra, G., Koivisto, J., & Wolff, J. (2018). CT image segmentation methods for bone used in medical additive manufacturing. *Medical engineering & physics, 51*, 6–16.

Virtanen, S., Milošev, I., Gomez-Barrena, E., Trebše, R., Salo, J., & Konttinen, Y. T. (2008). Special modes of corrosion under physiological and simulated physiological conditions. *Acta Biomaterialia, 4*(3), 468–476.

Wang, A. (2001). A unified theory of wear for ultra-high molecular weight polyethylene in multi-directional sliding. *Wear, 248*(1–2), 38–47.

Webster, T. J., Patel, A. A., Rahaman, M. N., & Bal, B. S. (2012). Anti-infective and osteointegration properties of silicon nitride, poly (ether ether ketone), and titanium implants. *Acta biomaterialia, 8*(12), 4447–4454.

Willert, H. G., Brobäck, L. G., Buchhorn, G. H., Jensen, P. H., Köster, G., Lang, I.,... Schenk, R. (1996). Crevice corrosion of cemented titanium alloy stems in total hip replacements. *Clinical Orthopaedics and Related Research (1976–2007), 333*, 51–75.

Williams, S., Isaac, G., Hatto, P., Stone, M. H., Ingham, E., & Fisher, J. (2004). Comparative wear under different conditions of surface-engineered metal-on-metal bearings for total hip arthroplasty. *The Journal of Arthroplasty, 19*(8), 112–117.

Williams, S., Schepers, A., Isaac, G., Hardaker, C., Ingham, E., van der Jagt, D., & Fisher, J. (2007). The 2007 Otto Aufranc Award: Ceramic-on-metal hip arthroplasties: A comparative *in vitro* and *in vivo* study. *Clinical Orthopaedics and Related Research®, 465*, 23–32.

Yao, S. H., Su, Y. L. & Lai, Y. C. (2017). Antibacterial and tribological performance of carbonitride coatings doped with W, Ti, Zr, or Cr deposited on AISI 316L stainless steel. *Materials, 10*(10), 1189.

Zienkiewicz, O. C., Taylor, R. L., & Zhu, J. Z. (2005). *The Finite Element Method: Its Basis and Fundamentals*. Elsevier.

ZumGahr, K. H. (1987). *Microstructure and Wear of Materials* (Vol. 10). Elsevier.

22 Aqueous Lubrication

Manjesh Kumar Singh
Indian Institute of Technology Kanpur

CONTENTS

22.1 INTRODUCTION

Traditionally, all the lubricants are based on mineral oils as the base fluid. This is because of good technical properties and reasonable price of mineral oils. But, the poor biodegradability of mineral oils is a big disadvantage associated with it and hence may cause long-term pollution to the environment [1,2].

As per the Fourth Assessment Report 'Climate Change 2007' of Inter-governmental Panel on Climate Change (IPCC), it is very likely that human activities are mainly responsible for the global warming in the past 50 years. It has also been mentioned that global greenhouse gases (GHG) have increased since industrialization and a 70% increment is reported between 1970 and 2004 [3].

Tribology is very important for reducing the GHG emissions. A suitable lubricant reduces friction and wear. So it will reduce energy consumption and will have a direct impact on GHG emissions [2].

The main objective of using lubricants in the metalworking process is to improve its efficiency. Lubricants control friction, reduce wear of tool, improve the surface quality of the product and in addition to this they control the energy requirement for the process. Generally, various additives are added to the base oils to improve their performance. Oil and additives are also emulsified in water for many applications. Oil-in-water emulsion is used as lubricant for metalworking processes. Different lubricants are applied in different metalworking processes due to the varying tribological conditions [4].

However, the lubricants used for metalworking processes are not good from the environmental standpoint. The metalworking lubricants have a detrimental effect on the environment. Traditional lubricating oils are generally characterized by a low biodegradation rate [2,4].

Following are the main environmental challenges before metalworking tribology [5]:

a. health and well-being of workers/people,
b. effect on capital goods,
c. destruction and/or disposal of hazardous chemicals and other waste generated.

Improvement activities are directed towards

1. eliminating hazardous chemicals, like chlorinated additives or phosphates with (heavy) metal sludge, and
2. limiting waste generation, improving the durability of tools and lubricants life,
3. recycling lubricants and
4. minimal quantity lubrication.

22.2 OIL-IN-WATER EMULSIONS

An emulsion is a heterogeneous mixture of two or more liquids. The liquids are normally immiscible. The emulsions have at least one liquid dispersed in form of droplets in the continuous phase of other liquid. The droplets of the dispersed phase are of minimum 0.1 μm diameter [6]. Metalworking emulsions are mostly oil-in-water (O/W) emulsions. In O/W emulsions, oil-droplets form the dispersed phase and water is the continuous phase. Such emulsions are generally very unstable due to the difference in the density of the dispersed and continuous phases. Buoyancy causes segregation of the phases. To increase the stability: (1) emulsions (mixture of oil and water) are aggressively agitated to break the oil-droplets to smaller sizes and to also achieve better distribution of the oil-droplets in water, (2) surfactants are added [7].

Emulsifiers or surfactants generally have two distinct parts: (1) water-loving polar part, also called hydrophilic end and (2) oil-soluble lipophilic part, also referred to as

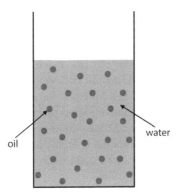

FIGURE 22.1   Oil-in-water emulsions.

hydrophobic end. When O/W emulsion is formed, emulsifiers get arranged in such a manner that the hydrophilic end orients towards the water phase and the hydrophobic end orients towards the oil phase. Such an arrangement stabilizes the interface between dispersed oil and continuous water phase in the emulsion (Figure 22.1).

Wilson et al. proposed the underlying mechanism for lubrication with O/W emulsion during metalworking. The inlet wedge formed between the slider (inclined or curved) and substrate can be divided into four well-defined zones: (1) supply zone, (2) concentration zone, (3) pressurization zone and (4) contact zone. When the O/W emulsion transits through these zones, it gets converted from oil-in-water emulsion to water-in-oil emulsion. Then, oil with very little water enters the contact zone to control the friction. Oil reduces friction and water acts as a coolant to take away the heat released as a result of plastic deformation of the metal workpieces [8].

22.3   WATER AS LUBRICANT

There are few specific applications where only water-based lubricants are used. For example, in the mining industry, oil-free water-based lubricants are used to lubricate pumps because the flammability of oil is dangerous in this application. This is, however, a relatively special application in the world of lubricated contacts, and besides this, oil-free, water-based lubricants are used very scarcely.

There are some properties of water which makes it a potential candidate as lubricants:

- Readily available
- Non-toxic
- Environmentally friendly
- High heat capacity which gives it excellent heat-transfer properties.

At the same time there are some properties of water which are disadvantageous:

- Low boiling point
- Low pressure-viscosity index which results in poor load carrying capacity.[9]

Researchers have been investigating water-based lubrication in recent times with an aim to substitute or reduce the use of oil in tribological applications. This will be useful in providing solutions to the environmental problems we are facing due to the use of oil in lubrication. Many industries like manufacturing, automotive, biomedical, food processing and ship-building are exploring the possibilities of using water as lubricants in their applications. In addition to the economic advantages, the use of water for lubrication will have a positive impact on environment conservation. There will be no need for the disposal and transportation of toxic polluting oils normally used for lubrication. Aqueous lubrication is highly desirable for a green and sustainable future. Higher resource allocations and research activities are need of the hour to develop this green technology further.

Polymer brushes – Polymers can be used for aqueous lubrication. Here, one end of the polymer chains is grafted to a surface and the other end is kept free. If the grafting density is sufficiently high, the polymer chains will stretch out in presence of a good solvent due to entropic reasons to form a brush-like structure known as polymer brushes. One can use hydrophilic polymers to exploit polymer-brush-based aqueous lubrication. Such kind of lubrication is abundant in nature e.g. in cartilage and other joints. In polymer-brush-based lubrication, polymer supports the normal load, and the fluid layer in between the opposing brush layers helps in reducing friction [10–15]. One can also use co-solvency and co-nonsolvency behaviour of polymers to design smart lubricants [16,17].

22.4 NANOPARTICLES

Nanoparticles have several advantages in comparison to organic molecules, which makes it a potential candidate to be used as additive [18]:

- The nanometre size makes it possible for them to move into the contact zone without any difficulty.
- As these particles work efficiently at the ambient temperature, no induction period is required to get desired lubrication behavior.

However, it is difficult to disperse nanoparticles in water.

In recent times, there have been attempts to disperse nanoparticles of inorganic materials in base liquids, e.g. oil or water. Most of these nanoparticles have a layered structure that shear easily to give lower friction. Additionally, the particles can also support the applied normal load and have very good thermal stability [19,20]. These nanoparticle-based suspensions are good boundary lubricants [21–24].

The most commonly known low friction layered materials are metal dichalcogenides of the form MX_2 (M=W, Mo; X=S, Se). The excellent tribological properties of these layered materials are of high scientific interest especially in applications where fluid lubrication is not an option to control friction and wear, such as space technology, ultra-high vacuum and automobiles. When put under traction, these materials undergo basal plane slip [25,26]. The structure of these materials is such that the M–X atoms in layers have covalent bonding in hexagonal arrays and the inter-layer binding is with weak van der Waals interaction. In addition to hexagonal symmetry,

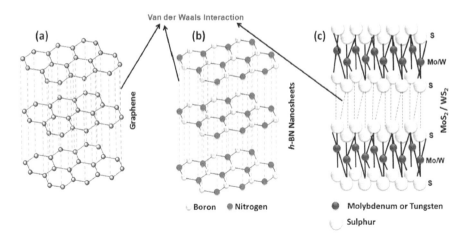

FIGURE 22.2 Representative atomic structure of (a) graphene, (b) h-BN, (c) MoS_2 and WS_2 nanosheets along with the illustration of van der Waals interaction between their atomic/molecular lamellae [27]. (Reproduced after permission.)

the unit cell has two adjacent lamellae (2H arrangement). Since very low shear force is required for inter-crystalline slip in the weak interlayer region, the coefficient of friction is very low when these materials are used as lubricants (Figure 22.2).

The friction mechanism when nanoparticles are used as additives for lubrication is quite different from when organic-surfactant-based additives are used. In the nanoparticle-dispersed-liquid lubrication, the liquid transports the particles to the contact region of the mating surfaces [28]. These solid particles surface get adhered to the asperities of the mating surfaces. A plane of weak shear develops within the particles under the relative motion of the mating surface [29–31]. The applied normal load is supported by the elastic or plastic strength of the solid particle. While the particle material is transferred and back transferred, already formed thin films of these nano-sized particulate materials grow on contacting surfaces [32]. It results in the formation of a system of low friction steady-state tribology. This tribo-mechanism is even more interesting when the particles are dispersed in water [33–36]. Because of the high pressure–viscosity index of oil, it is able to support the normal loads. But, water, with a poor pressure–viscosity index, is unable to support a high normal load. In nanoparticle-based aqueous dispersion, the particles support the normal load, and additionally, their layered structure helps in reducing friction.

These nano-sized particles have a very high surface-area-to-volume ratio. As surface forces dominate the body forces, these particles are attracted towards each other to form agglomerates. So, agglomeration is a major problem when nanoparticles are used as additives in liquid. As these agglomerates do not easily move into the contact zone of mating surfaces, the low friction effect of the layered structure of particles is not effective. The agglomerates of nanoparticles result in poor tribology [37–40]. If these agglomerates are transited to the contact by the liquid, the bulk isotropic

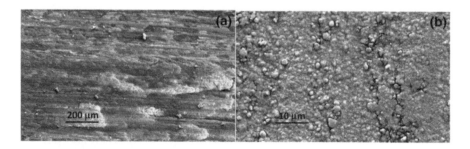

FIGURE 22.3 SEM micrograph of (a) disc and (b) pin surface used in the POD macrotribological test for the monolithic MoS_2 particles under ambient conditions. (Reproduced after permission.)

properties rather than the weak interlayer shear properties are invoked to account for dissipation.

MoS_2 and some white inorganic compounds such as calcium hydroxide and zinc pyrophosphate were added to oils and greases, and their friction behaviour was studied by J. Ghanseimer et al. [20] The white solid lubricants were found to exhibit similar frictional behaviour as by MoS_2 when added to greases or oils. They formed solid deposits on the mating surfaces during sliding and help in reducing wear. But the mechanism of formation of these films was found to be very different from that of MoS_2.

Sahoo et al. studied the frictional behaviour of MoS_2 particle suspension in water and oil. They observed that medium and agglomeration affect the deformation mode. They also found that the particle size as in the contact is also an important parameter. A thick deposit of MoS_2 was observed on the disc and pin surfaces (Figure 22.3) [41].

22.5 EXPERIMENT

Tribological experiments presented in this chapter were carried out using a ball-on-disc tribometer (Make: CSM Instruments, Switzerland). The 6-mm-diameter balls were used for friction experiments. The steel balls were cleaned in an ultrasonic bath of acetone prior to the experiments. The experiments were carried on a stainless steel (SS 304) sample.

The experiments were done under both dry and lubricated contact. The following lubricants were examined: water, base oil and aqueous suspension of different particles. The particles MoS_2, nanoclay and Kaoline of concentrations 1 mg/ml were suspended in water and sonicated for 2 mins using Ultrasonicator.

The experiments were carried out for a range of speeds and loads. To analyse the effect of load on friction behaviour of various lubricants, the experiments were done for a load of 1–6N at a speed of 0.5 cm/s. The Hertzian contact pressure was calculated to be 440–800 MPa which is very close to contact pressure in rolling. The speed variations are maintained between 0.005 and 5 cm/s at a load of 6N. All tribological experiments were carried out at room temperature and humidity ($T = 301$ K and RH $= 35\%$).

22.5.1 MATERIALS

For the ball-on-disc tribometer experiments, the ball material is EN-31 steel (bearing steel) with the following composition:

Constituents	% by wt
C	0.980
Si	0.260
Mn	0.300
S	0.025
P	0.028
Cr	1.540

and the disc material was SS-304 (stainless steel) with the following composition:

Constituents	% by wt
C	0.027
Si	0.487
Mn	1.094
S	<0.005
P	0.023
Cr	18.077
Ni	8.168

The layered MoS_2 particles were procured from M. K. Impex, Canada (MK-MoS$_2$-N3050, agglomerates consists of 50 nm MoS_2 crystallites). The halloysite nanoclay was bought from Sigma-Aldrich, USA, and Kaoline from Thomas Baker (Chemicals) Pvt. Ltd, Mumbai. The surfactant SDS was supplied by S. D. Fine-Chemical Ltd, Mumbai, India. All the water used in the preparation of emulsions was distilled water, processed using Millipore purification (Milli-Q, USA) system. The base oil used to benchmark the frictional results of emulsion was supplied by Bharat Petroleum Corporation Ltd, India.

22.5.2 STEEL SAMPLE PREPARATION

SS 304 and EN 31 steel were used as the substrates. The SS-304 disc samples were machined and ground to give desired sample dimensions of 25-mm diameter and 5-mm thickness. The samples were polished using abrasive silicon carbide papers of different grit sizes 220, 320, 400, 600, 800, 1,000, 1,200 and 1,500. After polishing, the samples were rinsed using the organic solvent acetone. Thereafter, the discs were diamond polished using a paste of 1–3 μm grade. The substrates were then sonicated in acetone for 15 min to get rid of all polishing debris. The steel discs were then flushed with a stream of dry nitrogen and preserved in a desiccator to avoid oxidation of substrate surfaces. The steel balls (EN 31) used in tribometer were used as received (no polishing but sonicated with acetone prior to use).

22.5.3 Emulsion Preparation

Emulsions were prepared by using particles (either of MoS_2, nanoclay or Kaoline) and water. The particles (conc. 1 mg/ml) were mixed with water. The mixture was sonicated using high-intensity ultrasonic processor (Sonics 500 Watt, Connecticut, USA) for 2 mins. To see the effect of surfactant on the emulsion, the surfactant SDS (conc. 1 mM) was also added to the mixture of water and particles and sonicated using Ultrasonicator for 2 mins.

22.5.4 Particle Size and Zeta Potential Measurement

The emulsions prepared as stated above were characterized, using dynamic light scattering (DLS), based on two parameters; droplet size distribution (multimodal size distribution, MSD, or logarithmic size distribution, LSD) and zeta potential (ζ). The droplet-size distribution of the emulsions was measured using a 90 Plus™ particle size analyzer (Brookhaven Instruments Corp., Holtsville, NY).

Zeta potential (surface charge) of the suspended particles in the emulsion was measured with Zeta Plus™ zeta potential analyzer. Zeta potential indicates the stability of dispersion. A higher zeta potential means higher stability of emulsions and resistance to the formation of agglomerates of the nanoparticles.

22.5.5 Nanohardness

Young's Modulus and hardness of nanoparticles were measured using nanoindentation. Diamond Berkovitch tip of 100 nm radius was used in a nanoindenter (Hysitiron Triboindenter, Hysitron Inc., Minneapolis, USA) for estimation of mechanical properties of different nanoparticles. The nanoparticles were immobilized on silicon wafers for the indentation experiments. The measured mechanical properties are reported in the Table 22.1.

22.6 RESULTS

22.6.1 Dry Contact

In the absence of any lubrication, the coefficient of friction of steel–steel contact is found to be 0.5–0.65 which has been reported previously in many literatures (Figure 22.4).

TABLE 22.1
Properties of the Nanoparticles

Sl. No.	Particles	Hardness (GPa)	Elastic Modulus (GPa)
1.	MoS_2	0.2, 0.3	20, 30
2.	Nanoclay	3.77, 4.27	192.2, 104.04
3.	Kaoline	2.16, 2.06	98.1, 79.2

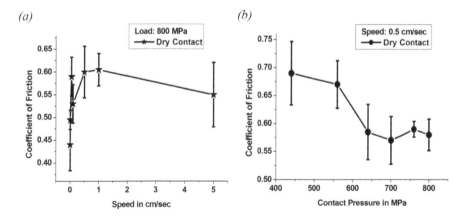

FIGURE 22.4   Coefficient of friction against (a) speed and (b) for dry contact.

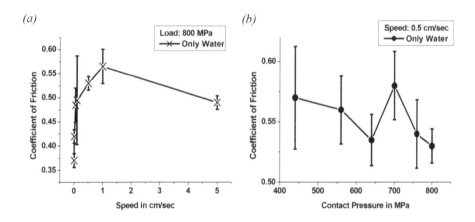

FIGURE 22.5   Coefficient of friction against (a) speed and (b) load for water as lubricant.

22.6.2   Only Water as Lubricant

In Figure 22.5a and b, the coefficient of friction is plotted against speed and load respectively. The coefficient of friction is found to be 0.35–0.6. It is known that the pressure–viscosity index of water is very low. That is why water can not support a high normal load. In present experiments, pressure is maintained at 400–800 MPa which is sufficiently high, so a high coefficient of friction is observed.

22.6.3   Oil as Lubricant

The oil used for the tribo test is an aromatic base oil supplied by a commercial oil company.

In Figure 22.6a and b, the coefficient of friction is plotted against speed and load with oil as the lubricant. At low speed, a comparatively high coefficient of friction

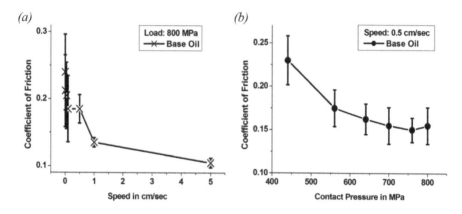

FIGURE 22.6   Coefficient of friction against (a) speed and (b) load for oil as lubricant.

is observed. With an increase in speed, the coefficient of friction reduces. This suggests the interfacial boundary condition is most likely to be in the boundary lubrication regime. Similarly, at low load, a high coefficient of friction is observed, but a decrease in friction coefficient is observed as load increases.

22.6.4   AQUEOUS SUSPENSION OF MoS₂ (PARTICLE CONC.: 1 MG/ML, SDS CONC.: 1 MM)

MoS₂ is a black crystalline inorganic compound and occurs as the mineral molybdenite. In comparison to the other transition metal chalcogenides, MoS₂ is lesser reactive. It is unaffected by dilute acids. It is one of the most commonly used solid lubricants having good friction properties even at relatively high temperatures (Figures 22.7 and 22.8, Table 22.2).

Observations:

- At low speeds (typical of boundary lubrication regime) the friction coefficient of oil-lubricated contact is high (~ 0.25, Figure 22.6).
- Use of MoS₂ nanoparticle-based water lubrication brings down the low-speed friction dramatically (from that of typical oil lubrication) to 0.1.
- Unlike in the case of oil lubrication, nanoparticle-based water lubrication gives the coefficient of friction that is insensitive to load and speed.
- The use of surfactant does not have a dramatic effect on friction in aqueous lubrication.

22.6.5   AQUEOUS SUSPENSION OF NANOCLAY (PARTICLE CONC.: 1 MG/ML, SDS CONC.: 1 MM)

Clay is a naturally occurring material. It is mainly composed of a finely grained phyllosilicate group of minerals. Nanoclay is a multiphase solid material having high aspect ratios. At least one of the dimensions of particles in nanoclay is less than

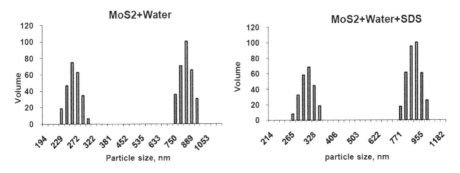

FIGURE 22.7 Particle size distribution for aqueous suspension of MoS$_2$.

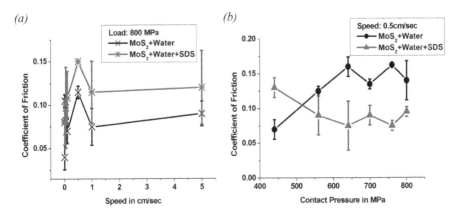

FIGURE 22.8 Coefficient of friction against (a) speed and (b) load for aqueous suspension of MoS$_2$ as lubricant.

TABLE 22.2
Particle Size and Zeta Potential of Aqueous Suspension of MoS$_2$

Sample ID	Particle Size(nm)	Zeta Potential
MoS$_2$+Water	430.8	−40.13
MoS$_2$+SDS+Water	409.8	−43.61

100 nanometers (nm). Nanoclay finds application in various industries such as oil drilling, catalysis, medicine, pet food, skin and body care (Figures 22.9 and 22.10, Table 22.3).

Only nanoclay-in-water gives a high particle size, indicating the formation of agglomerates and a high coefficient of friction. When surfactant SDS is added, a reduction in particle size and coefficient of friction is observed. It can be concluded that the surfactant helps in the better distribution of nanoparticles in water. The coefficient of friction changes from 0.5–0.6 to 0.1–0.15.

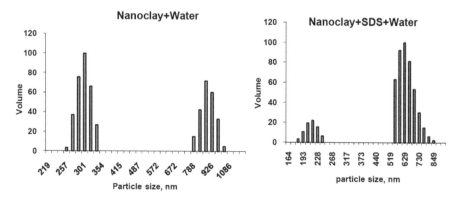

FIGURE 22.9 Particle size distribution for aqueous suspension of nanoclay.

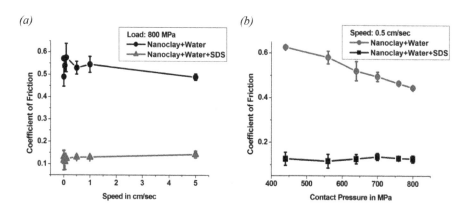

FIGURE 22.10 Coefficient of friction against (a) speed and (b) load for aqueous suspension of nanoclay as lubricant.

TABLE 22.3

Particle Size and Zeta Potential of Aqueous Suspension of Nanoclay

Sample ID	Particle Size (nm)	Zeta Potential (mV)
Nanoclay+Water	457.2	−21.53
Nanoclay+Water+SDS	367.6	−24.56

22.6.6   AQUEOUS SUSPENSION OF KAOLINE (PARTICLE CONC.: 1 MG/ML, SDS CONC.: 1 MM)

Kaoline ($Al_2Si_2O_5(OH)_4$) is a layered silicate mineral. It is composed of clay minerals of the kaolinite group. It is usually white but seldom it has red, blue or brown tints due to impurities (Figures 22.11 and 22.12, Table 22.4).

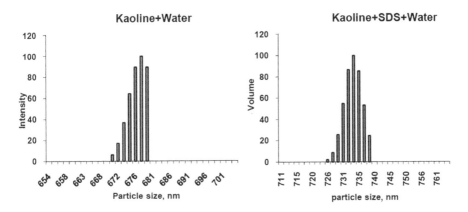

FIGURE 22.11 Particle size distribution for aqueous suspension of Kaoline.

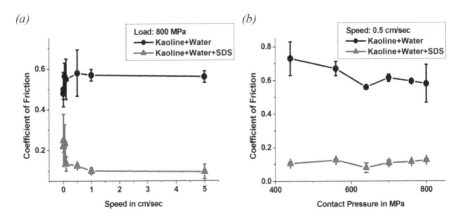

FIGURE 22.12 Coefficient of friction against (a) speed and (b) load for aqueous suspension of Kaoline as lubricant.

Only Kaoline and water gives a high particle size and high coefficient of friction, but when surfactant SDS is added, a reduction in the coefficient of friction is observed. The coefficient changes from 0.5–0.6 to 0.1–0.15.

22.6.7 Aqueous Suspension of SDS (SDS Conc.: 1 mM)g

SDS ($C_{12}H_{25}SO_4Na$) is an anionic emulsifier. This organosulphate salt has a 12-carbon tail attached to a sulfate group. This structure gives it the required amphiphilic properties of a surfactant. It has a critical micelle concentration (CMC) value of 8.2 mM. This detergent is very commonly used in cleaning and hygiene products (Figure 22.13 and Table 22.5).

Since SDS did not effect the friction behaviour of MoS_2 particles but changed the frictional behaviour of nanoclay and Kaoline particles, we decided to study the

TABLE 22.4
Particle Size and Zeta Potential of Aqueous Suspension of Kaoline

Sample ID	Particle Size (nm)	Zeta Potential (mV)
Kaoline+Water	790.20	−17.80
Kaoline+SDS+Water	844.00	−45.33

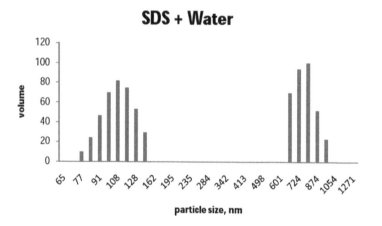

FIGURE 22.13   Particle Size distribution for Aqueous suspension of SDS.

TABLE 22.5
Particle Size and Zeta Potential of Aqueous Solution of SDS

Sample ID	Particle Size (nm)	Zeta Potential (mV)
SDS+Water	306.29	−3.29

frictional behaviour of only SDS in water. It was observed that it gave friction value as low as 0.1.

The role played by the surfactant SDS in aqueous lubrication is not clear. Figure 22.14 shows that SDS in aqueous suspension alone generates low friction. The presence of SDS on the other hand prevents agglomeration of nanoclay (Figure 22.9) and reduces the friction of nanoclay-in-water and Kaoline-in-water to the same level as done by SDS alone.

The reasons for this are not clear and require further investigation.

22.6.8   COMPARISON OF DIFFERENT LUBRICANTS

We studied the suspension of different nanoparticles in water with or without surfactant. We also studied tribological behaviour of the dry contacts, only water as

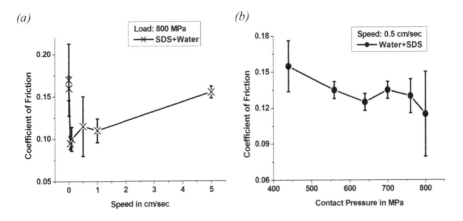

FIGURE 22.14 Coefficient of friction against (a) speed and (b) load for aqueous suspension of SDS as lubricant.

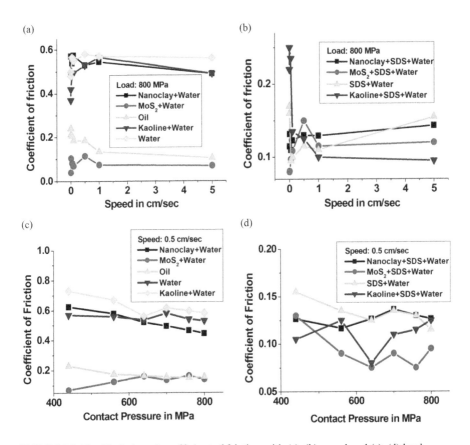

FIGURE 22.15 Variation of coefficient of friction with (a), (b) speed and (c), (d) load.

lubricant and included commonly used lubricants like oil and O/W emulsion in our study. The tribological behaviour of different lubricants were studied over a range of speeds and loads.

Figure 22.15 presents the comparison of tribological behaviour of different lubricants.

22.7   CONCLUSIONS

The present study on using nanoparticle-dispersed aqueous lubrication in rolling process leads to the following conclusions:

- Nanoparticle-based aqueous lubricants can bear high contact pressure. In the current study, a contact pressure as high as 800 MPa was applied during experiments.
- Unlike in the case of oil lubricants, nanoparticle-dispersed aqueous lubricants give a coefficient of friction which is insensitive to load and speed.
- The use of surfactant does not have a dramatic effect on friction in an aqueous suspension of MoS_2.
- Kaoline and nanoclay particles in an aqueous medium yield a high coefficient of friction, but when SDS surfactant is added, the coefficient of friction values fall dramatically.
- The role played by the surfactant SDS in aqueous lubrication is not clear. SDS in aqueous suspension alone generates low friction. The reason for this behaviour of SDS is not understood and requires further investigation.
- Of all the particles studied, MoS_2-based aqueous lubrication gives the best results.

ACKNOWLEDGEMENT

The author would like to thank Prof. (Late) Sanjay Biswas and Prof. Vikram Jayaram from the Indian Institute of Science Bangalore, India, for their help in conducting the research presented in this chapter.

REFERENCES

1. Lenard, J. G. *Primer on Flat Rolling*; Elsevier Ltd, 2007.
2. Habereder, T.; Moore, D.; Lang, M. Eco requirements for lubricant additives. In *Lubricant Additives*; CRC Press, 2009; p 647.
3. Netz, B.; Davidson, O. R.; Bosch, P. R.; Dave, R.; Meyer, L. A. *Climate Change 2007: Mitigation. Contribution of Working Group III to the Fourth Assessment Report of the Intergovernmental Panel on Climate Change. Summary of Policy Makers.* 2007.
4. Tahir, M. *Some Aspects on Lubrication and Roll* Wear in Rolling Mills, KTH Sweden, 2003.
5. Bay, N. *Lubrication Aspects in Cold Forging of Aluminium and Aluminium Alloys*; Meisenbach GmhH, 1995.

6. Keife, H.; Sjögren, C. A Friction Model Applied in the Cold Rolling of Aluminum Strips. *Wear* **1994**, *179* (1–2), 137–142.

7. Wilson, B. Lubricants and Functional Fluids from Renewable Sources. *Ind. Lubr. Tribol.* **1998**, *50* (1), 6–15.

8. Wilson, W. R. D.; Sakaguchi, Y.; Schmid, S. R. A Mixed Flow Model for Lubrication with Emulsions. *Tribol. T.* **1994**, *37* (3), 543–551.

9. Müller, M. T. *Aqueous Lubrication by Means of Surface-Bound Brush-like Copolymers*, ETH Zurich, **2005**.

10. Singh, M. K.; Kang, C.; Ilg, P.; Crockett, R.; Kröger, M.; Spencer, N. D. Combined Experimental and Simulation Studies of Cross-Linked Polymer Brushes under Shear. *Macromolecules* **2018**.

11. Singh, M. K. *Simulation and Experimental Studies of Polymer-Brushes under Shear*, ETH Zurich, **2016**.

12. Singh, M. K.; Ilg, P.; Espinosa-Marzal, R. M.; Kröger, M.; Spencer, N. D. Polymer Brushes under Shear: Molecular Dynamics Simulations Compared to Experiments. *Langmuir* **2015**, *31* (16), 4798–4805.

13. Singh, M. K.; Ilg, P.; Espinosa-Marzal, R. M.; Kröger, M.; Spencer, N. D. Effect of Crosslinking on the Microtribological Behavior of Model Polymer Brushes. *Tribol. Lett.* **2016**, *63* (2), 17.

14. Singh, M. K. Polymer brush based tribology. In *Tribology in Materials and Application*; Katiyar J., Ramkumar P., Rao T., D. J., Ed.; Springer, 2020; pp 15–32.

15. Singh, M. K.; Ilg, P.; Espinosa-Marzal, R. M.; Spencer, N. D.; Kröger, M. Influence of Chain Stiffness, Grafting Density and Normal Load on the Tribological and Structural Behavior of Polymer Brushes: A Nonequilibrium-Molecular-Dynamics Study. *Polymers (Basel).* **2016**, *8* (7), 254.

16. Mukherji, D.; Marques, C. M.; Kremer, K. Smart Responsive Polymers: Fundamentals and Design Principles. *Annu. Rev. Condens. Matter Phys.* **2020**, *11* (1), 271–299.

17. Zhao, Y.; Singh, M. K.; Kremer, K.; Cortes-Huerto, R.; Mukherji, D. Why Do Elastin-Like Polypeptides Possibly Have Different Solvation Behaviors in Water-Ethanol and Water-Urea Mixtures? *Macromolecules* **2020**, *53* (6), 2101–2110.

18. Martin, J. M.; Ohmae, N. *Nanolubricants*; John Wiley & Sons, 2008.

19. Black, A. L.; Dunster, R. W. Comparative Study of Surface Deposits and Behaviour of MoS2 Particles and Molybdenum Dialkyl-Dithio-Phosphate. *Wear* **1969**, *13* (2), 119–132.

20. Gänsheimer, J.; Holinski, R. A Study of Solid Lubricants in Oils and Greases under Boundary Conditions. *Wear* **1972**, *19* (4), 439–449.

21. Winer, W. O. Molybdenum Disulfide as a Lubricant: A Review of the Fundamental Knowledge. *Wear* **1967**, *10* (6), 422–452.

22. Holinski, R.; Gänsheimer, J. A Study of the Lubricating Mechanism of Molybdenum Disulfide. *Wear* **1972**, *19* (3), 329–342.

23. Killeffer, D. H.; Linz, A. Molybdenum Compounds, Their Chemistry and Technology. *Soil Sci.* **1953**, *75* (1), 83.

24. Black, A. L.; Dunster, R. W.; Sanders, J. V.; McTaggart, F. K. Molybdenum Bisulphide Deposits-Their Formation and Characteristics on Automotive Engine Parts. *Wear* **1967**, *10* (1), 17–32.

25. Dickinson, R. G.; Pauling, L. The Crystal Structure of Molybdenite. *J. Am. Chem. Soc.* **1923**, *45* (6), 1466–1471.

26. Bragg, W. The Investigation of the Properties of Thin Films by Means of X-Rays. *Nature* **1925**, *115*, 266–269.

27. Chouhan, A.; Mungse, H. P.; Khatri, O. P. Surface Chemistry of Graphene and Graphene Oxide: A Versatile Route for Their Dispersion and Tribological Applications. *Adv. Colloid Interface Sci.* **2020**, *283*, 102215.

28. Bakunin, V. N.; Suslov, A. Y.; Kuzmina, G. N.; Parenago, O. P. Synthesis and Application of Inorganic Nanoparticles as Lubricant Components-a Review. *J. Nanopart. Res.* **2004**, *6* (2), 273–284.

29. Singer, I. L. Mechanics and Chemistry of Solids in Sliding Contact. *Langmuir* **1996**, *12* (19), 4486–4491.

30. Wahl, K. J.; Singer, I. L. Quantification of a Lubricant Transfer Process That Enhances the Sliding Life of a MoS2 Coating. *Tribol. Lett.* **1995**, *1* (1), 59–66.

31. Wahl, K. J.; Belin, M.; Singer, I. L. A Triboscopic Investigation of the Wear and Friction of MoS2 in a Reciprocating Sliding Contact. *Wear* **1998**, *214* (2), 212–220.

32. Scharf, T. W.; Singer, I. L. Quantification of the Thickness of Carbon Transfer Films Using Raman Tribometry. *Tribol. Lett.* **2003**, *14* (2), 137–145.

33. Peng, Y.; Hu, Y.; Wang, H. Tribological Behaviors of Surfactant-Functionalized Carbon Nanotubes as Lubricant Additive in Water. *Tribol. Lett.* **2007**, *25* (3), 247–253.

34. Chiñas-Castillo, F.; Lara-Romero, J.; Alonso-Núñez, G.; Barceinas-Sánchez, J. D. D. O.; Jiménez-Sandoval, S. MoS$_2$ Films Formed by In-Contact Decomposition of Water-Soluble Tetraalkylammonium Thiomolybdates. *Tribol. Lett.* **2008**, *29* (2), 155–161.

35. Jia, J.; Lu, J.; Zhou, H.; Chen, J. Tribological Behavior of Ni-Based Composite under Distilled Water Lubrication. *Mater. Sci. Eng. A* **2004**, *381* (1–2), 80–85.

36. Radice, S.; Mischler, S. Effect of Electrochemical and Mechanical Parameters on the Lubrication Behaviour of Al$_2$O$_3$ Nanoparticles in Aqueous Suspensions. *Wear* **2006**, *261* (9), 1032–1041.

37. Gubarevich, A. V.; Usuba, S.; Kakudate, Y.; Tanaka, A.; Odawara, O. Diamond Powders Less than 100 Nm in Diameter as Effective Solid Lubricants in Vacuum. *JPN. J. Appl. Phys. Lett.* **2004**, *43* (7A), L920.

38. Shafiei, M.; Alpas, A. T. Friction and Wear Mechanisms of Nanocrystalline Nickel in Ambient and Inert Atmospheres. *Met. Mater. Trans. A* **2007**, *38* (7), 1621–1631.

39. Gubarevich, A. V.; Usuba, S.; Kakudate, Y.; Tanaka, A.; Odawara, O. Frictional Properties of Diamond and Fullerene Nanoparticles Sprayed by a High-Velocity Argon Gas on Stainless Steel Substrate. *Diam. Relat. Mater.* **2005**, *14* (9), 1549–155.

40. Moshkovith, A.; Perfiliev, V.; Lapsker, I.; Fleischer, N.; Tenne, R.; Rapoport, L. Friction of Fullerene-like WS$_2$ Nanoparticles: Effect of Agglomeration. *Tribol. Lett.* **2006**, *24* (3), 225–228.

41. Sahoo, R. R.; Biswas, S. K. Deformation and Friction of MoS$_2$ Particles in Liquid Suspensions Used to Lubricate Sliding Contact. *Thin Solid Films* **2010**, *518* (21), 5995–6005.

Index

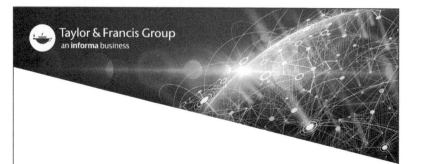